Structural Design of Buildings

Structural Design of Buildings

EurIng Paul Smith DipHI, BEng(Hons), MSc, CEng, FCIOB, MICE, MCIHT, MCMI

WILEY Blackwell

This edition first published 2016

Registered Office
John Wiley & Sons, Ltd, The Atrium, Southern Gate, Chichester, West Sussex, PO19 8SQ, United Kingdom.

Editorial Offices
9600 Garsington Road, Oxford, OX4 2DQ, United Kingdom.
The Atrium, Southern Gate, Chichester, West Sussex, PO19 8SQ, United Kingdom.

For details of our global editorial offices, for customer services and for information about how to apply for permission to reuse the copyright material in this book please see our website at www.wiley.com/wiley-blackwell.

Library of Congress Cataloging-in-Publication Data applied for.

ISBN: 9781118839416

A catalogue record for this book is available from the British Library.

Wiley also publishes its books in a variety of electronic formats. Some content that appears in print may not be available in electronic books.

Set in 10/12pt TrumpMediaeval by SPi Global, Pondicherry, India

Printed in Singapore by C.O.S. Printers Pte Ltd

1 2016

To Tracy, Harriet, George, Henry and Niamh

Contents

Acknowledgements

I would like to acknowledge and thank the following people and organisations for their contribution to this book.

Tracy P. Smith for drawing the plans, figures and illustrations. Also for her endless proof-reading and tireless support and encouragement.

Mr J. and I. Richardson at Richardson's Botanical Identifications for providing information on the close proximity of trees to foundations, and a particular thank you to James Richardson for his assistance in obtaining this information.

Amy Baker for assistance in the workplace during the compilation of this book.

Simon Coyle for assistance in producing drawings.

Tony Gwynne for his support and encouragement.

Tekla (UK) Ltd for the use of their structural engineering software packages to provide examples of structural calculations demonstrating key structural principles.

Tŷ Mawr Lime Ltd. Martin Tavener, *Hydraulic Lime Mortar Application Guide.*

Building Research Establishment, *BRE British stone testing and assessment – stone list.*

Mrs R.W. Brunskill for the use of diagrams demonstrating traditional timber frame construction.

Permission to use extracts derived from British Standards is granted by BSI. British Standards can be obtained in PDF or hardcopy formats from the BSI online shop: www.bsigroup.com/shop or by contacting BSI Customer Services for hardcopies only: tel. +44(0)2089969001; e-mail cservices@bsigroup.com.

York Press for the use of Figure 12.1, photograph of a sinkhole in Magdalen's Close, Ripon.

About the Author

Paul Smith, BEng (Hons) Civil Engineering, MSc Engineering Management, CEng Chartered Engineer, MICE – Member of the Institution of Civil Engineers, FCIOB – Fellow of the Chartered Institute of Building, MCMI – Member of the Chartered Institute of Management, EurIng – Member of Fédération Européenne d'Associations Nationales d'Ingénieurs, MCIHT – Member of the Chartered Institute of Highways and Transportation.

Paul has worked for over 20 years in the public and private sectors, mainly on infrastructure projects. He now runs his own company, Geomex – Structural Engineers & Architectural Design Consultants, which specialises in architectural design, surveying, project management and structural design.

Introduction

The intention of this publication is to embark on a journey taking the reader through a brief history of buildings, how the construction of buildings has evolved over the years and then examining in more detail the structure of buildings and their principal elements. We also examine other factors which affect the stability and structure of buildings, including ground investigations and environmental factors, and detail the materials used in their construction. Finally, we examine structural failures of buildings, their likely causes and common remedies.

This book explains some of the structural engineering principles in the design of residential dwellings and their various structural elements. Some structural theory has been included to demonstrate and reinforce understanding of the comments made. In addition, structural calculations have been included to demonstrate the key points. Diagrams and photographs add clarity.

The theoretical concepts contained in this book are equally applicable to all building structures, whether commercial, traditional or modern. To emphasise some of the issues raised, large examples such as castles and churches are used, which clearly demonstrate the building science and technology.

It must be understood from the onset that specialist structural advice should be sought before undertaking any alterations, or in the identification of structural failures and defects. This book attempts to provide some guidance on understanding the behaviour and construction of buildings, but should not be taken as an exhaustive text.

Health and Safety

It must be recognised that the building and construction industry can be a hazardous environment in which to work, and each individual has responsibility to minimise the risks to both themselves and others.

There is legislative framework which ensures that everyone involved in the commissioning of works, the design, construction and maintenance of building structures has clearly defined responsibilities for health and safety. It is essential that you are aware of your responsibilities under the legislation to reduce risks and prevent accidents. The Health and Safety at Work Act 1974 places responsibilities on contractors, members of the public, clients and construction workers, and is enforced by the Health and Safety Executive.

The Construction Design and Management Regulations 2015 (CDM) place further responsibilities on clients, contractors and designers. Under Regulation 9, a designer must not commence works in relation to a project unless the client is aware of his/her responsibilities. The responsibilities are different for commercial and domestic clients. For domestic clients, unless the designer has a written agreement, the responsibilities must be carried out by the contractor and if more than one contractor is engaged on a project then the client must appoint a principal contractor. This does not mean that the regulations do not have to be carried out, but merely places the responsibilities on another duty holder. The client always has the responsibility for ensuring all pre-construction information is available.

Commercial clients have responsibility for ensuring a construction phase plan is drawn up by the contractor, and that the principal designer prepares a health and safety file for the project. This is undertaken by another duty holder if the client is domestic.

Commercial clients also ensure that management arrangements are in place for health, safety and welfare. The regulations make it clear that clients are accountable for their decisions and the approach they have in regard to the health, safety and welfare of the project.

The client is responsible for the submission of a notice to the Health and Safety Executive subject to the responsibility being undertaken by another duty holder, and with the criteria set out below.

Projects are notifiable to the Health and Safety Executive if the construction work on a construction site is scheduled to:

- *Last longer than 30 working days and have more than 20 workers working simultaneously at any point in the project. Or*
- *Exceed 500 person-days.*

Works may include alterations, maintenance, construction and demolition.

The activities that are defined as domestic works require consideration as the client dictates the classification, for example works can be undertaken on a residential dwelling but if these are undertaken by a private landlord or someone engaged in property development, they would be defined as commercial activity and not domestic works because they relate to a trade or business.

Although in the case of a domestic client regulations 4(1) to (7) and 6 must be carried out by another duty holder, the client still has responsibilities under the regulations. If the domestic client fails to make the necessary appointments under regulation 5, the client's responsibilities are then passed on to other duty holders.

Ignorance of the legislation is no protection against prosecution, and professionals have been prosecuted for not informing clients of their responsibilities under this legislation. Further information is available from the Health and Safety Executive website. Clients, contractors and designers who are in any doubt about their responsibilities are strongly advised to check with the appropriate body or seek professional advice.

Building Regulations, Listed Buildings and Planning Consent

It is also important to recognise that works undertaken on buildings may be subject to other conditions and restraints.

All works must be compliant with the Building Act 1984 and Building Regulations 2010, Local Authority Planning Conditions, Listed Building Consents and Building Regulation Approvals. These should be checked via your Local Authority before embarking on any works. Other legislation may also apply to the proposed works, such as the Wildlife and Country Act 1981 and the Conservation of Habitats and Species Regulations 2010. This legislation makes it an offence to disturb certain species such as bats, and licensed ecologists are required to provide advice.

There are many factors and considerations that may affect the proposed works, and for this reason it is always wise to seek competent professional advice.

Chapter 1 The History of Buildings

The development of building knowledge

In order to understand the construction of buildings it is necessary to determine the age of the building and the technologies likely to be included in the construction and design of that period. For this reason, this first chapter briefly explains the construction and features of buildings over the years and this is further developed in chapter three where the construction is discussed in more detail.

Since the beginning of time man has been engaged in building structures and it is remarkable that many of the early structures still exist. The Neolithic period as early as 6500–10 200 BC saw the first structures being made which may have been simple huts and bridges but nevertheless commenced mankind's quest to construct buildings. Buildings continued to develop through the Mesopotamian, Ancient Greek and Ancient Egyptian periods, which ranged from 6000 BC until 146 BC, and some of these structures – such as the pyramids – are a lasting legacy to the ingenuity and understanding of building construction principles. Following this, the period of the Ancient Romans from around 753 BC until 476 AD saw large-scale buildings become more commonplace. As techniques and materials became better understood, more adventurous structures were constructed.

The Medieval period of the 12th century until the 18th century saw timber frame houses being constructed and some of the early timber frame houses of this era still exist, such as the Medieval Merchants House in Southampton, Hampshire. The development of these structures is intrinsically linked to the understanding of materials and the behaviour of structures which carpenters gained over these centuries.

Masons involved in the construction of churches would travel across the east and west, refining techniques and applying them to new and larger structures. One such example is the development of the arch from a circular arch to a gothic pointed arch, which improved its ability to carry loads, thus resulting in larger-scale and more impressive structures. This is evident in the late 16th century when large glass windows became fashionable in churches to provide light, which also had a significant theological meaning.

The understanding of flying buttresses to resist large lateral and horizontal loads meant that vaulted ceilings could be constructed which accommodated large spans. The first example in England was in Durham Cathedral, which was commenced in 1093. Other early examples include the apse of the Basilica of Saint-Remi in Reims dating from 1170.

Although some of the structural principals were understood, many were based on trial and error and then carried through as tried and tested means of developing structures.

Structural Design of Buildings, First Edition. Paul Smith.

Such scholars as Marcus Vitruvius Pollio wrote some of the earliest books on architecture, and his work *De architectura* (known as *Ten Books of Architecture*) is the only surviving book from the classical period. This provided dimensions for columns based on the number and type of column used and the style of temple required. The height of the column was expressed as a multiple of the diameter. This work was not discovered until 1414 in a library in Switzerland, and interestingly there had been no other printed works prior to this time.

During the Renaissance period, in 1450, Leon Battista Alberti published *De re aedificatoria*, which translates as *The Art of Building*. This was one of the first printed books on architecture. Later, Sebastiano Serlio (1475–1554) published *Regole generali d'architettura*, which translates as *General Rules of Architecture*. Then, in 1570, Adrea Palladio published *I quattro libra dell'architettura*, which translates as *Four Books of Architecture*. This final publication carried many of the Renaissance ideas into Europe.

Prior to these publications there were very few books for architects and masons to reference how structures were constructed. Following the Renaissance period (15th–17th centuries), more information became available.

During the years 1100–1200, fire was the major concern and a hazard in built-up cities. The construction of houses during this time was predominantly in timber, and densely populated areas resulted in accommodation being provided by extending existing properties and adding additional storeys.

In 1666 the Great Fire of London transformed building control and regulation in the UK. The following year the London Building Act banned the use of timber and insisted on the use of brick and stone in the construction of houses. In 1694, following another major fire in Warwick, more major cities were prompted to introduce Building Acts based upon that introduced in London. By the 18th century, most cities had a Building Control Authority and had adopted a Building Act.

The Building Act of 1858 meant that plans had to be deposited with the authorities for new buildings and alterations. This makes it easier – after this period – to ascertain the history and construction of properties throughout the UK.

Styles of architecture and building construction

It is remarkable that today we still dwell in houses constructed as far back as Medieval times, and it is at this point that we begin our analysis of the structures of buildings based on the techniques used in the past.

Medieval

The majority of the remaining residential dwellings of this period are of timber frame. Predominantly these were of cruck construction or box frame, where the roof is a separate structure to the walls. Medieval buildings tended to have thick timber members which were irregular in shape, and the timber posts were placed directly onto or inserted into the ground. The floor joists were generally large and laid flat rather than upright, typically these would be 200 mm × 150 mm timbers. Figure 1.1 shows a photograph of a typical cruck frame construction used in a house in Herefordshire.

In their simplest form, Medieval buildings were four-bay cruck frame structures with a large hall occupying at least two of the bays. The open-plan design centred around a large

Figure 1.1: Photograph of typical cruck construction in Herefordshire.

fire, which was the only means of heating. Access was gained through two large doors normally located on opposite sides, which formed cross passages. Of the remaining bays, one would form a parlour which would create some privacy for its occupants and the other would be split as a pantry and buttery for storing food and drink. Over 4000 cruck frame buildings remain in the UK today.

Other forms of construction existed at this time, and stone cottages have been constructed from materials close to hand from a very early period in history. Cob construction is another form of construction with the main component being mud, earth or clay. This form of construction can be traced back to the 14th century and was particularly evident in the south-west and central-southern England.

Tudor (1485–1560)

With the exception of churches, most buildings in the Tudor period were also of timber frame construction with box frame construction being dominant. Houses tended to be one-room deep with a limited span, as the walls did not have sufficient load-bearing capacity to support the heavy roof structure. Some masonry brick construction was used to fill the timber panels and some stone construction for windows and quoins.

Bricks were a luxury product and found only in the homes of the wealthy, and generally in the east and south parts of the country. This was predominantly because the people who knew how to make and use bricks were Flemish immigrants who settled on the east coast.

Most large houses were constructed around a central hall, with wings containing private chambers at one end and kitchens and service rooms at the other. As today, space within the towns and cities was valuable and the timber frame houses were generally

Figure 1.2: Photograph showing the close proximity of timber frame housing in Leominster, increasing the risk of fire.

owned by rich merchants. Plots in the cities tended to be long and narrow, and houses often had a rectangular form with the gable end facing onto the street. The ground floor was used for commercial enterprise, with the living accommodation being above. To gain additional space, jetties were introduced to extend over ground floors and create additional storeys.

The jetties extended the higher storeys forward of the building line into the street, reducing the distance between the facing properties. Consequently, this resulted in an increased fire risk, as fire could travel easily from one building to another. Figure 1.2 is a photograph showing the close proximity of timber frame properties in Leominster, increasing the risk of fire. This was a prominent reason for the spread of the Great Fire of London in 1666.

Initially little consideration was given to external appearance, but towards the end of the Tudor period the finest timber frame houses featured close timber studding, decorative panels and brick panels with diagonal patterns.

Glazing was not generally used in properties at this time, and only the finest properties enjoyed this privilege. Timber frame houses at this time had shutters and mullions rather than glass.

Floorboards above the floor joists were usually left exposed and the joists were chamfered and did not have plastered ceilings. Joists were laid flat and were usually 125 mm × 100 mm in size.

During the Medieval and Tudor periods thatch was the most likely roof covering, and the pitch of the roof had to be steep (generally 45°–60°) to dispel water from the roof. The overhang was deep so that the water was thrown clear of the walls.

Elizabethan and Jacobean (1560–1660)

During the Elizabethan and Jacobean periods, timber frame houses were still the most popular type of building construction but masonry and brick buildings began to appear and transcend down the social scale.

In larger houses the house footprint moved away from a large central hall into smaller rooms; fireplaces reduced in size and chimneys were introduced. These were a status symbol and often tall and topped with decorative chimney pots. The buildings often had an E- or H-shaped floor plan.

Glazing was still only for the privileged, but more modern methods of construction allowed larger areas to be glazed. Windows with patterned glass were being introduced, but more common was a diamond-shaped pane with lead casing.

Buildings became even more decorative, with some containing hidden messages showing allegiance to the Queen by having an E-shaped floor plan or containing the forbidden Trinity Triangle. During this period of religious division Catholic houses often had secret chambers constructed, known as priest holes.

Internally, large fireplaces with decorative mantles and elaborate panelling to the walls further demonstrated the wealth of the owner.

Large columns were introduced to the finest properties, but their proportion did not always match the property. Symmetry became important during this era and properties faced outwards rather than inwards towards courtyards.

Bricks were becoming more popular and ranged in size from 210 mm to 250 mm in length and 100 mm to 120 mm in depth, with a height of 40 mm to 50 mm. An example of a Jacobean property can be seen in the photograph in Figure 1.3.

Figure 1.3: Photograph of a Jacobean house of brick construction.

The commoner houses were still constructed using a timber frame with steeply pitched roofs having thatching or slate tiles. Cruck frame and box frame were still the main methods of construction. Jetties were becoming more common to achieve second storeys in market towns and cities, with the beams becoming more decorative and the introduction of carved finishes. Doors and windows were tall and narrow.

Restoration (1660–1714)

The Restoration period brought about great change to architectural style and building construction. Initially with the restoration of Charles II to the throne, many exiled royalists returned home to reclaim their lands and with them came the European influence on architectural style. In addition during this period was the Great Fire of London, the aftermath of which brought sweeping changes with the introduction of the Building Act of London – legislation controlling the structure and materials used in new houses. This began the new era of Building Control.

New properties were now constructed using mainly masonry and, although some timber frame structures were still being constructed, masonry became more prevalent. There were rapid developments, with properties becoming wider through the introduction of a second room at the rear. Thus, properties became two rooms deep. Quoins became fashionable, and these were highlighted on properties. Steep roofs were introduced, and hips were necessary to achieve this style of roof structure. Dormer windows along the roof line of grand and terraced houses also became noticeable during this period. A Restoration-style property can be seen in the photograph in Figure 1.4.

Figure 1.4: Photograph of a Restoration-period house in Herefordshire.

The brick bond changed from English to Flemish, and lintels were typically brick with stone keystones.

By the end of the 17th century window styles also changed, from casement windows to sash windows. There was a desire to maintain the flat façade of the building and not break the front line when the windows were opened. Single mullions were also popular, with the transom set slightly above the centre – thus the top part of the window was smaller than the bottom.

Although available in the 13th century, clay tiles became widespread during the 17th century (mainly due to the need to protect properties from fire, thus replacing thatch).

Window tax

In 1696 a property tax was introduced on all properties in England and Wales. The window tax was introduced in Scotland in 1748, which was some time after the Union in 1707. Those properties having more than ten windows were subject to an additional tax depending on the number of windows. Interestingly, in Scotland this was later reduced to seven windows.

This additional property tax was known as the window tax, and at the time windows were in-filled or properties were being built with in-filled windows with the intention of the windows being glazed or re-glazed at a later date (on anticipation of the tax being lifted). The in-filled windows meant that the tax was not payable, but the legislation contained no definition of a window and the smallest opening could be included as a window. The tax was repealed in England, Wales and Scotland in 1851 following the argument that the tax was a tax on health and a tax on light and air. Consequently, properties with in-filled windows are likely to have been constructed before 1851.

Georgian (1714–1790)

During the Georgian period houses became much more substantial and focused on symmetry and larger rooms. The architecture was based on Greek and Roman architecture, constructed using uniform stone or brick with Corinthian, Doric and Ionic capitals on columns. The architecture was elegant and based on ancient worlds and temples, with smaller houses using the same approach on a reduced scale. A photograph of a large Georgian-period Grade II listed building can be seen in Figure 1.5.

Two-up, two-down properties were introduced in urban areas, with most terraces being constructed of brick with sloping shallow-pitched slate roofs hidden behind parapets. New regulations on fire introduced standards for compartmentalisation of terraced houses. Party walls were built to prevent the spread of fire and carry the weight of the chimney.

The Building Act 1774 tried to reduce the risk of fire by improving the quality of construction, and houses were rated based on their value and floor area; each category had its own set of structural requirements in terms of foundations, external and party walls.

Windows were predominantly sash windows with thin wooden glazing bars. On the ground floor the windows were kept smaller to ensure the stability of the building, on the first floor the windows were tall and elegant but these reduced in height on successive floors, with the top-floor windows being almost square. Front doors traditionally contained six panels.

Terraced houses also had basements with the front door approached at road level, but later in this period the basement protruded above ground level and formed a half

Figure 1.5: Photograph of a Grade II listed Georgian property.

basement. Thus, the elevated front door was approached via steps, sometimes spanning the void between the basement and the road.

Wall construction was achieved by using thin mortar joints, and darker mortars were covered with lime putty to lighten the colour to match the surrounding brickwork.

As the population grew in rural areas, huge numbers of stone cottages were constructed. These were usually of one or two bays and single storey, with an end chimney.

Regency (1790–1830)

The Regency period saw the introduction of stucco, which is render made to look like stone. Elegant buildings became the fashion and towns such as Brighton and Cheltenham displayed fashionably elegant houses. Figure 1.6 is a photograph of a typical terraced property of this era, taken in Malvern.

At this time cement was also used in mortars, which meant that taller, more robust structures could be constructed in masonry. The half basement was still used during this period, and rear extensions accommodating servants became popular. Gothic style began to replace the Greek and Roman styles of the Georgian period.

Bow windows became fashionable on the finest houses of the wealthy. For terraced properties the first floor accommodated French doors leading to balconies comprised of decorative ironwork. Sash windows were set further back and small timber strips were used, often reinforced with metal strips.

Roof lines became much shallower with the introduction of lightweight Welsh slate, and this was used in abundance. Roof valleys became popular which were hidden behind the parapets, and Mansard roofs were also used to achieve low pitches. In the suburbs detached and semi-detached houses, known as villas, were also being constructed.

Figure 1.6: Photograph of a typical Regency-period terrace in Malvern.

In 1833, at the end of the Regency period, John Claudius Loudon published a book which contained over 2000 designs for houses: the *Encyclopaedia of Cottage, Farm and Villa Architecture and Furniture*. This demonstrates the wide variety of house designs available at the time.

Victorian (1830–1900)

Following this came the Victorian era and the Industrial Revolution, which led to many advances in the use of materials and the construction of buildings. The improvements in iron production led to the economic production of pig and wrought iron. Structural-grade iron was now achievable, and this meant that structures began to appear constructed using iron. The introduction of the railways also meant that bridge engineering and huge retaining structures had to be constructed, increasing our understanding of the materials used and the structures' behaviour.

The quality and availability of materials improved as goods could be transported all over the country by rail. Towards the end of the Victorian era machine-made bricks became widely available, although these were still expensive and as such were only used to construct the façade of buildings.

Britain's population doubled during this time and towns expanded dramatically. Owing to work places being predominantly the mills, collieries and factories, housing was concentrated around these, leading to the densely populated terraced houses known as back-to-back terraces, which can still be seen today.

Middle-class terraced houses were of substantial construction and contained quite a number of rooms, sufficient to accommodate servants. The basement area was usually given over to the kitchen, larder and scullery, with the servants' quarters being accommodated in the attic. These properties also had the benefit of a garden.

The Gothic style was prevalent at this time, but houses were also influenced by other styles of architecture. Buildings were often asymmetrical, with steep-pitched roofs and

Figure 1.7: Photograph of the Cotford Hotel in Malvern.

forward-facing gables. Decorative brickwork was commonplace, and this was complemented by ornate bargeboards. Figure 1.7 is a photograph of the Cotford Hotel in Malvern, which demonstrates the Gothic-style architecture with steep-pitched roofs and ornate bargeboards. The walls are constructed using Malvern stone with ribbon pointing, which is characteristic of this area.

Doors typically had four panels as opposed to the previous six-panel Georgian style. In terraces, doors were often recessed and set in pairs rather than along the same side of the row.

With improvements made in the manufacture of glass windows, increased size and number with larger panes and, by the end of the era, coloured glass was a common decorative feature.

During the Industrial Revolution stone cottages became commonplace and were used for industrial and rural workers alike. Generally constructed using rubble stone walls and lime mortar, the accommodation typically comprised two-up, two-down properties and was used across the country. This method of construction has changed little over time and this type of construction can be found in the south west, Derbyshire, Yorkshire and any area where stone is readily available.

The use of damp-proof courses was introduced by the Victorians, but it was later when they became used throughout the UK. Early damp-proof courses included asphalt, bitumen, tar, three courses of engineering brick or even such materials as lead and copper.

Post-1900

Before the 1900s, local materials were used in the construction of buildings. Walls were of solid brick in towns and cottages were of stone construction with walls some 450–500 mm thick. In later years many of the walls have been rendered in an attempt to improve protection against the infiltration of damp. Parking was not a consideration at this time, since people did not own cars, and consequently properties in the countryside could be constructed in locations some distance from main highways.

At the turn of the century balconies became fashionable in town houses and looked towards the street. Between 1900 and the 1920s the quality of building materials improved. Decorative brickwork panels using coloured bricks such as buff and blue were seen over window openings, and at eaves and first-floor level. Porches and hallways from the entrance comprised decorative tiles. It was not uncommon for the density of housing to be 20 to 30 houses per acre, particularly in terraced rows in cities and towns. Figure 1.8 is a photograph of a typical semi-detached house of this period.

During the 1920s the cavity wall was used more extensively, but it should be noted that cavity walls had been in use since the 1870s in the west of England and parts of Ireland in an attempt to prevent the penetration of damp. The density of housing was reduced to six to ten houses per acre, and larger gardens were provided as minimum distance rules on the close proximity of back-to-back housing were introduced.

The accommodation became more spacious and the concept of housing estates began to become established. During this era social housing was introduced, and this new concept meant that large, spacious estates began to grow.

1930s housing saw the widespread implementation of the cavity wall and hipped roofs, with large overhangs over fashionable bay windows that extended over both storeys.

Figure 1.8: Photograph of a typical brick semi-detached house of the 1900–1920s era.

Figure 1.9: Photograph showing a typical 1930s property.

The gables extended over the bay windows and could be mock Tudor or tiled with timber framing. Another characteristic of this period was the recessed front door, usually with windows either side and approached through a brick arch. Housing estates became more densely populated than in the 1920s. Figure 1.9 is a photograph of a typical 1930s property.

In the 1940s the construction of houses was halted as a result of the war. During the war the housing stock was reduced considerably as a result of bomb damage, particularly in the major cities. This led to a post-war housing shortage and to cope with demand approximately 156 000 prefabricated houses were constructed, which were low cost and could be built rapidly. Figure 1.10 shows a photograph of a Cornish dwelling which was a particular type of prefabricated concrete house.

These prefabricated houses were only intended as a short-term solution but still endure today. Chapter 3 explains this type of housing in more detail, and how it has been adapted to overcome problems with corrosion of the concrete. In addition, flats and maisonettes became popular, which were low-rise constructions up to four storeys in height.

During the 1950s metal-casement Crittal windows became fashionable, and large overhanging cantilevered porches supported on metal posts became characteristic of the period. The London Brick Company mass produced bricks, and walls were constructed without decorative panels as in the 1900–1920s period. Social changes meant that open-plan estates evolved, which removed the concept of front boundaries thus

Figure 1.10: Photograph of a Cornish dwelling which was a particular type of prefabricated concrete house.

leading to open-plan gardens. Transport policies at the time favoured the increased use of the motor car rather than trains, and this led to more garages being constructed.

Bungalows also became fashionable during this period, and this trend continued into the next decade.

During the 1950s timber was in short supply after the war, and the TRADA truss was introduced to complement a cut roof by supporting a 50 mm × 150 mm purlin at mid-span. This reduced the size of the timber rafters and negated the need for a wall extending into the roof space to support the purlin. Typically, the TRADA trusses were spaced at between 1.8 m and 2.4 m, but were soon superseded by the complete trussed roof as seen in modern houses. Typically, TRADA trusses are Fink trusses with double-tie and rafter members laid side by side. A photograph of a typical TRADA truss can be seen in Figure 1.11.

During the 1950s roofing materials also saw a change, with the increased use of concrete tiles.

The decade of the 1960s saw the development of town planning and led to a new approach to inner-city development, with high-rise tower blocks being constructed to increase the density of housing – for example Park Hill, later known as the Hyde Park Estate, in Sheffield. One of the tower blocks in this development has 19 storeys and contains 1160 dwellings.

Even the bungalow was developed to have additional rooms in the roof space, effectively increasing the accommodation on the same floor area. Bungalows grew in popularity and were constructed with dormer windows and rooms in the roof. These are still evident on housing estates across the country. In some cases these dormers have been added

Figure 1.11: Photograph of a TRADA truss.

retrospectively or widened. Chapter 17 explains some of the issues to examine in these circumstances. A typical 1960s bungalow can be seen in the photograph in Figure 1.12.

The trussed roof, introduced during the 1960s, was becoming popular and this removed the need for labour-intensive cut roof construction. A series of single Fink trusses was placed at approximately 600 mm centres along the length of the roof, making the so-called trussed roof.

Between the 1930s and the 1960s clinker or aggregate blockwork was used for the internal leaf of cavity walls. These blocks were cheaper to manufacture than bricks and were larger in size, so larger areas could be constructed in less time.

Figure 1.12: Photograph of a typical 1960s bungalow.

In the 1970s most estates became more densely populated, often using a mix of different standardised housing types to give the impression of individuality. Garages were replaced with off-road parking or car ports, and gardens became enclosed. Larger double-glazed aluminium windows were introduced and became fashionable. The early skylight, such as Velux™ windows, was also introduced to roof structures. The increased use of natural gas as a heating fuel meant that new houses of this period were being connected to mains gas supplies, hence the requirement for the chimney was now obsolete.

Until this time most cavities were 50 mm in width and the introduction of insulation in the cavity was being implemented in recognition of the need to insulate houses.

By the 1980s estates became even more densely populated in an attempt to meet housing shortages. Open-plan gardens took on a revival and estates accommodated a number of similar property types but with a range of styles to give the appearance of variety, but ensuring that mass construction could be undertaken. Figure 1.13 shows a photograph of a typical 1980s house using brick with a rendered gable.

During this time calcium silicate bricks became more popular, but these were more prone to reversible and irreversible moisture movements than their clay brick counterparts. Consequently, walls experienced contractions in the brick material. For this reason BS 5628 Part 3 recommends that movement joints be placed at intervals of not more than 7.5 m–9.0 m in calcium silicate blocks (as opposed to 15.0 m in clay bricks). Where this criterion has not been introduced, many of the houses have experienced vertical cracking due to the natural formation of the movement joints. Figure 1.14 is a photograph showing a typical 1980s house using calcium silicate bricks.

Figure 1.13: A typical 1980s house on a residential estate using clay bricks with a rendered gable.

All walls were constructed using cavity wall insulation, since this was enshrined in the Building Regulations. Although first introduced in the 1970s, by the 1980s the use of Thermalite lightweight blocks had become widespread to improve the thermal qualities of walls. Cavity widths increased to 75 mm to ensure the additional thicknesses of insulation could be accommodated. Flues began to appear on roofs to accommodate solid-fuel burners.

During the 1990s larger estates were being constructed, creating densely populated areas of housing. Decorative brickwork became fashionable, but this was in an attempt to disguise the similarity of properties through subtle differences in their appearance. The introduction of extractor fans to bathrooms and kitchens, and trickle vents to windows, was used to improve ventilation to properties.

In the 21st century the use of sustainable materials, the development of timber frame houses and the introduction of structurally insulated panels (SIPs) have become widespread. The use of renewable energy sources and green technologies has also become commonplace. There have been a number of changes to the Building Regulations in the last two decades, with particular emphasis on sustainable development. Figure 1.15 shows a photograph of SIPs being used in construction.

Figure 1.14: A typical 1980s house on a residential estate using calcium silicate bricks.

Figure 1.15: A photograph of SIPs being used in construction.

In summary, the 20th century saw advances in technology and improvements in the thermal qualities of houses. This resulted in construction methods changing, and the increased use of non-breathable construction was more evident than the traditional breathable construction. The difference is explained further in Chapter 3. Cavities in walls had been used as early as the Roman period, but traditional houses were of solid brick until 1911. By 1920 the use of cavity walls had become widespread. Since the 21st century, the cavity width has increased to 90 mm to 100 mm to improve the thermal qualities of housing.

Chapter 2 Loadings and Aspects of Structural Theory Relating to Buildings

Before launching into the practical aspects of structural engineering in relation to buildings, the reader should spend a few moments on some of the more commonly used terms and concepts in relation to structural engineering. This will enable the reader to understand some of the theory behind the practical aspects and appreciate the context of the practice.

Weight and mass

It is worth noting the difference between mass and weight, since both can be mistakenly referred to as the same thing. However, there is a difference. Mass is measured in kilograms (kg), whereas the weight of an object is measured in newtons (N). By definition, newton is a force and therefore weight is a force.

The weight is derived from the mass and is dependent on gravitational acceleration (9.80665 m/s^2):

$$\text{Weight} = \text{Mass} \times \text{Acceleration (gravitational acceleration)} \quad (2.1)$$

To identify the behaviour of a material or structural element it is necessary to identify the load acting on it. To obtain the weight or force on a structural element, the mass has to be multiplied by 9.80665 (usually rounded up to a value of 10) as described in equation (2.1).

For example, the mass of a tile may be 55 kg/m^2.

$$\text{The load or weight is therefore } 55 \times 10 = 550\,\text{N/m}^2 \quad (2.2)$$

There are three basic types of load in building, and these are defined depending on whether British standards or Eurocodes are employed for the purposes of design. In the Eurocodes, loads are referred to as "Actions".

Permanent actions or dead loads

As the name suggests, these loads are the permanent loads acting on a structural element and are the self-weight of materials such as a roof, a floor, a wall, etc. For example a roof may comprise tiles, baton, felt and rafters, all of which contribute to the dead load or permanent load of the roof.

Structural Design of Buildings, First Edition. Paul Smith.

Table 2.1: Based on BS 648: 1964 showing mass of materials

Material	Unit mass
Asbestos 25.4 mm thick	3.9 kg/m^2
Batons (slating and tiling 38.1 mm × 19.1 mm) softwood 100 mm gauge	3.4 kg/m^2
Bitumen roofing felts type 1E	3.4 kg/m^2
Blockwork (per 25 mm thick)	55 kg/m^2
Brickwork (per 25 mm thick)	55 kg/m^2
Concrete natural aggregate	2307 kg/m^3
Flagstones natural stone (50.8 mm thick)	137 kg/m^2
Lead (sheet BS 1178) (3.0 mm thick)	34.2 kg/m^2
Plaster two coat (12.7 mm thick)	22 kg/m^2
Plywood (per mm thick)	0.7 kg/m^2
Rendering (12.7 mm thick)	29.3 kg/m^2
Tiling (concrete)	48.8±7.3 kg/m^2

Publications are available, such as BS 648: 1964 Schedule of Weights of Building Materials and product literature, which provide units of mass per unit area, length or volume. As detailed in equation (2.1) above, to obtain the weight or force on a structural element these values have to be multiplied by 9.80665 (usually rounded up to a value of 10; see equation (2.2)). An extract from BS 648 can be seen in Table 2.1.

Variable actions or imposed loads

These loads are the loads on structural elements that can vary, for example people or furniture. Buildings have different usages and occupancies thus, depending on their use, the imposed or variable loads will change. Snow is also considered as a variable or imposed load.

BS 6399-1: 1996 provides minimum imposed load values for different occupancies and usages of buildings, and covers such buildings as warehousing, storage, offices, garages and industrial buildings. A residential building is generally understood to have an imposed or variable floor load of 1.5 kN/m^2.

Wind load

The wind load varies across a country depending on topology, wind speed and other factors. BS 6399-2 can be used to calculate wind loading. Further information on wind loading is available in Part A of the Building Regulations. EN 1991-1-1-4: 2005 also provides design values for wind loading. Under the Eurocodes the wind load is considered as a variable action.

The following example is for a building in Bristol. The calculations have been undertaken using structural engineering software and we gratefully acknowledge Tekla (UK) Ltd for their approval in the use of this software. As can be seen, the calculations for wind loading are complex and fortunately we have computer programmes to assist in this process.

The location will provide a basic wind speed by using figure 6 of BS 6399: Part 2. The site altitude and upward distance from the sea are also known.

WIND LOADING (BS 6399).

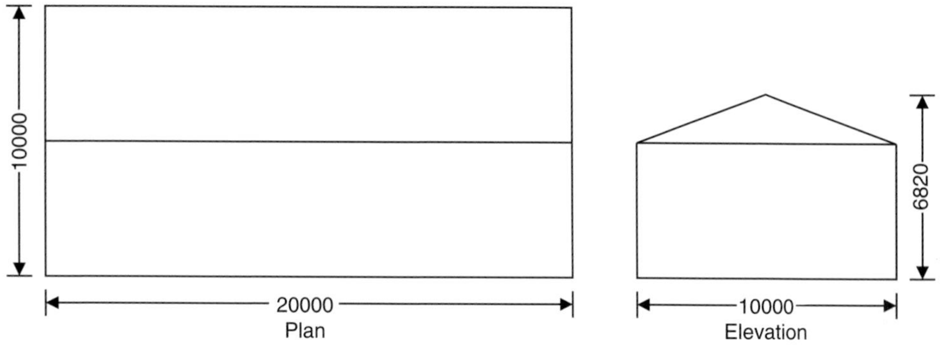

Building data

Type of roof;	Duopitch
Length of building;	L = **20000** mm
Width of building;	W = **10000** mm
Height to eaves;	H = **5000** mm
Pitch of roof;	α_0 = **20.0** deg
Reference height;	H_r = **6820** mm

Dynamic classification

Building type factor (Table 1);	K_b = **1.0**
Dynamic augmentation factor (1.6.1);	$C_r = [K_b \times (H_r/(0.1\ \text{m}))^{0.75}]/(800 \times \log(H_r/(0.1\ \text{m}))) = \mathbf{0.02}$

Site wind speed

Location;	Bristol
Basic wind speed (figure 6, BS 6399: Pt 2);	V_b = **20.4** m/s
Site altitude;	Δ_S = **72** m
Upwind distance from sea to site;	d_{sea} = **90** km
Direction factor;	S_d = **1.00**
Seasonal factor;	S_s = **1.00**
Probability factor;	S_p = **1.00**
Critical gap between buildings;	g = **5000** mm
Topography not significant	
Altitude factor;	$S_a = 1 + 0.001 \times \Delta_S/1\ \text{m} = \mathbf{1.07}$
Site wind speed;	$V_s = V_b \times S_a \times S_d \times S_s \times S_p = \mathbf{21.9}$ m/s
Terrain category;	Town
Displacement height (sheltering effect excluded);	H_d = **0** mm

The velocity pressure for the windward face of the building with a 0 degree wind is to be considered as 1 part, since the height h is less than b (cl. 2.2.3.2)

The velocity pressure for the windward face of the building with a 90 degree wind is to be considered as 1 part, since the height h is less than b (cl. 2.2.3.2)

Dynamic pressure – windward wall – wind 0 deg and roof

Reference height (at which q is sought);	H_{ref} = **5000** mm
Effective height;	$H_e = \max(H_{ref} - H_d, 0.4 \times H_{ref})$ = **5000** mm
Fetch factor (Table 22);	S_c = **0.889**
Turbulence factor (Table 22);	S_t = **0.192**
Fetch adjustment factor (Table 23);	T_c = **0.754**
Turbulence adjustment factor (Table 23);	T_t = **1.630**
Gust peak factor;	g_t = **3.44**

Terrain and building factor;	$S_b = S_c \times T_c \times (1 + (g_t \times S_t \times T_t) + S_h) = \mathbf{1.39}$
Effective wind speed;	$V_e = V_s \times S_b = \mathbf{30.4}$ m/s
Dynamic pressure;	$q_s = 0.613$ kg/m$^3 \times V_e^2 = \mathbf{0.567}$ kN/m^2

Dynamic pressure – windward wall – wind 90 deg and roof

Reference height (at which q is sought);	$H_{ref} = \mathbf{6820}$ mm
Effective height;	$H_e = \max(H_{ref} - H_d, 0.4 \times H_{ref}) = \mathbf{6820}$ mm
Fetch factor (Table 22);	$S_c = \mathbf{0.932}$
Turbulence factor (Table 22);	$S_t = \mathbf{0.187}$
Fetch adjustment factor (Table 23);	$T_c = \mathbf{0.781}$
Turbulence adjustment factor (Table 23);	$T_t = \mathbf{1.583}$
Gust peak factor;	$g_t = \mathbf{3.44}$
Terrain and building factor;	$S_b = S_c \times T_c \times (1 + (g_t \times S_t \times T_t) + S_h) = \mathbf{1.47}$
Effective wind speed;	$V_e = V_s \times S_b = \mathbf{32.1}$ m/s
Dynamic pressure;	$q_s = 0.613$ kg/m$^3 \times V_e^2 = \mathbf{0.633}$ kN/m^2

Size effect factors

Diagonal dimension for gable wall;	$a_{eg} = \mathbf{12.1}$ m
External size effect factor gable wall;	$C_{aeg} = \mathbf{0.920}$
Diagonal dimension for side wall;	$a_{es} = \mathbf{20.6}$ m
External size effect factor side wall;	$C_{aes} = \mathbf{0.872}$
Diagonal dimension for roof;	$a_{er} = \mathbf{20.7}$ m
External size effect factor roof;	$C_{aer} = \mathbf{0.871}$
Room/storey volume for internal size effect factor;	$V_i = \mathbf{0.125}$ m^3
Diagonal dimension for internal size effect factors;	$a_i = 10 \times (V_i)^{1/3} = \mathbf{5.000}$ m
Internal size effect factor;	$C_{ai} = \mathbf{1.000}$

Pressures and forces

Net pressure;	$p = q_s \times c_{pe} \times C_{ae} - q_s \times c_{pi} \times C_{ai}$
Net force;	$F_w = p \times A_{ref}$

Roof load case 1 – wind 0, c_{pi} 0.20, $-c_{pe}$

Zone	External pressure coefficient, c_{pe}	Dynamic pressure, q_s (kN/m^2)	External size factor, C_{ae}	Net pressure, p (kN/m^2)	Area, A_{ref} (m^2)	Net force, F_w (kN)
A (-ve)	-0.90	0.63	0.871	-0.62	19.80	-12.32
B (-ve)	-0.70	0.63	0.871	-0.51	9.23	-4.73
C (-ve)	-0.33	0.63	0.871	-0.31	77.39	-24.01
E (-ve)	-1.17	0.63	0.871	-0.77	19.80	-15.23
F (-ve)	-0.77	0.63	0.871	-0.55	9.23	-5.07
G (-ve)	-0.50	0.63	0.871	-0.40	77.39	-31.11

Total vertical net force;	$F_{w,v} = \mathbf{-86.90}$ kN
Total horizontal net force;	$F_{w,h} = \mathbf{3.54}$ kN

Walls load case 1 – wind 0, c_{pi} 0.20, $-c_{pe}$

Zone	External pressure coefficient, c_{pe}	Dynamic pressure, q_s (kN/m^2)	External size factor, C_{ae}	Net pressure, p (kN/m^2)	Area, A_{ref} (m^2)	Net force, F_w (kN)
A	-1.44	0.63	0.920	-0.96	14.99	-14.46
B	-0.85	0.63	0.920	-0.62	44.11	-27.31
w	0.77	0.57	0.872	0.27	100.00	26.57
l	-0.50	0.57	0.872	-0.36	100.00	-36.08

Overall loading

Equivalent leeward net force for overall section; $F_l = F_{w,wl} = \mathbf{-36.1}$ kN

Net windward force for overall section; $F_w = F_{w,ww} = \mathbf{26.6}$ kN

Overall loading overall section; $F_{w,w} = 0.85 \times (1 + C_r) \times (F_w - F_l + F_{w,h}) = \mathbf{57.2}$ kN

Roof load case 2 – wind 0, c_{pi} –0.3, +c_{pe}

Zone	External pressure coefficient, c_{pe}	Dynamic pressure, q_s (kN/m^2)	External size factor, C_{ae}	Net pressure, p (kN/m^2)	Area, A_{ref} (m^2)	Net force, F_w (kN)
A (+ve)	0.40	0.63	0.871	0.41	19.80	8.12
B (+ve)	0.30	0.63	0.871	0.36	9.23	3.28
C (+ve)	0.27	0.63	0.871	0.34	77.39	26.06
E (+ve)	−1.17	0.63	0.871	−0.45	19.80	−8.97
F (+ve)	−0.77	0.63	0.871	−0.23	9.23	−2.15
G (+ve)	−0.50	0.63	0.871	−0.09	77.39	−6.64

Total vertical net force; $F_{w,v} = \mathbf{18.51}$ kN

Total horizontal net force; $F_{w,h} = \mathbf{18.89}$ kN

Walls load case 2 – wind 0, c_{pi} –0.3, +c_{pe}

Zone	External pressure coefficient, c_{pe}	Dynamic pressure, q_s (kN/m^2)	External size factor, C_{ae}	Net pressure, p (kN/m^2)	Area, A_{ref} (m^2)	Net force, F_w (kN)
A	−1.44	0.63	0.920	−0.65	14.99	−9.72
B	−0.85	0.63	0.920	−0.30	44.11	−13.36
w	0.77	0.57	0.872	0.55	100.00	54.95
l	−0.50	0.57	0.872	−0.08	100.00	−7.71

Overall loading

Equivalent leeward net force for overall section; $F_l = F_{w,wl} = \mathbf{-7.7}$ kN

Net windward force for overall section; $F_w = F_{w,ww} = \mathbf{54.9}$ kN

Overall loading overall section; $F_{w,w} = 0.85 \times (1 + C_r) \times (F_w - F_l + F_{w,h}) = \mathbf{70.4}$ kN

Roof load case 3 – wind 90, c_{pi} 0.20, –c_{pe}

Zone	External pressure coefficient, c_{pe}	Dynamic pressure, q_s (kN/m^2)	External size factor, C_{ae}	Net pressure, p (kN/m^2)	Area, A_{ref} (m^2)	Net force, F_w (kN)
A (−ve)	−1.47	0.63	0.871	−0.93	5.32	−4.97
B (−ve)	−1.37	0.63	0.871	−0.88	5.32	−4.68
C (−ve)	−0.60	0.63	0.871	−0.46	42.57	−19.46
D (−ve)	−0.43	0.63	0.871	−0.37	159.63	−58.31

Total vertical net force; $F_{w,v} = \mathbf{-82.16}$ kN

Total horizontal net force; $F_{w,h} = \mathbf{0.00}$ kN

Walls load case 3 – wind 90, c_{pi} 0.20, –c_{pe}

Zone	External pressure coefficient, c_{pe}	Dynamic pressure, q_s (kN/m^2)	External size factor, C_{ae}	Net pressure, p (kN/m^2)	Area, A_{ref} (m^2)	Net force, F_w (kN)
A	−1.60	0.57	0.872	−0.90	10.00	−9.05
B	−0.90	0.57	0.872	−0.56	40.00	−22.35
C	−0.90	0.57	0.872	−0.56	50.00	−27.93
w	0.60	0.63	0.920	0.22	59.10	13.16
l	−0.50	0.63	0.920	−0.42	59.10	−24.67

Overall loading

Equivalent leeward net force for overall section; $F_l = F_{w,wl} = $ **−24.7** kN

Net windward force for overall section; $F_w = F_{w,ww} = $ **13.2** kN

Overall loading overall section; $F_{w,w} = 0.85 \times (1 + C_r) \times (F_w - F_l + F_{w,h}) = $ **32.7** kN

Roof load case 4 – wind 90, c_{pi} −0.3, $-c_{pe}$

Zone	External pressure coefficient, c_{pe}	Dynamic pressure, q_s (kN/m^2)	External size factor, C_{ae}	Net pressure, p (kN/m^2)	Area, A_{ref} (m^2)	Net force, F_w (kN)
A (−ve)	−1.47	0.63	0.871	−0.62	5.32	−3.29
B (−ve)	−1.37	0.63	0.871	−0.56	5.32	−3.00
C (−ve)	−0.60	0.63	0.871	−0.14	42.57	−6.00
D (−ve)	−0.43	0.63	0.871	−0.05	159.63	−7.83

Total vertical net force; $F_{w,v} = $ **−18.91** kN

Total horizontal net force; $F_{w,h} = $ **0.00** kN

Walls load case 4 – wind 90, c_{pi} −0.3, $-c_{pe}$

Zone	External pressure coefficient, c_{pe}	Dynamic pressure, q_s (kN/m^2)	External size factor, C_{ae}	Net pressure, p (kN/m^2)	Area, A_{ref} (m^2)	Net force, F_w (kN)
A	−1.60	0.57	0.872	−0.62	10.00	−6.21
B	−0.90	0.57	0.872	−0.27	40.00	−11.00
C	−0.90	0.57	0.872	−0.27	50.00	−13.75
w	0.60	0.63	0.920	0.54	59.10	31.85
l	−0.50	0.63	0.920	−0.10	59.10	−5.98

Overall loading

Equivalent leeward net force for overall section; $F_l = F_{w,wl} = $ **−6.0** kN

Net windward force for overall section; $F_w = F_{w,ww} = $ **31.8** kN

Overall loading overall section; $F_{w,w} = 0.85 \times (1 + C_r) \times (F_w - F_l + F_{w,h}) = $ **32.7** kN

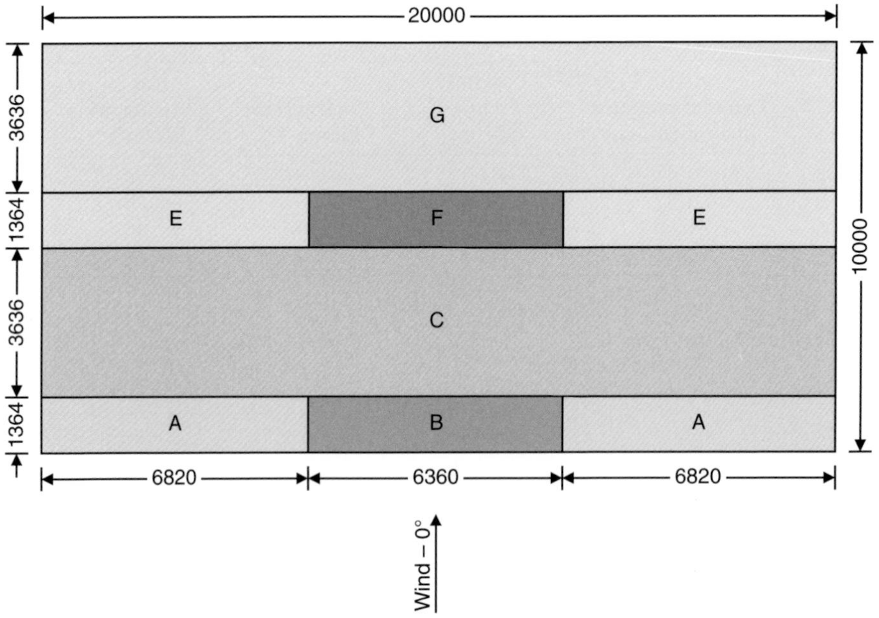

Plan view – Duopitch roof

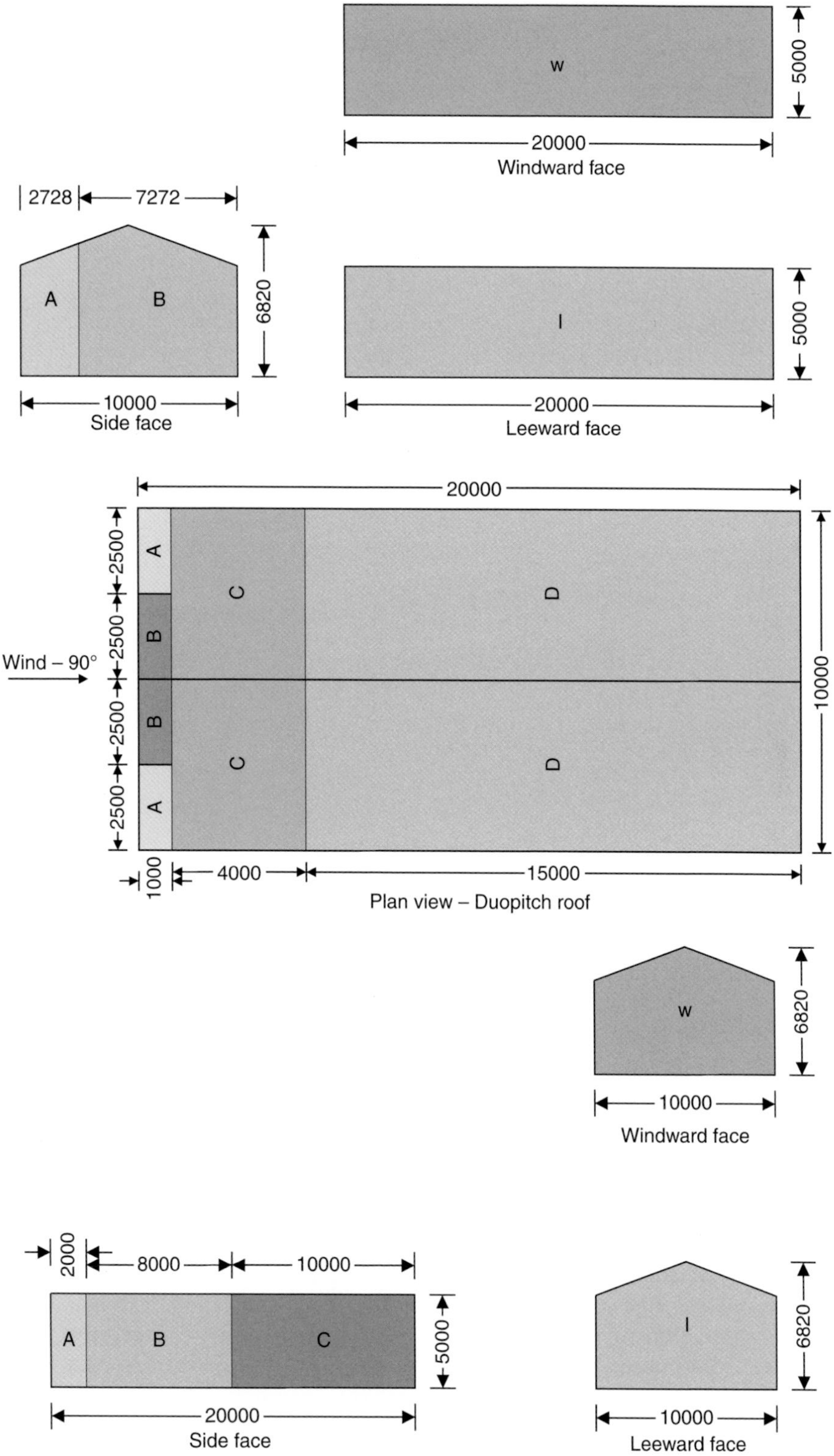
w
5000
20000
Windward face
2728
7272
A
B
6820
10000
Side face
I
5000
20000
Leeward face
20000
A
B
B
A
C
C
D
D
2500
2500
2500
2500
10000
Wind – 90°
1000
4000
15000
Plan view – Duopitch roof
w
6820
10000
Windward face
2000
8000
10000
A
B
C
5000
20000
Side face
I
6820
10000
Leeward face

As can be seen, the calculations provide a series of loads for different wind loading scenarios for the walls and roof of the structure. These loadings can then be used in conjunction with the other loads to design the components of the building.

In the Eurocodes there are further actions or loads to consider.

Accidental actions

These actions include such actions as fire, explosions or vehicle impact and are considered under the Eurocode design standards.

Seismic action

This action is as a result of earthquake forces.

Under Eurocode EN 1990, permanent, variable, accidental and seismic actions are classified depending on their variation over time.

BS EN 1991: Actions on structures EC1

This publication includes the weights of materials and loadings for occupancy and usage previously found in BS 648 and BS 6399. This latest publication under the new Eurocodes also provides information on actions for the structural design of buildings and civil engineering structures.

Combinations of load and factors of safety

There are different combinations of the above loads which can be applied to a structure, and these are usually factored as required under the relevant code of practice depending on the load type and combination. Under the British Standard design codes dead loads are factored by 1.4 and imposed loads by 1.6. If, however, wind load is also experienced, the factor of safety is 1.2 across the dead, imposed and wind loadings.

Under the Eurocodes the factors of safety are different and the maximum values for permanent and variable loads are 1.35 and 1.5, respectively, compared with 1.4 and 1.6 under the British Standards. Furthermore, other aspects can influence the partial safety factors to be used in the permanent loads – for example if the designer is checking for equilibrium and strength the partial safety factor is 1.1 and 1.35, however if the designer is checking only for equilibrium the partial safety factor is 1.1 or 0.9 depending on the risk of instability. If the risk is decreased this is said to be a favourable action but if the risk is increased it is said to be unfavourable.

Manufacturers of lintels and other products will provide load capacities for their specific product. This means that the loads which the product can sustain have already been calculated and include a factor of safety. These will often be referred to as the safe working load. When specifying these products the loadings need to be calculated and compared against the safe working load for the product to ensure the structural strength is not compromised.

Stress

Stress (σ) is the intensity of a force measured in N/mm^2 and forces can be compressive or tensile, when caused by bending known as bending stresses. Shear stresses are covered later.

$$\text{Stress is defined as } \sigma = \frac{\text{Force applied to an object}}{\text{Cross-sectional area of the object}} \tag{2.3}$$

Strain

When a stress acts on an object it will cause a deformation and the object is said to be under strain (ϵ). This strain is measured in the same direction as the stress which causes the strain, and is defined by the following formula:

$$\epsilon = \frac{\text{Change in length of object}}{\text{Original length of object}} \tag{2.4}$$

Young's modulus or modulus of elasticity

This is a measure of a material's resistance to deformation in compression, tension or bending. The member under examination has to be within its elastic limit; this means that once the load is removed, the member returns to its original shape or length.

If a point is reached where the load applied to a member is such that it does not return to its original shape, the member is said to have reached its plastic limit and the member becomes "plastic" in that it continues to deform or the deformation is irreversible. At this point the section has passed its yield point. If a graph of stress versus strain is plotted, the point of yield can be seen as in Figure 2.1.

$$E = \text{Young's modulus} = \frac{\sigma}{\epsilon} \tag{2.5}$$

The graph in Figure 2.1 shows that the yield stress is typically 250 N/mm^2 and the Young's modulus – which is the gradient of the straight line, also known as the modulus of elasticity – is 200 kN/m^2.

Plastic deformation

From the graph in Figure 2.1 we can see following the yield point that the material is plastic, and plastic deformation will occur. The graph plateaus between the elastic and plastic regions and there is an area where the strain increases under the same stress.

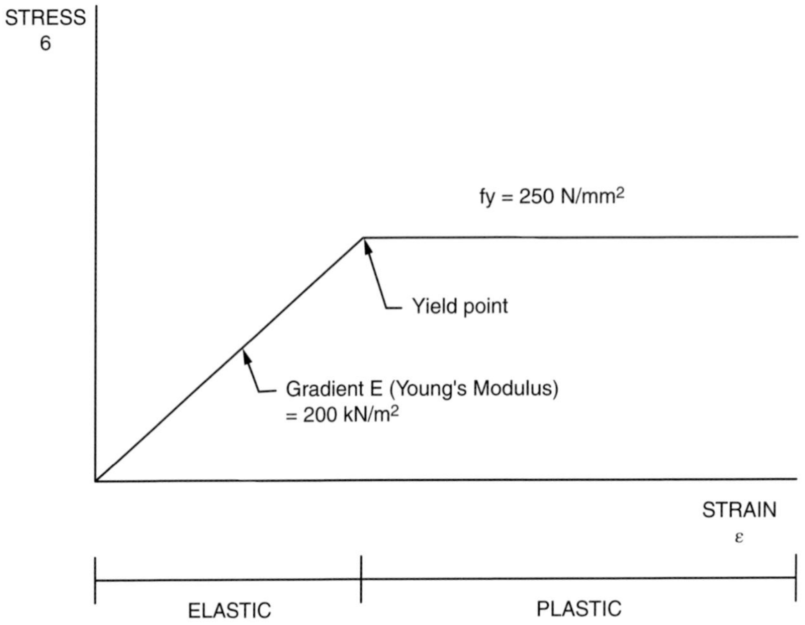

Figure 2.1: Graph showing idealised stress vs. strain relationship for steel.

After this point the material will continue to deform and absorb the stress until it reaches a point of ultimate strength. This period is called the strain hardening.

Following the point of ultimate strength the material is said to undergo necking and is characterised by a reduction in cross-section. The strain quickly increases since the material can no longer cope with the maximum stress applied. Eventually the cross-sectional area decreases and the material fractures.

The elastic range is the basis for steel design, however, plastic design can be undertaken which is based on the section of the graph between the yield point and where the strain hardening commences. This design utilises the strength of the material beyond the elastic limit due to the redistribution of internal forces.

Interestingly, plastic design has some advantages in the design of buildings against explosions. The area under the stress–strain curve is the amount of energy that can be absorbed per unit volume and is known as the toughness or resilience of a material. If we consider the plastic design, the area is much greater than the elastic envelope and consequently the material will be able to sustain a greater amount of energy. In 1938 William Peterson and Oscar Karl Kerrison designed the Anderson shelter in readiness for World War II. This comprised galvanised corrugated steel sheets secured together to form a robust and ductile structure. This shelter had advantages over concrete bunkers in terms of absorbing energy from blasts, since the steel absorbed energy through plastic deformation.

Buckling

This mode of failure will be discussed later in more detail but is a measure of the slenderness of the member under consideration. For example, a short stocky member will be able to resist a larger load and will eventually fail in compression. A slender thinner member will fail through bowing or bending before failing by being crushed, and this is known as buckling.

The buckling length is the length between the pins or points of contra-flexure of the member under consideration. In simple terms the buckling length is the length between the points at which it is pinned. However, the lateral restraint and end conditions will also affect the effective buckling length – for example a beam or column which has a pinned connection will have an effective buckling length equal to its length, but a beam or column that is fixed will have a buckling length equal to half its length as having a fixed connection makes the beam or column much stronger (the slenderness ratio is less). See "connections and restraints" later in this chapter.

Local buckling

This is the same as buckling, but where a member under compressive stress fails in a localised area.

Second moment of area

This is a geometric property of a section and is a measure of its resistance to bending and deflection. For example, for a rectangular section of breadth b and depth d the second moment of area about the x-axis through the centroid is given by $bd^3/12$ and for a circular section of diameter d the second moment of area about the x-axis is $\frac{\pi d^4}{64}$. See Figure 2.2.

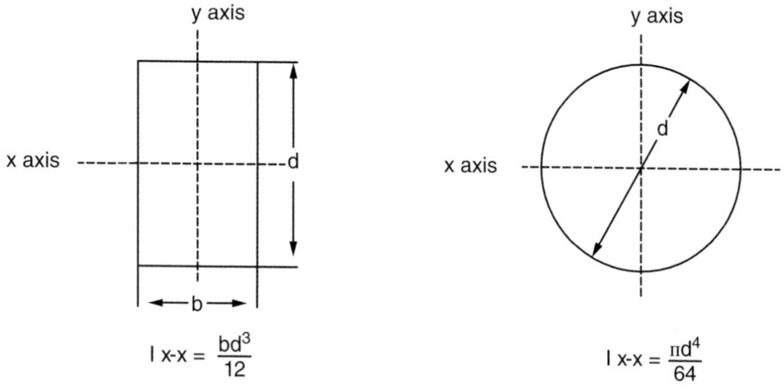

Figure 2.2: Second moment of area for a rectangular and circular section.

Centre of gravity

This is the point from which the weight of the body is considered to be acting. It is the point where a single force could support the body, or where a body would remain in equilibrium if supported at this point.

Lateral torsional buckling

Let us consider a beam under compression along its length and across the depth of the beam, consider it failing like a column. Thus, the top of the beam is under compression and will buckle, however, the underside of the beam remains in tension and stays straight. Eventually the top of the beam that is in compression will buckle sideways, causing twisting of the beam. This is known as lateral torsional buckling.

Neutral axis

This is the axis of the cross-section of a member at which the bending stress and bending strain are zero. The stress and strain diagrams are directly proportional to each other. The neutral axis will pass through the centroid of the section. For example, if a rectangular beam is loaded transversely across the top, the top fibres will be in compression and the bottom fibres will be in tension. At some point across the section of the beam the fibres change from being in compression to being in tension, and the net effect is zero. This point is known as the neutral axis. See Figure 2.3.

Bending force

Bending is a force which causes an object to bend. For example, if a plastic ruler spans between two clamps and a load is applied to the centre it will bend. The force causing the ruler to bend is called bending force.

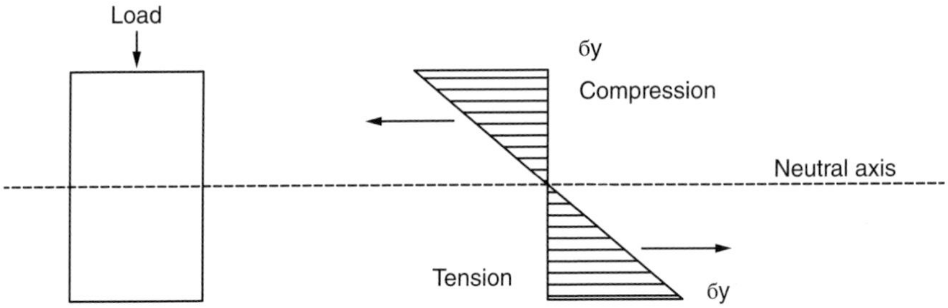

Figure 2.3: Neutral axis along a rectangular section under a transverse load.

Shear force and bending moment

Shear force

The shear force is defined as the sum of the vertical forces to the right- or left-hand side of the section being considered.

In examining the behaviour of structures and the structural members that make up a building structure we must understand bending moments, shear forces and deflections. These can be calculated and, for standard load cases, formulae are available to determine these parameters. Once calculated this allows us to analyse such structures.

With this in mind let us examine these parameters in more detail.

Bending moment

A moment is a force multiplied by the perpendicular distance from the fulcrum. The bending moment is an internal moment or reaction when a force is applied to a structural element causing it to bend. This can be calculated as the sum of all the moments of the forces to the left or right of a section.

Deflection

The deformation or deflection of a structure caused by the addition of a load is an important consideration, and the codes of practice place limits on the deflection of members such as beams, usually expressed as a fraction of the length of the member under consideration. The deflection is important for three reasons:

- Excessive deflection could cause damage to finishes such as plaster.
- Deflections may cause excessive movement in the frames or elements of a building which accommodate glazing, for example.
- Finally, the users of the building may find the deflection uncomfortable if these are allowed to be excessive.

Simple calculations can be undertaken to calculate the bending moment, shear force and deflection within a structural element provided the element under consideration is statically determinate. To be statically determinate there has to be a number of equations able to determine the unknowns, and this will invariably involve the solution of simultaneous equations. In the case of a beam the unknowns are the shear force and the bending moment, which can be solved using the three equations of equilibrium as shown below. Hence, the number of unknowns is less than the number of equations of equilibrium.

Statically determinate structures will be covered later in this book.

Static equilibrium

There are three equations of static equilibrium:

$$\textbf{The sum of the vertical forces } \sum \mathbf{FV} = \mathbf{0} \quad (+\mathbf{ve})\uparrow \tag{2.6}$$

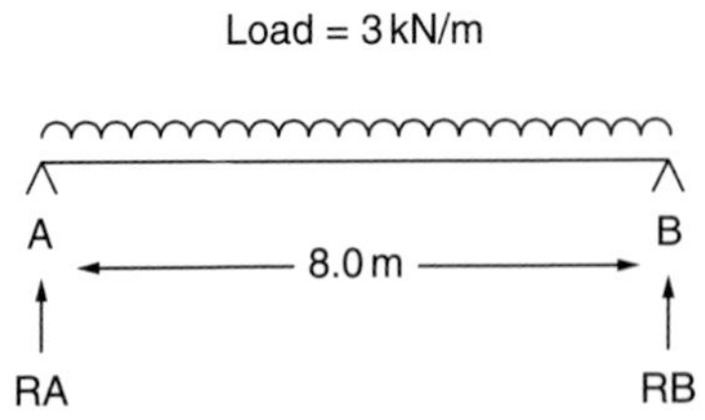

Figure 2.4: A simply supported beam, 8 m long and loaded with a uniformly distributed load of 3 kN/m.

$$\textbf{The sum of the horizontal forces} \sum \mathbf{FH} = \mathbf{0} \quad (+\mathbf{ve}) \rightarrow \tag{2.7}$$

$$\textbf{The sum of the moments} \sum \mathbf{M} = \mathbf{0} \quad (+\mathbf{ve}) \circlearrowright \tag{2.8}$$

(+ve) means positive direction. Hence, vertically upwards is positive and downwards is negative.

These equations allow us to determine the reactions acting on a member if we consider the simply supported beam in Figure 2.4 with a uniformly distributed load of 3 kN/m and a length of 8.0 m.

Applying the equilibrium equations (2.6), (2.7) and (2.8):

The sum of the vertical forces using equation (2.6)

$$\sum \mathbf{FV} = \mathbf{0} \quad (+\mathbf{ve}) \uparrow$$

Hence $RA + RB - (3 \times 8) = 0$

The sum of the moments using equation (2.8)

$$\sum \mathbf{M} = \mathbf{0} \quad (+\mathbf{ve}) \circlearrowright$$

If we take moments about A or B, this eliminates one of the reactions and thus the moments about A give

$$-RB \times 8 + [(3 \times 8) \times (8/2)] = 0$$

$$RB = 12\,\text{kN}$$

Clearly the beam under analysis is symmetrical, hence one would expect RA = RB. However, for completeness, if we put the value determined for RB into equation (2.6) for vertical equilibrium, we can see that RA = 12 kN as expected.

Internal forces

To determine the internal forces, the beam is cut and this is achieved by using a section line. A section line can be taken at any position along the length of the member under consideration and this cuts the member into a left and a right component – giving two free bodies from which the forces and moments can be determined. This is illustrated in Figure 2.5.

The internal shear force at 4 m along the beam is denoted by V_{4m}.

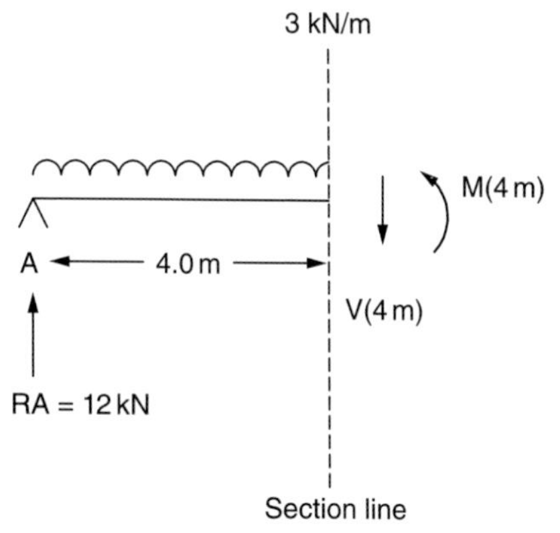

Figure 2.5: Section line through beam.

The bending moment at 4 m along the beam is denoted by M_{4m}.

Considering the left-hand side of the beam and applying the equations of equilibrium, using equation (2.6)

$$\sum \mathbf{FV} = \mathbf{0} \quad (+\mathbf{ve}) \uparrow$$

$$RA - (3 \times 4) - V_{4m} = 0$$

We know RA = 12 kN. Hence, substituting into equation (2.6) the above provides $V_{4m} = 0$.

Hence the shear force (internal force) at 4.0 m along the beam measured from A is zero.

Therefore, substituting into the equation above:

$$\sum \mathbf{M} = \mathbf{0} \quad (+\mathbf{ve}) \circlearrowright$$

$$RA \times 4 - (3 \times 4 \times 2) - M_{4m} = 0$$

We know that RA = 12 kN. Therefore, substituting into equation (2.8):

$$48 - 24 = M_{4m}$$

Hence the bending moment 4.0 m from A is 24 kN·m.

We can take a section at any distance along the beam, as shown in Figure 2.5, and the bending moment and shear force can be determined for each distance. If this is undertaken at 1.0 m intervals we can compile the results in Table 2.2.

A positive moment is called a sagging moment. The values found can be plotted on a graph, with the distance along the section under analysis being the x-axis and the y-axis being the bending moment and the shear force, respectively. Figures 2.6 and 2.7 plot these graphs, which are known as bending moment diagrams and shear force diagrams.

Interestingly, we can see that as a rule where the shear force is zero the bending moment is a maximum and vice versa. Hence, in the centre of the beam, 4.0 m from A, the bending moment is a maximum at 24 kN·m and the shear force is zero.

Table 2.2: Shear force and bending moment at 1.0 m intervals along the beam

Distance from A (m)	Shear force (kN)	Bending moment (kN·m)
0	12	0
1	9	10.5
2	6	18
3	3	22.5
4	0	24
5	-3	22.5
6	-6	18
7	-9	10.5
8	-12	0

Distance	Bending
0	0
1	10.5
2	18
3	22.5
4	24
5	22.5
6	18
7	10.5
8	0

Distance	Shear
0	12
1	9
2	6
3	3
4	0
5	−3
6	−6
7	−9
8	−12

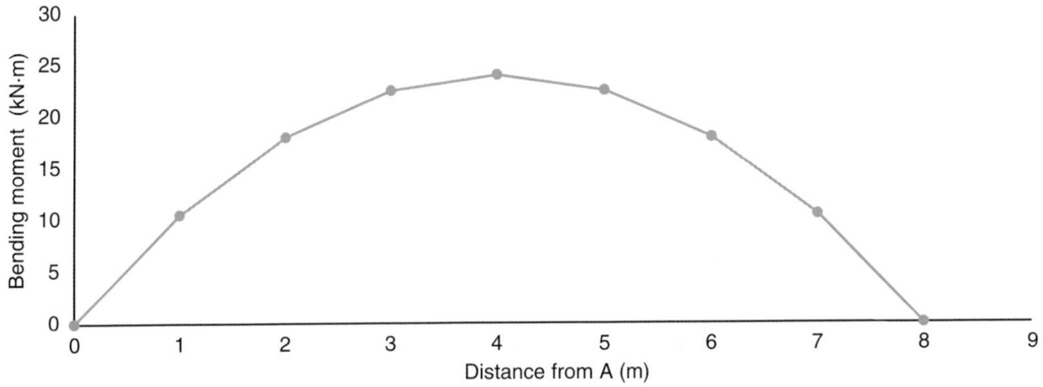

Figure 2.6: The bending moment graph.

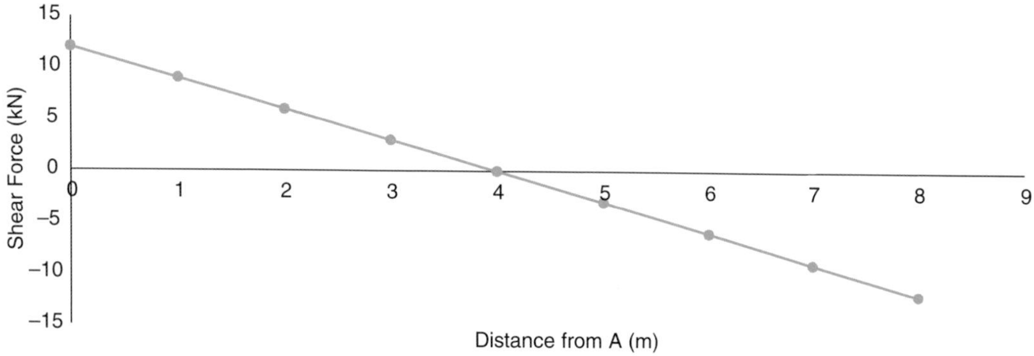

Figure 2.7: The shear force graph.

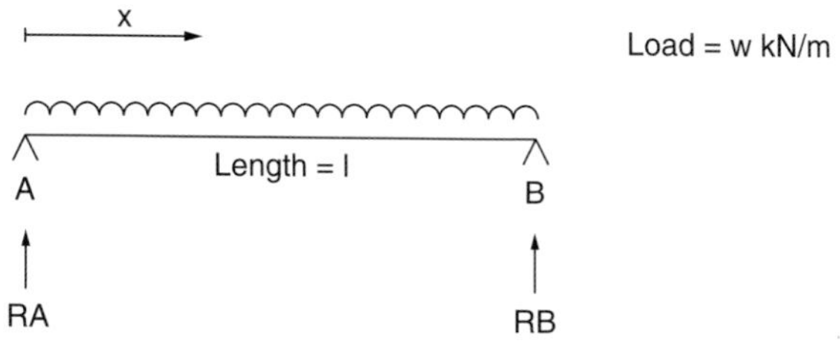

Figure 2.8: Derivation of shear force diagram.

Knowing the above, let us examine a simply supported beam considering the shear force, bending moment and deflection at any point along the beam and examine the formulae in some more detail.

Derivation of shear force

The derivation of shear force can be undertaken by considering a beam of length L with a uniformly distributed load of w (kN/m). A diagram of this can be seen in Figure 2.8.

From our knowledge of shearing force gained from the above example, we can see that the reaction and maximum shear force for a uniformly distributed load along a beam would be wL/2.

Derivation of bending moment

As stated above, the bending moment is defined as the sum of all the moments of the forces to the left or right of a section. In the above example, and as seen in Figure 2.8, the moment M at a distance x along the beam becomes

$$M = RAx - wx\tfrac{1}{2}x \tag{2.9}$$

$$M = wL\tfrac{1}{2}x - wx\tfrac{1}{2}x \tag{2.10}$$

where wL/2 is the reaction at RA.

The bending moment is maximum at x = L/2, thus substituting into the above equation (10):

$$M_{max} = \frac{wL^2}{4} - \frac{wL^2}{8} \tag{2.11}$$

$$M_{max} = \frac{wL^2}{8} \tag{2.12}$$

This is the maximum bending moment for a uniformly distributed load on a beam.

Derivation of deflection

Taking the moment above, where the moment at any point x along the beam is provided by the following:

$$M = \frac{wx(L-x)}{2} = w/2\left(Lx - x^2\right) \tag{2.13}$$

the maximum bending moment occurs when

$$d^2y/dx^2 = M/EI = \frac{w}{2EI}\left(Lx - x^2\right) \tag{2.14}$$

Here:
E = Young's modulus (or modulus of elasticity)
I = second moment of area

Integrating the above:

$$dy/dx = w/2EI\left(\frac{Lx^2}{2} - \frac{x^3}{3}\right) + c \tag{2.15}$$

The boundary conditions are
dy/dx = 0 when x = L/2 and C = $-L^3/12$.

$$\text{Therefore } dy/dx = \frac{w}{2EI}\left(\frac{Lx^2}{2} - \frac{x^3}{3} - \frac{L^3}{12}\right) \tag{2.16}$$

Integrating again

$$y = \frac{w}{2EI}\left(\frac{Lx^2}{6} - \frac{x^4}{12} - \frac{L^3x}{12}\right) + c \tag{2.17}$$

The boundary conditions are that $y = 0$ when $x = 0$ and $C = 0$.
Therefore, substituting into the above equation (2.17) provides
y = a maximum when $x = L/2$
Replacing these values in equation (2.17) provides us with

$$y = \frac{w}{2EI}\left(\frac{L^4}{48} - \frac{L^4}{192} - \frac{L^4}{24}\right) \tag{2.18}$$

Rearranging provides us with the following:

$$y = \frac{5wL^4}{384EI} \tag{2.19}$$

where wL = total load along the beam, W. Thus this can be rewritten as

$$y = \frac{5WL^3}{384EI} \tag{2.20}$$

Thus the deflection for a beam with a uniformly distributed load is solved.

Formulae are provided in Figure 2.9 for the more common load cases showing the deflection, bending moments and shear forces for various loading situations and end restraints. These are calculated in a similar manner to the above analysis for a uniformly distributed load on a simply supported beam.

Basic theory of bending

Let us consider a beam in cross-section rather than along the length of a beam.

If a beam is subjected to a pure bending moment – that is a constant bending moment along the length of the beam with no shear force – then the fibres in the lower section of the beam are extended and those in the upper part of the beam are compressed. See Figure 2.10.

Tensile and compressive stresses, known as bending stresses, are induced across the beam section, producing a moment of resistance. This moment of resistance is equal and opposite to the applied bending moment.

At some depth across the section the fibre lengths are unchanged, as they move from compression to tension, and this is known as the neutral plane. The transverse axis lying in the neutral plane is the neutral axis of bending of the section.

The following formula applies to this situation:

$$M/I = E/R = \sigma/Y \tag{2.21}$$

M = moment of resistance

I = second moment of area (a geometrical property based on the cross-section; for a rectangular section it is $bd^3/12$, where b = breadth and d = depth)

E = Young's modulus

Beam/Loading	Bending	Shear force	Deflection
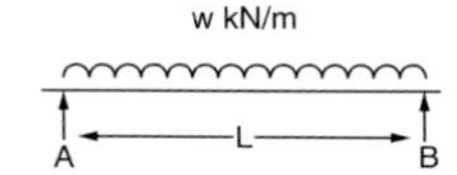	$\frac{wL^2}{8}$	$\frac{wL}{2}$	$\frac{5wL^4}{384EI}$
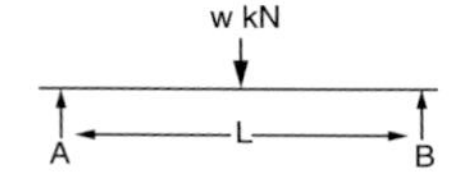	$\frac{wL}{4}$	$\frac{L}{2}$	$\frac{wL^3}{48EI}$
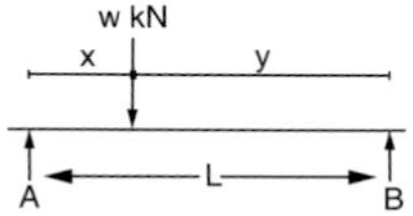	$\frac{wxy}{L}$	Va $\frac{wb}{L}$ Vb $\frac{wa}{L}$	$\frac{wL^3 [(3x/L) - (4x^3/L)]}{48EI}$
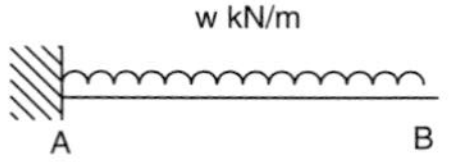	$\frac{-wL^2}{2}$	wL	$\frac{wL^4}{48EI}$
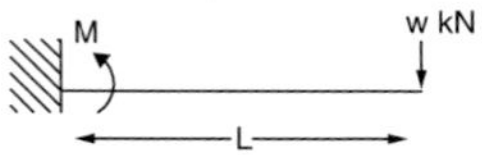	$-wL$	w	$\frac{wL^3}{3EI}$
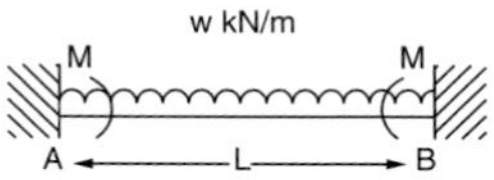	Maximum negative bending moment MA $= \frac{-wL^2}{12}$; Maximum negative moment at mid span $= \frac{-wL^2}{24}$	$Va = \frac{wL}{2}$	$\frac{wL^4}{384EI}$
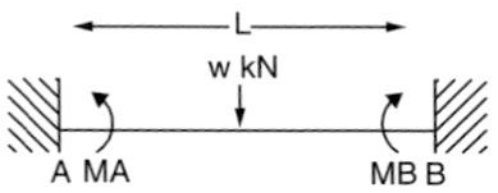	Support bending moment $MA = \frac{-wL}{8}$, $MA = \frac{+wL}{8}$; Maximum positine bending moment at mid span $\frac{+wL}{8}$	$\frac{w}{2}$	$\frac{wL^3}{192EI}$

Figure 2.9: Bending moments, deflection and shear force for standard load cases.

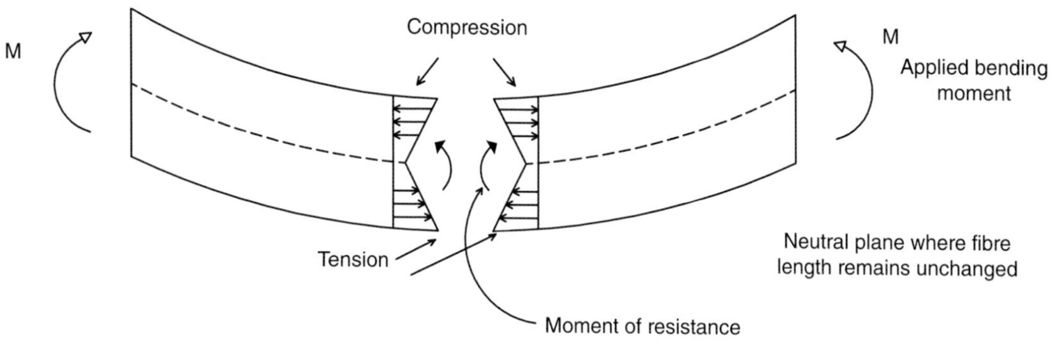

Figure 2.10: Compressive and tension stresses through a beam cross-section.

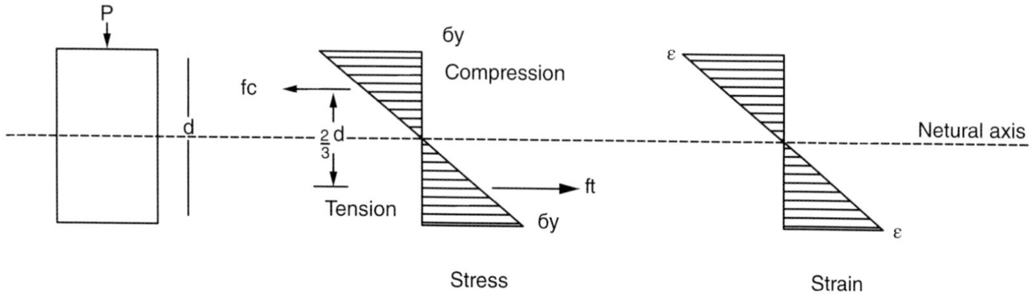

Figure 2.11: Bending stress and strain across a rectangular beam section.

R = radius of curvature of the bend section
Y = distance from the neutral axis of bending
σ = stress induced in the section
Further definitions of the above have been provided earlier in this chapter.

If we consider a rectangular beam section with an applied load, the resultant bending stress and strain can be seen in Figure 2.11: it is a maximum at the extreme fibres and zero at the neutral axis. This can be seen using the formula of equation (2.21).

This shows that not only do we have a bending moment and shear force distribution along the length of the beam, but also a stress and strain distribution across the depth of the beam. The strain is dimensionless, hence has no units, and is proportional to the stress.

Moment of resistance

It can be seen from Figure 2.11 that if the compressive and tension forces are a distance apart and acting against each other, they form a moment which resists the applied moment. Owing to the triangular stress distribution, the distance between the tension and compressive forces Ft and Fc is taken as 2d/3. The moment is equal to the force multiplied by the distance. In this case the equation for the moment of resistance can be derived as follows:

$$\text{Force} = \text{Stress} \times \text{Area} = \frac{\sigma_y}{2} \times \frac{bd}{2} = \frac{\sigma_y bd}{4} \tag{2.22}$$

$$\text{Thus the moment of resistance becomes } Mr = \frac{\sigma_y bd}{4} \times \frac{2d}{3} \tag{2.23}$$

$$\text{Thus the moment of resistance } Mr = \frac{bd^2\,\sigma_y}{6} \tag{2.24}$$

Since at equilibrium Mr = M (i.e., the moment of resistance = applied moment),

$$\text{we can derive that } M = \sigma_y Z \tag{2.25}$$

$$\text{where } Z \text{ is known as the section modulus and is equal to } \frac{bd^2}{6} \tag{2.26}$$

$$\text{This can be rewritten as } M = \frac{\sigma_y I}{Y} \tag{2.27}$$

Therefore, if we know the yield strength of a material σ, we can determine the dimensions of a section b and d that are required to resist a particular moment M.

Combined bending and direct stress

A structural member subject to a pure bending moment along its length, with zero shear force, means that the fibres in the lower part of the section are extended and those in the upper part are compressed. As we have seen above, compressive and tensile stresses are induced which produce a moment called the moment of resistance. This is equal and opposite to the applied bending moment.

Direct load and moment will create a stress distribution through the depth of the section. The formula for the combined stress analysis is as follows:

$$\sigma = \frac{P}{A} \pm \frac{M_x}{Z_x} \tag{2.28}$$

External and internal statically determinate structures

In the case of external statically determinate structures, the number of equilibrium equations has to be equal to the number of unknowns. If there are more unknown forces than equilibrium equations, then the structure is statically indeterminate and additional equations have to be found by considering the displacement or deformation of the body. If the number of unknowns is matched by the number of equilibrium equations, the structure is said to be externally statically determinate and the equations of static equilibrium seen earlier in the chapter can be used to solve the unknown reactions. The degree of indeterminacy is the number of unknown forces in excess of the equations of statics.

However, we also have to consider whether the overall structure – for example a steel frame – is internally statically determinate and this determines if the overall structure contains redundant members and if the structure is stable. This is found from the following equation:

$$m + r = 2j \tag{2.29}$$

where
m = number of members
r = number of reactions
j = number of joints

If equation (2.29) holds true for the structure, that structure is stable.

When m + r > 2j the structure is considered to have redundant members.

When m + r < 2j the structure is considered unstable.

The above describes the internal statical determinacy of the structure.

Consider the structures in Figure 2.12, where the structural determinacy has been shown. Taking example (a), there are three external reactions: VA, VB and HA. There are four members and four joints. Therefore, applying equation (2.29) the sum becomes 2 × 4 = 3 + 4 and the structure is considered unstable.

If we examine Figure 2.12(c), the structure is statically determinate externally since there are three unknown reactions HA, VA and VB and three equations of equilibrium. However, internally the forces in the members cannot be determined by the equations of equilibrium alone and the truss is statically indeterminate to first degree since

$$m + r = 9 > 2 \times j = 8 \tag{2.30}$$

Connections and restraints

There are a number of ways in which a member can be restrained, and this affects how the structural element behaves. Let us consider the different restraint types and provide some examples of what this means in practical terms.

Roller

An example of where a roller connection may be employed is at the end of bridges to allow for expansion and contraction. These are effectively rubber bearings, rockers or rollers and allow movement in the horizontal plane. Consequently they do not provide resistance to lateral forces such as wind or earthquake forces. The roller support has one reaction and in the example shown in Figure 2.13 this is a vertical reaction. The connection can support a vertical load and remains in place supporting itself. If there is a horizontal force, this type of connection will move in response to the force. A similar analogy is someone standing on roller skates.

Pinned

The pinned connection is the type of connection usually employed in a truss. The pinned connection can resist vertical and horizontal forces but not moments. This type of

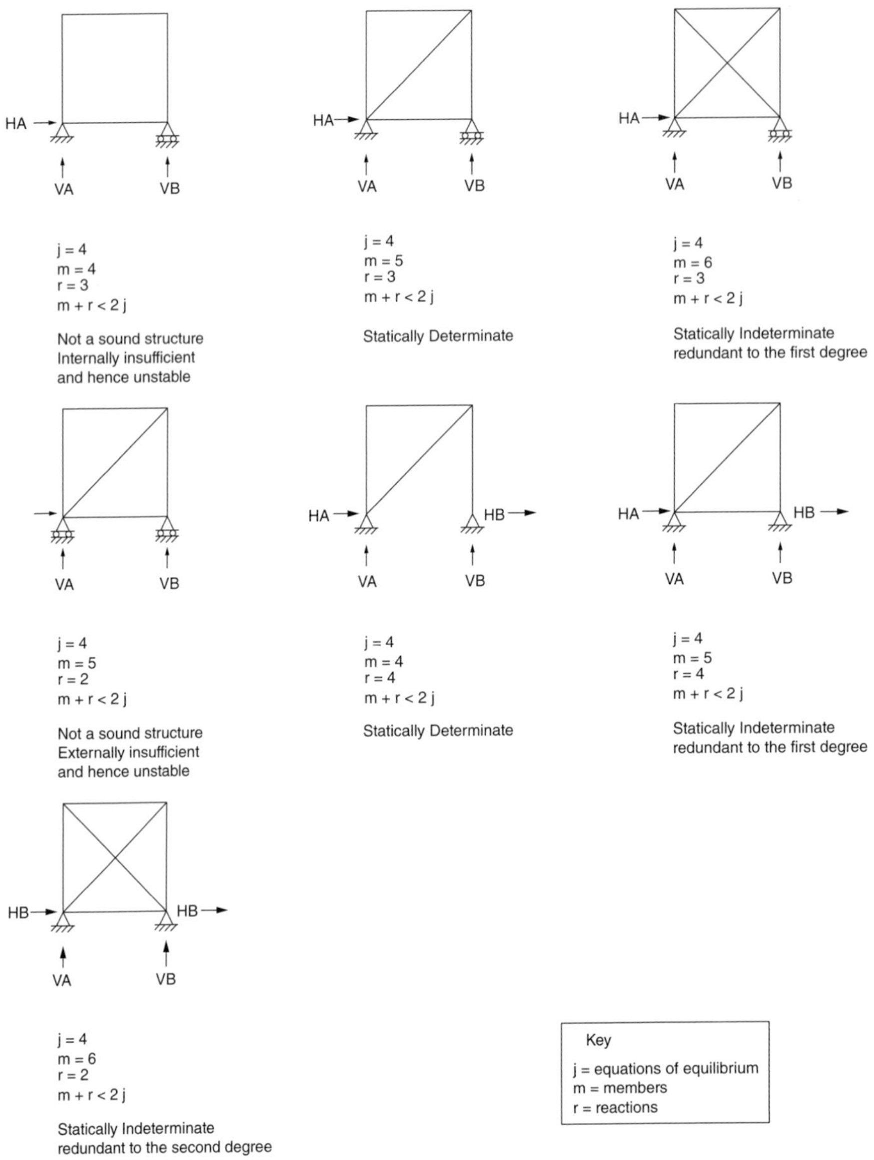

Figure 2.12: Statically determinate structures.

connection allows the connection to rotate in one direction but provides resistance in another. A pinned connection will have no resistance to a moment and consequently there will be no bending moment at a pinned connection. Take a door hinge for example, which is a pinned connection and when it is opened this allows rotation; the moment which causes the opening of the door is not resisted by the pin or hinge, thus there is no moment at the pin.

A pin connection is not usually sufficient to make a structure stable and another support must be provided to prevent rotation. An example of a pinned connection is seen in Figure 2.13.

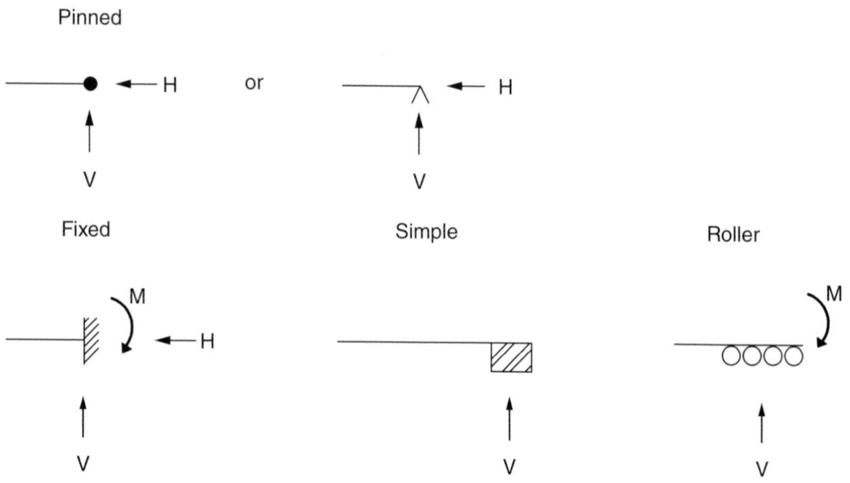

Figure 2.13: Representation of roller, pinned, fixed and simple connection.

Fixed or rigid

The fixed connection can be found in concrete structures that are cast in situ or on welded connections. These types of connection require careful consideration since they allow no rotation or translation, and therefore are often the source of most building failures.

The vertical, horizontal and moment forces are resisted and consequently this type of connection has three reactions: a force in the horizontal and vertical direction and a moment. An example of a fixed or rigid connection is seen in Figure 2.13.

Simple connection

An example of a simple connection is a plank lying across a surface. It is idealised by a frictionless surface. This type of connection is not usually found in buildings. Figure 2.13 shows a diagram of this arrangement.

The type of connection determines the type and magnitude of the load that can be resisted by the support or connection. For example, a pinned connection cannot resist a moment load.

The type of connection will determine the buckling length between the points of contra-flexure of the structural member under consideration. Chapter 8 demonstrates this and explains it in further detail with the aid of a simple experiment for a pinned-end and fixed-end timber dowel.

The Euler critical load is a measure of the maximum axial load a member can sustain and is dependent on the buckling length, thus if this is reduced the Euler critical load will increase – hence making the structural member stronger. If we consider a structural column of length L and assume it is connected via a pinned connection at each end, the effective length or effective buckling length is considered to be L.

$$\text{Thus, for a pinned structure the Euler critical load } Pe = Pc = \frac{EI\pi^2}{L^2} \quad (2.31)$$

where Pe is the Euler critical load and Pc is the critical load for the member.

If we consider the same structural member which has a fixed end, the effective buckling length reduces to 0.5L. This is an effective buckling length compared with that of a pinned-end member. The ratios of effective lengths compared with a pinned-end member can be shown by theoretical analysis outside the scope of this book.

$$\text{Thus, for a fixed-end member the critical load becomes } Pc = \frac{EI\pi^2}{(0.5L)^2} \quad (2.32)$$

$$\text{Equation (2.32) is equal to } P = \frac{4EI\pi^2}{L^2} \text{ or } 4Pe \quad (2.33)$$

Thus, within the elastic range the end connection makes a significant difference to the strength of the structural member. In the case of a fixed-end column the critical load is four times that if the same column has a pinned-end connection. There are more details and information about the Euler load and how it is applied to the analysis of walls in Chapter 8.

It should be noted that this is a theoretical analysis and although it shows that the end conditions and connections have an impact on the strength of a column or beam, a number of assumptions have been made that are not practical. For example, the analysis assumes that the column is perfectly elastic, the material is completely homogeneous and the load is applied centrally to the column.

Stiffness

Stiffness is given for a member by the term bd^3, where d is the depth of the section and b is the breadth. It is defined as the load which is required to cause a displacement, or in other words a measure of the resistance offered to displacement. Thus, the stiffness can be increased significantly by increasing the depth of a section. Timber laminate structures work on this principle by increasing the depth or width by gluing additional members to the structure. Composite members are also possible, such as flitch beams which comprise a steel plate sandwiched between timber sections.

If we take a structural beam and apply a point load at the midpoint along the beam, the stiffness can be found to be inversely proportional to the span cubed. Therefore, we have the relationship $S \alpha 1/L^3$, where S = stiffness and L = span of the beam. Thus, it can be seen that by halving the span the stiffness of the beam is increased eight times.

Taking this a step further, it can be shown that moments will redistribute themselves through a structure such as a building or a frame depending on the relative stiffness of the adjacent members; this is called moment distribution.

There is more information and an explanation on flexural stiffness in Chapter 8, which also provides details of a simple experiment showing how the end connection of a structural member can affect its strength. An example of the application of stiffness, end

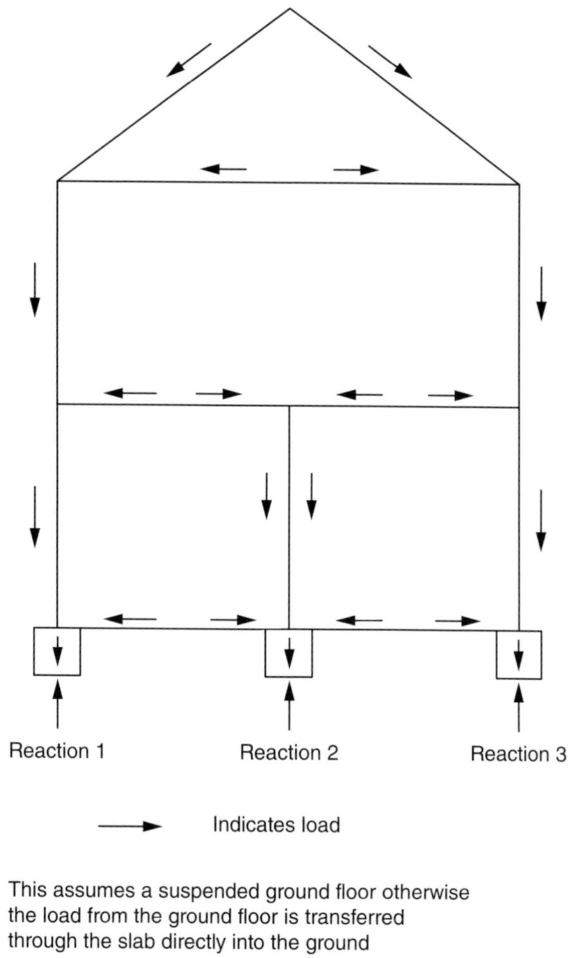

Figure 2.14: Load path through a dwelling.

restraints and the Euler load to a structural element is also demonstrated in Chapter 8, where we show the effect of this theory on a wall.

Buildings and load paths

Throughout history, mankind has realised that the manner and orientation in which building materials and components are employed will determine the strength and load-bearing capacity of the materials. Our knowledge has evolved through trial and error, and later through the advent of mathematical analysis and software packages that provide detailed structural models on which complex analysis can be undertaken. This has resulted in fantastic structures being constructed across the world.

To understand this basic concept, one has to have a basic understanding of load paths and how loads are transferred through a structure. Although we will not delve into moment distribution, suffice to know that the moment can be redistributed depending

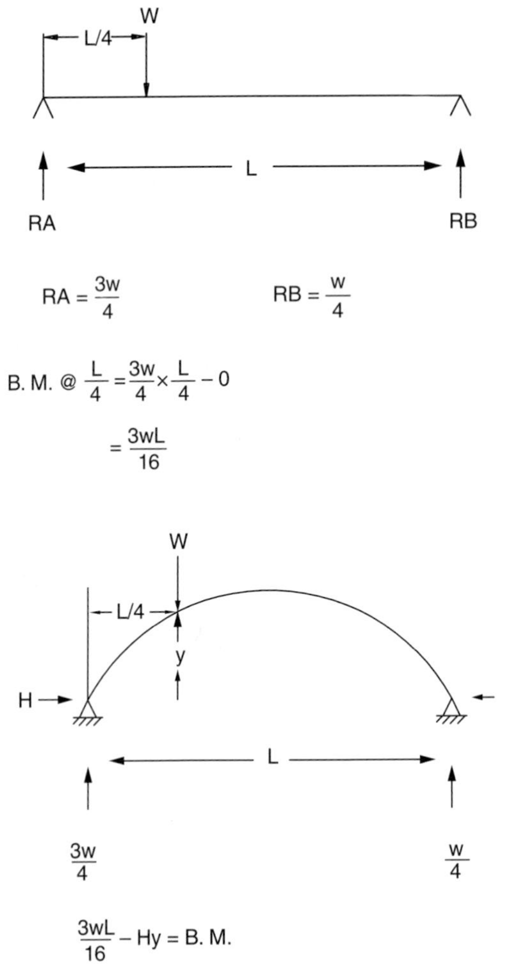

Figure 2.15: Arch analysis.

on the relative stiffness of adjoining structural members. For now, let us examine a basic load path of a house.

Loads are transmitted through the house from their point of impact, eventually leading to the foundation, where loads are transmitted to the ground. The loads from the roof will be distributed via the tiles to the rafters. The ceiling joists will transfer loads to prevent the roof from spreading. The wall will transfer ceiling and roof loads to the first floor and then "collect" the floor loads to the walls on the ground floor and on to the foundations. Eventually the load is transferred to the ground. Figure 2.14 illustrates this load path, following the arrows through the property. The loads may be wind, imposed (variable) or dead (permanent).

As mentioned above, the use and orientation of materials can influence the strength of a particular structural element. For example, if we analyse a piece of flat plate or a beam supporting a load as shown in Figure 2.15, it will experience the bending moment increase along its length. However, if the same thickness of material is used in an arch, horizontal reactions are incurred which reduce the design bending moment.

If we take the moment at L/4 along the arch or plate, we can compare the results.

$$\text{For a simply supported beam the bending moment is } 3wL/16 \tag{2.34}$$

$$\text{Considering a two-hinged symmetrical arch of span L, the bending moment is } 3wL/16 - Hy \tag{2.35}$$

Thus, by comparing equations (2.34) and (2.35) we can see that the bending moment for the arch is less than that of a simply supported beam.

Consequently, the manner in which we have used the plate has reduced the bending moment and allowed the structure to carry a heavier moment. It has been shown that the manner in which materials and structures are used, and how these structural elements are jointed and orientated, can have a huge impact on their strength. This is important since a structural member used in a certain way, or connected differently, can determine if the structure will sustain the given loads or fail.

Chapter 3 The Construction of Buildings

Anyone undertaking alterations to a property should understand the methodology and materials used in the construction of the building. In traditional properties, careful consideration should be given to the architectural heritage of the building to be renovated or altered to ensure the work reflects the character of the original building.

Breathable and non-breathable construction

Traditionally, buildings were constructed of breathable materials but in more recent times, during the past 100 years, building materials have been employed to seal buildings and prevent them from breathing.

In breathable buildings stone and brick were bonded together using lime mortars, which allowed damp and rain to pass a few millimetres through the walls; through the advent of wind and sun, the damp would then evaporate from the masonry. Internally, open fires and natural ventilation such as draughts would allow the dampness arising from the habitation of the rooms to disperse. Lime mortars and renders allowed for the passage of water and moisture possibly more so than the stone. In this scenario a house is said to breathe through the passage and removal of moisture.

In more recent constructions, houses have been sealed to prevent draughts – mainly to improve the insulating qualities – and damp is prevented from entering the building through damp-proof membranes and damp-proof courses. This is said to be non-breathable, since the damp is prevented from entering the fabric of the building.

When more recent technology is introduced to traditional construction, problems can arise and in the extreme, structural distress and failure can occur. One example is the extensive use of cement. This has been used to replace lime mortars and in addition applied as a render to improve the appearance of stone and brick masonry. This has been commonplace over the last 40–50 years, in some cases through the provision of grants to improve cottages and houses.

The problems associated with using cement renders are that cement is not breathable and consequently traps damp in the stone or brick masonry. This can result in the trapped damp freezing, leading to expansion of the water and subsequent spalling of the stone or brickwork. The introduction of cement mortars used in re-pointing also traps damp, causing structural damage to the stone and brickwork since moisture cannot easily escape from the masonry. If we take the example of sandstone, although it is porous, the pores are densely spaced and thus most of the evaporation of moisture is via the lime joints. If this is a cement mortar the moisture becomes trapped, causing structural

Structural Design of Buildings, First Edition. Paul Smith.

deterioration of the stone. Characteristically the stone will begin to powder and break down. Figure 3.1 shows a photograph of damage caused by cement pointing.

Other examples are where traditional stone flags are replaced by a damp-proof membrane and a concrete floor slab. This approach prevents the damp from escaping from the ground through the stone slabs and traps it in the ground. The damp migrates towards the shallow foundations and can in extreme cases cause a softening of the subsoil, loss of bearing under the walls and consequently subsidence.

Traditional timber frame houses can also suffer a similar fate when applying insulation to the walls and brick panels. This causes condensation to be trapped against the wall, which in turn can cause deterioration of the timbers through wet and dry rot.

Cob construction comprises walls constructed from clay subsoil mixed with straw. Clearly, if this material suffers from the advent of damp, then the clay in the wall will soften and can be destroyed in a relatively short space of time. In some areas it has been the practice to render such walls, which also traps damp and causes the constituents of the wall to degrade and collapse.

As can be seen, the construction of a property needs to be understood and fully appreciated prior to commencing any extensions, remedial works or alterations, and the impact of such activities considered and fully understood prior to embarking on such projects.

At this juncture it is worth considering some of the more common types of construction that are likely to be encountered.

Figure 3.1: Damage caused by cement pointing.

Timber frame

Historically, timber has been used in the construction of houses since the earliest dwellings were produced. The cruck frame was introduced in the 13th century and continued until the 16th century, when this was superseded by the box frame. Four types of traditional timber frame house construction exist, and these are the post and truss, box frame, cruck frame, and aisled construction. Figures 3.2–3.5 show the different types of construction.

Figures 3.2–3.5 are reproduced by kind permission of Mrs R.W. Brunskill (see Brunskill, 1994, p. 46).

The cruck frame comprises a tree, split to form a pointed bow. The timber was usually obtained from a bank where the tree had grown in a curve towards the sunlight.

Timber frame properties continued for a number of centuries and, although the material remained the same, the construction changed with the advancement of knowledge. The box frame continued to be used up until the 18th century.

The earlier Medieval houses have irregular timbers, whereas the much later properties of the 16th and 17th centuries have more regular sections of timber. More decorative panels were introduced in the 16th century to symbolise the wealth of the property owner. As the understanding of timber frames progressed into the 17th and 18th centuries, the sections became thinner. In the 17th century, although still in construction, timber frame houses began to decline in urban areas. In the 18th century, steel straps were introduced to replace the traditional peg joint construction. These are not to be confused with straps that have been added retrospectively as a repair. Other features, such as

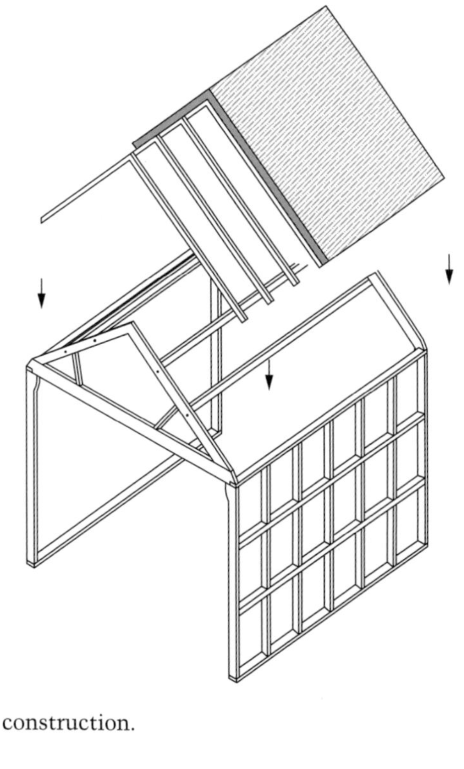

Figure 3.2: Post and truss construction.

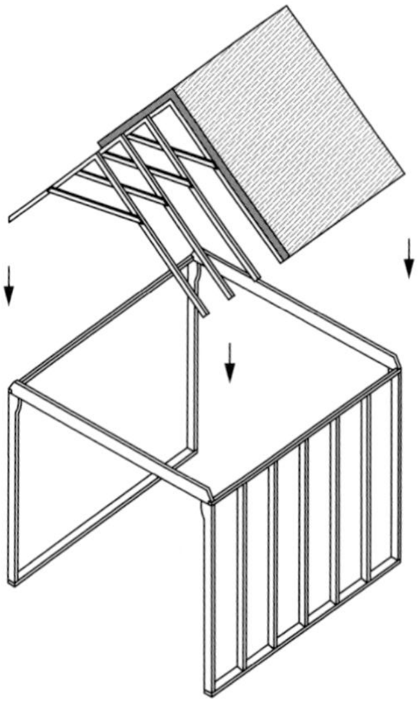

Figure 3.3: Box frame construction.

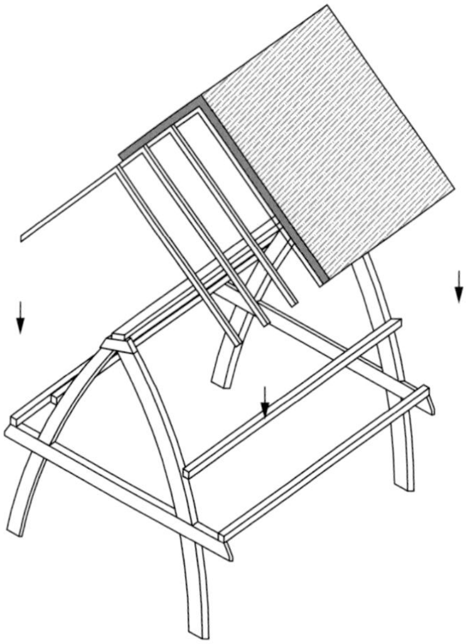

Figure 3.4: Cruck frame construction.

timber weather board and brick which covered the timber frame, are typical of this era. Another fashion during the 18th century was to encase the timber frame in stone or brick, which was sometimes rendered, creating a façade around the original timber frame. In some cases this practice has been undertaken at the front of the property

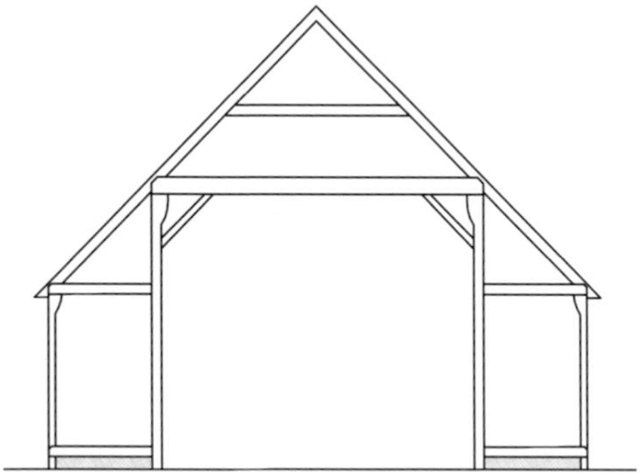

Figure 3.5: Aisled construction.

and the side of the property still exhibits the original timber frame. This can be seen in some properties in Worcester.

Figures 3.6–3.11 are photographs showing traditional timber frame houses of different ages, as listed below.

- Figure 3.6 is a photograph of the internal cruck frame of Leigh Court Barn in Worcestershire, dated 1300.

Figure 3.6: Internal cruck frame of Leigh Court Barn in Worcestershire, dated 1300.

- Figure 3.7 is a photograph taken in Greyfriars, Worcester and shows a building dating from 1480.

Figure 3.7: A building in Greyfriars, Worcester dating to 1480.

- Figure 3.8 is a photograph of a timber frame building circa 1500s with an extension to the right-hand side dated 1718.

Figure 3.8: A timber frame building circa 1500s with an extension to the right-hand side dated 1718.

- Figure 3.9 is a photograph showing a timber frame construction in Shrewsbury circa 1500s with decorative timber panelling.

Figure 3.9: A timber frame construction in Shrewsbury circa 1500s with decorative timber panelling.

- Figure 3.10 is a photograph of a property constructed circa 1600s; note the convex knee brace and queen post truss.

Figure 3.10: A property constructed circa 1600s with convex knee brace and queen post truss.

• Figure 3.11 is a photograph of a traditional timber frame from the 1700s.

Figure 3.11: A traditional timber frame from the 1700s.

The upright columns in traditional frame construction are referred to as posts and the beams as rails. These were constructed in frames held in position by a sole or sill plate, invariably supported on a dwarf stone or brick wall. However, early examples consisted of posts directly into the ground without the use of a sill beam. Between the roof members and the roof structure there is a wall plate or head beam, which in turn supports the rafters. The main posts and corner posts support trusses, which support purlins. A diagram showing a typical traditional timber frame, identifying the main parts of the building, can be seen in Figure 3.12.

The panels would originally be in-filled with wattle and daub, as can be seen in Figure 3.13, which is a photographic illustration of wattle and daub taken in Herefordshire. Sticks would be placed at regular intervals between the rails, inserted into holes in the timber. Wattle was weaved between the sticks and a render of mud and straw, known as daub, would be applied to the mesh to complete the panel construction. Figure 3.14 illustrates the construction using this type of material.

During the 16th and 17th centuries panels were sometimes filled with stone or brick; in such cases the brick would be in a herringbone pattern and was mainly used in the south and south-east of England. In other parts of the country wattle and daub continued until the 18th century when it was replaced by more fashionable brick panels.

Lime wash was used to cover the timber frame and walls, which added some protection from moisture. Interestingly, it was the Victorians who found it fashionable to paint these types of property black and white.

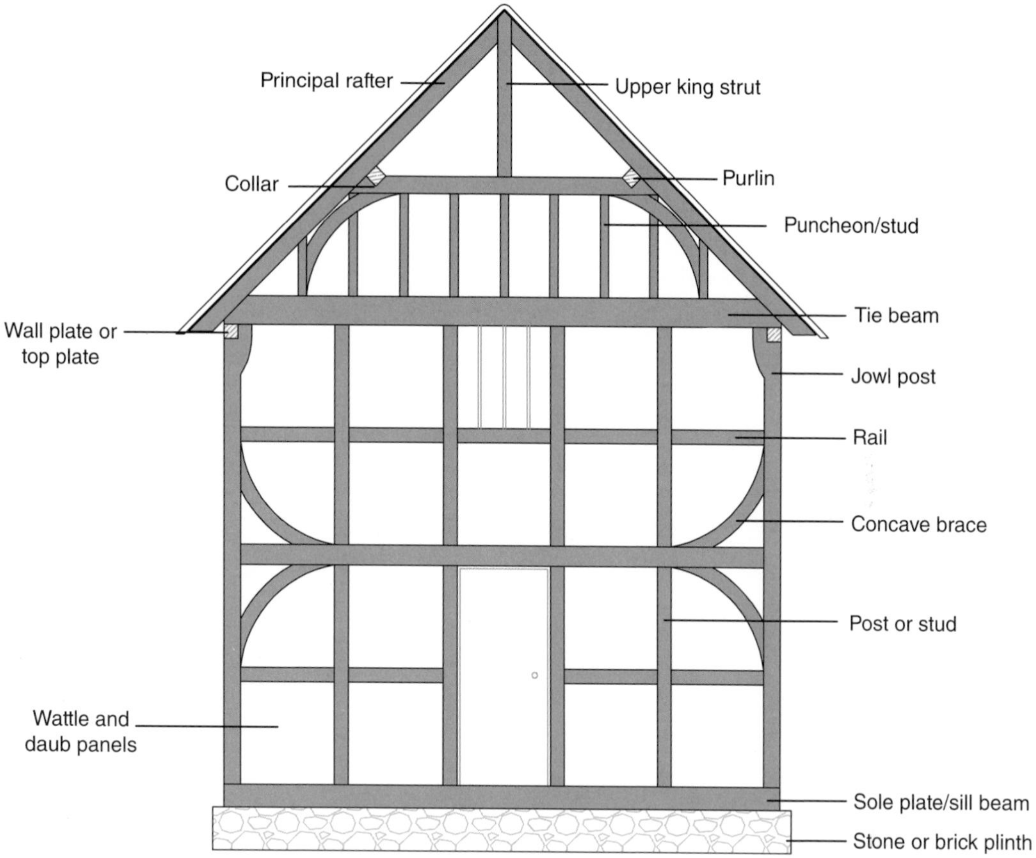

Figure 3.12: The named parts of a traditional timber frame constructed house.

Figure 3.13: Photograph showing wattle and daub exposed on a damaged property in Herefordshire.

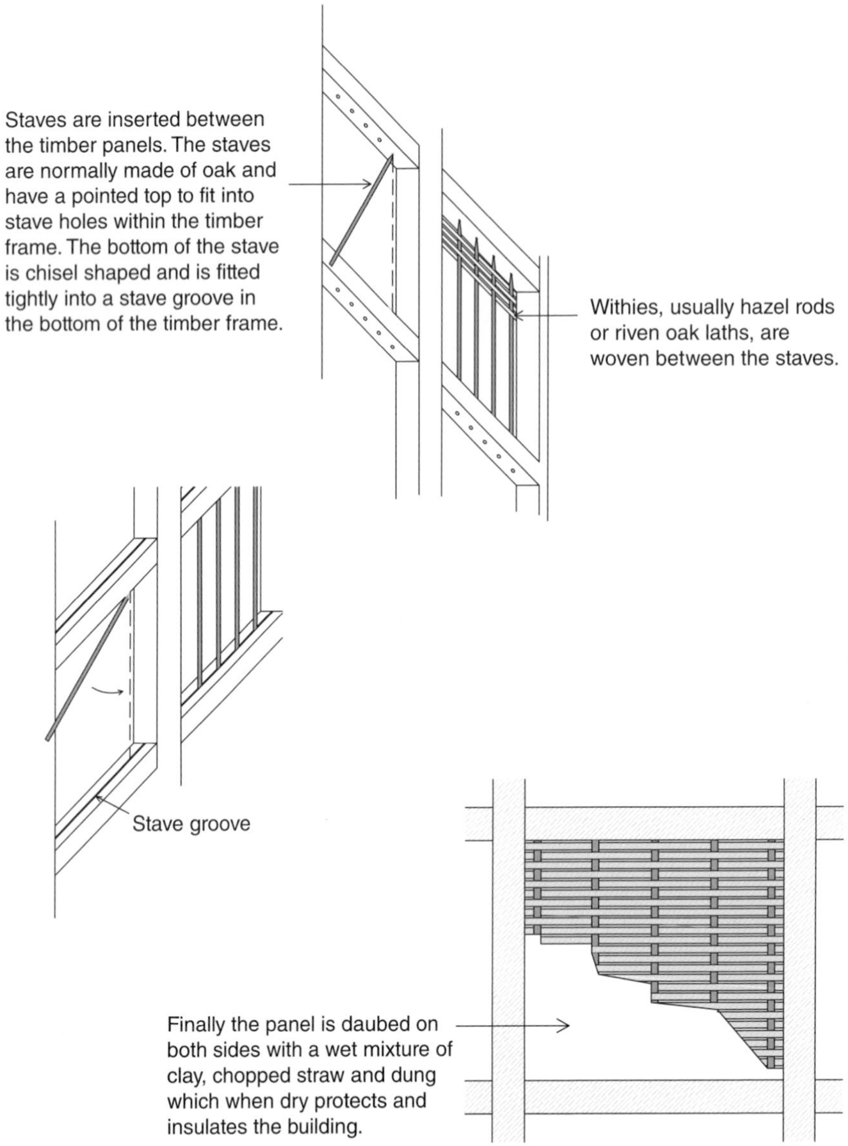
Staves are inserted between the timber panels. The staves are normally made of oak and have a pointed top to fit into stave holes within the timber frame. The bottom of the stave is chisel shaped and is fitted tightly into a stave groove in the bottom of the timber frame.
Withies, usually hazel rods or riven oak laths, are woven between the staves.
Stave groove
Finally the panel is daubed on both sides with a wet mixture of clay, chopped straw and dung which when dry protects and insulates the building.

Figure 3.14: Illustration showing wattle and daub construction.

Other notable changes through the history of timber frame buildings are that glazing was introduced to timber frame construction circa the 1600s; prior to this the windows were left open or had timber shutters. By the 18th century, rather than floor joists being exposed, ceilings were becoming plastered.

Structural problems

In the earlier timber frame buildings, as the elevations became exposed to the weather, the wattle and daub experienced wet-rot deterioration and consequently these elevations were replaced with brick panels but retaining the timber frame. If the brick panel is removed, exposing the timber rail, the original construction can be identified by the holes in the rail which accommodated the wattle. During the 17th and 18th centuries, as timber frames fell from fashion, in some cases the outside of the frame was encased in brickwork thus hiding the traditional construction.

Traditional timber frame buildings suffer from movement caused by lateral loading such as wind. The movement could be along the length of the building, sideways or in some cases twisting at a connection. Longitudinal movement is called raking and is not uncommon in traditional timber buildings that have little lateral restraint and suffer from wind loading. Figure 3.15 shows how raking affects a building.

Raking can occur in the roof separately from the walls, and this is evident where the gable exhibits a lean or displacement which is out of plumb. Figure 3.16 is a photograph showing raking of a gable wall on a house in Herefordshire.

In some cases, large chimneys offered the necessary lateral support along the length of the building. Therefore the removal of these large chimney structures is not advised unless professional structural advice has been sought. See the photograph in Figure 3.17.

Other problems that occur in timber frame buildings are the deterioration of the sole plate through wet rot, causing a failure of the foundation and a dropping of the gable or side walls, thus also resulting in sideways movement and raking. The combined effect of both these actions can cause a twisting of the building. An added complication is that the members which once sustained the loads then distribute those loads to adjacent members, placing additional stresses on these members which can lead to their failure.

We will discuss in more detail later how timber is very good in compression and how, where beams were thought likely to fail in deflection, the insertion of a brace, strut or diagonal member can act in compression to support the beam and redirect the load to a main strut. Using stocky short structures would reduce the slenderness ratio of the timber and thus the likelihood of buckling failure.

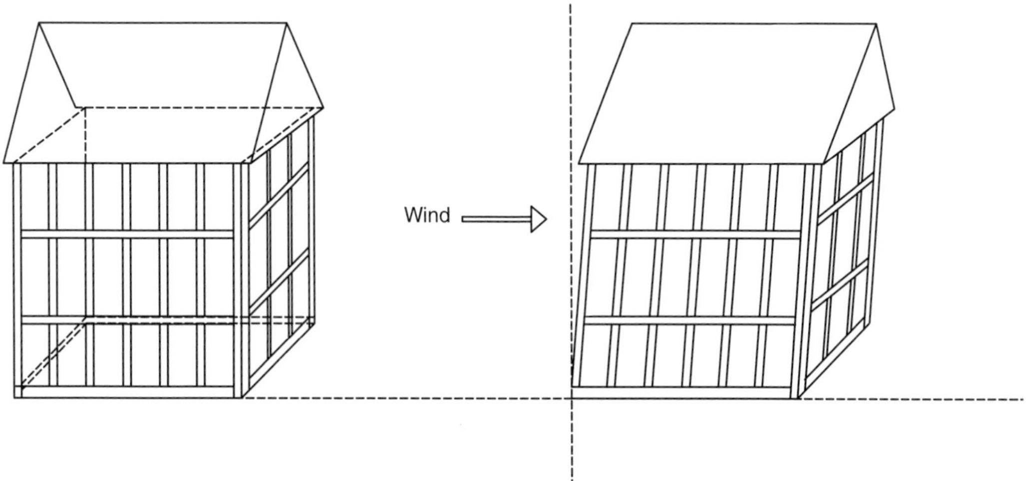

Figure 3.15: Diagram of raking to a building.

Figure 3.16: Photograph of raked gable wall of a house in Herefordshire.

Timber frame buildings have to maintain triangulation to ensure the overall stability of the building, and thus the use of triangles by employing diagonal braces reduces the chances of sideways movement or raking.

Construction

Buildings would be constructed as a series of frames with each joint numbered, usually in roman numerals, so that it could be identified. Often these marks can still be seen on traditional timber frame buildings. Figure 3.18 is a photograph of such markings on a timber truss.

The construction of timber frames would be undertaken by carpenters, and the property would be assembled by local people who gathered on the promise of a feast or food and ale following the lifting of the frames. The frames would be constructed on the ground and hoisted into position using levers and pulley systems.

Timber used in the construction of the buildings was felled locally and was green, thus as it dried a certain amount of movement would be experienced unless the timber was

Figure 3.17: Photograph showing how a chimney offers lateral strength to a timber frame property.

left exposed to the elements – such as in a barn construction where the sides were left open.

There was very little grading of the timber, but the timber employed in the construction of timber frame houses was traditionally oak, elm, sweet chestnut or hornbeam.

Jointing

Mortise and tenon joints were commonly used in the construction of timber frame buildings. These were strengthened by the use of pegs. In some circumstances the pegs were offset in the mortise joint to tighten the joint as the peg is driven into the pre-drilled hole. However, this can overstress the joint and in some cases shear the peg.

Scarf joints were also used, in particular to lengthen members. Lap joints were employed, usually between collars and rafters, and the more complicated lap dovetail joint was usually employed on the tie beam – connecting the wall plate to the tie beam of the truss and the stud. Owing to this jointing, cruck frame structures did not employ the beam lap dovetail joint but it is evident in almost all other timber frame buildings.

Figure 3.18: Photograph showing roman numbering of joint on a timber truss.

A jowl was an important enlargement at the head of a column or post, which ensured sufficient bearing was provided for the beams it was to carry and that the beams could be located and appropriately fixed to the column support. In some examples the jowl was a decorative piece of timber and in other cases it was merely a thickening of the head.

Second-storey extensions in timber frame buildings

First floors were not introduced to timber frame buildings until between 1485 and 1560. With the development of second and in some cases third storeys, this led to the introduction of jetties which expanded the floor area on the first and second floors. Figure 3.19 is a photograph showing a jetty on a building in Shrewsbury.

This method of expansion was used extensively in cities and towns, where accommodation was at a premium and space for expansion was undertaken by expanding vertically rather than horizontally to the side or by gable elevations. The spans across the building were usually too long for the joist spans, and additional beams were added to reduce deflections. These members were called binders or bridging beams and provided intermediate support to the floors by running to the side walls or posts. The joists either attached to the side rails built into the frame – known as a girding beam – or onto a beam attached to the frame – known as a ledge beam, which was traditionally pegged to the wall studs. Timber used for floorboards was either oak or elm.

Roof construction

Roof construction was traditionally a cut roof with purlins spanning from gable walls and trusses, which extended down through the property to the ground floor. As stated above,

Figure 3.19: Photograph showing the jetty floor of a timber frame building in Shrewsbury.

the trusses could be cruck frame but other types of construction were used which include aisled, crown post and hammer beam, and these can be seen in Chapter 10, Figure 10.11. Figure 3.20 shows a photograph of a barn with aisled construction. The timber posts would originally have extended to the floor but these have been replaced with brick walls, probably as a result of damp from the ground rotting the timber at the base.

During the 18th century construction began to change, and trusses were employed that rested on the walls and comprised queen post or king post trusses. The trusses supported the purlins and in some cases a ridge purlin which prevented roof spread. The purlin adjoined the principal rafter of the truss, either below the upper surface of the principal rafter – known as a "but purlin" and tenon jointed to the truss – or "through purlins" were carried over the principal rafter and may have been notched into the top of the principal rafter or held in position by a timber cleat.

The construction of timber frame buildings and the methods used are quite different from other types of construction, and specialist skilled knowledge is required when undertaking works or repairs to these types of structure. It is also important to realise that these buildings represent our architectural heritage, and careful consideration should be exercised before undertaking any works that may compromise the historical value of a building.

Figure 3.20: An aisled barn construction.

Stone

Stone walls in house construction are usually 450–500 mm thick. This type of construction originally used lime mortar to make the walls breathable. The strongest part of the wall was the masonry, hence the stones would be large and the joints small, since the mortar was the weakest part of the structure. Traditional stone walls were classified according to the general shape and tooling of the stones, as follows.

Rubble walls

Rubble walls comprise an inner and outer skin of roughly dressed stone with an in-fill of small stones grouted in mortar; larger stones are sometimes used to bridge between the two walls. Although such walls were solid, they could be quite weak structurally. Figure 3.21 is a photograph of a large rubble wall in cross-section at Pembroke Castle.

Ashlar walls

These walls have prepared and shaped rectangular stone consistently bonded, dressed and cut to form the wall.

Stone veneer

This type of wall provides a 25 mm thick stone fascia to a structural wall. The structural wall is constructed and the stone veneer is mortared to the face of the wall. Metal ties are mortared between the stones to tie it to the structural wall behind.

Slipform

Slipform is a concrete wall with a stone facing on one or both sides. Short forms of concrete are constructed using formwork up to a maximum of 1.0 m high at each

Figure 3.21: Photograph of a large rubble wall in cross-section at Pembroke Castle.

concrete pour. Stones are then placed against the formwork and reinforcement placed between. Concrete is then poured between the stonework. The stonework can be placed on one or both sides of the wall and the actual strength is provided by the reinforced concrete.

The strength of the construction of ashlar and rubble walls is dependent on the stone and the manner in which it is used. All stone has a grain, and it is important in the construction of walls that the strata are kept horizontal; if vertical, the stone will split. We will discuss this in more detail later in this chapter. In bridge construction the stone employed in the voussoirs of the arch should have the grain at right angles to the thrust.

Sandstone

Traditionally one of the best sandstones in the country is known as Blue Pennant, which originated at Fishponds in Bristol. The sandstone is a "moderately deep blue colour" and could be quarried in large blocks and slabs. The stone is hardwearing, strong and not prone to weathering. However, like all hard stones this does make it difficult to work and dress. The stone was used for wall construction and, if dressed, it was suitable for quoins, sills, lintels, hearths and a variety of other uses.

Forest of Dean sandstone is also good-quality stone which is hardwearing, resistant to weathering and available in large sizes. The main problem is that the colour was not as uniform as the Bristol Blue Pennant stone, ranging from blue to grey. Nevertheless, this stone was used for ashlar construction and a number of monuments have been constructed using this type of stone. It is still available and is quarried at Park End, Coleford, Gloucestershire.

There is also a Forest of Dean red sandstone, which was quarried at Mitcheldean and varies in strength. The dark red is a soft stone and can be used for general use but is prone to weathering. However, the very light red is arguably as strong as granite.

Quarella stone from Bridgend was available in large blocks and was obtained from three beds. One of the beds is light green stone, suitable for general building work. Another bed is the Ryder, which varies in colour and was used predominantly for the construction of cottages. Finally, there was a very hard white stone.

Sandstone was also available from Cheshire, Derbyshire, Yorkshire and Scotland, and was a general building stone.

Yorkshire and Liscsannor in Ireland are known for hard gritstones which are good for large structures.

Limestone

Predominantly limestone is a sedimentary rock which is quarried and can be used for road materials or burnt to produce lime. The exception to this is Bath limestone, which was mined. Limestone is softer than sandstone and easily worked by hand. Limestones are usually named after the locality from which they originate.

Flint

Flint is generally used in road construction and as a building stone, but full pointing is recommended with large mortar beds.

Portland stone

Portland stone is a decorative white limestone that originates from the Isle of Portland in Dorset. This is used particularly for large impressive public buildings such as Buckingham Palace in London.

Granites

In the UK, granites were mostly obtained from Aberdeen and Cornwall. Other granite was imported – due to the great demand – from Norway, Sweden and Russia. Blocks are usually obtained from quarries by blasting, and these are made into smaller blocks using wedges. The finished surface can be polished if desired.

Usage of stone

Table 3.1 is an extract of different text and tables taken from Middleton (1905, chapters 8 and 9) and Mitchell (1930, chapter 5). The table provides a description of the types of stone commonly found in the UK and their crushing loads. Primarily the table is taken from the BRE British stone testing and assessment – stone list and we kindly acknowledge the Building Research Establishment for their approval to use this table.

Table 3.1: Types of stone commonly found in the UK and their crushing loads

Stone	Description	Crushing resistance (MPa)
Limestones		
Ancaster Limestone-Weatherbed	Oolite limestone found in Lincolnshire, colour varies from cream to brown. A good dressing stone and many of the churches in Lincoln are constructed using this stone.	65.6
Beer Limestone	This stone is mined in the village of Beer near Seaton. The stone has a creamy greyish white colour with a fine texture. Sometimes the stone can have a greenish tint. This stone is used for moulding and has a limited resistance to weathering. The stone is used for new buildings, ecclesiastical work and restoration works.	21.8
Box Ground	This is a variety of the Bath Oolite limestone. Light cream to yellow in colour. This is a good weathering stone. Largely used for dressing, carved and moulded work. Good in all situations, external and internal, but is subject to penetrating damp so walls were generally constructed 450 mm thick.	
Broadway	The quarry is located on Broadway Hill in Worcester. The stone is an Oolithic limestone and has distinctive light and dark cream colours. The stone is used in buildings and has good resistance to weathering.	18.4
Caen	A fine Oolite quarried at Caen in Normandy. The stone weathers very badly but is much used for internal work, for carving and decorative purposes.	
Coombe Down	This is a variety of the Bath Oolite limestone. Colour light cream to yellow. This is used as a building stone.	16.4
Corsham Down	This is a variety of the Bath Oolite limestone. Varies in colour from light cream to yellow. Good for internal work and exterior work above ground level.	
Cotswold Hill Cream	The quarry is located at Stow-on-the-Wold and this is an Oolithic limestone. A traditional building stone. The different beds produce different quality stone. For example, the White Guiting stone (Bed 2) and the Honey stone (Bed 3) is a weaker stone used for walling.	36.6
Doulting Freestone	An Oolite limestone which has a uniform texture, quarried in Shepton Mallet, Somerset. The colour varies from cream to a brownish yellow. Very durable and suitable for general building work.	12.6
Elm Park	The mine is located near Garstad in Wiltshire and the stone is an Oolithic limestone, buff in colour. The stone weathers quickly and is used for walling or cladding.	28.3

(continued overleaf)

Table 3.1 (*continued*)

Stone	Description	Crushing resistance (MPa)
Freestone Purbeck	Quarried near Langton in Dorset and the stone is buff in colour with buff/cream-coloured shells. The stone is used in walling, door/window surrounds and flooring. The stone is considered to weather well. There is also a quarry in Swanage, which provides a blue/grey limestone with shells.	102.1 (93.7 Swanage quarry)
Golconda Dolomitic Limestone	Quarried at Bonehill near Matlock, Derbyshire. The stone is cream or gold in colour and is used in walling, architectural details such as columns, paving and flooring.	123.8
Little Casterton	An open Oolite limestone quarried near Stamford, used for general building.	
Hopton Wood Limestone	Quarried at Brassington in Derbyshire. The stone is used for architectural detailing, paving, flooring, walling and sculptures. The stone is cream/grey and contains fossils. The stone has good resistance to frost and salts.	119.8
Laning Vein Purbeck	Quarried at Swanage in Dorset, this limestone is a strong stone. The stone is a buff grey colour with shells. The stone is used for walling and paving.	217
Nailsworth	Quarried at Nailsworth in Gloucestershire. This is an Oolite limestone, brown in colour and similar to Portland limestone but much cheaper. Rocks are free from defects and can therefore be quarried in huge sizes. Recommended for internal work.	
Ham Hill	Quarried in Somerset. Yellow and grey in colour, being very bright when first quarried and then tones down. Used for facing and dressing work. Weathers well.	
Monks Park	Quarried at Corsham, Wiltshire. An Oolite limestone. Whitish cream colour. Compact and close grained. Has a relatively high crushing strength for a limestone and is amongst the best Bath stones. However, the stone is prone to weathering.	29.3
Orton Scar	Quarried near Orton, Penrith. The stone is grey/brown with colour variations. The stone is used in buildings and paving and is very durable with good frost resistance.	95
Portland Whitbed	Quarried at the Isle of Portland and is an Oolite limestone. White to light brown in colour. There are four beds of Portland limestone: True Roach, Whitbed, Roach and Basebed. Weathers better than all other Portland limestones in towns. There are three districts in Portland that quarry the stone. Wakeham: the stone from here is considered good to withstand atmospheres which contain sulphuric acid. Minesthay: lasts well in coastal regions and is used in London. Weston: good in moist conditions and forest areas, especially for the north of England and Ireland.	51

Table 3.1 (*continued*)

Stone	Description	Crushing resistance (MPa)
Chilmark	Oolite limestone which is light brown in colour, from Wiltshire. Used for general building and good for steps and paving.	131.5
Cap Purbeck	The quarry is in Swanage and the stone is buff – grey in colour but does vary. The stone is used for walling and paving.	226.7
Skerries	An Irish limestone quarried at Skerries County, Dublin. Generally a good building stone, grey to black in colour.	
Tullamore	An Irish limestone quarried at Tullamore, Kings County. Light blue in colour when chiselled but black when polished. Sometimes this rock is classed as a marble.	
Cairns Lodge	An Irish limestone quarried at Tyrrell's Pass, County West Meath. Considered one of the best Irish limestones. Grey in colour and is a good durable building stone.	
Crossdrum	An Irish limestone quarried at Old Castle, County Meath. Good working and durable building stone. Light grey in colour.	
Little Island	An Irish limestone quarried at Little Island, County Cork. Light grey in colour. A good working stone.	
Salterwath	Quarried near Orton, Penrith. The stone is a dense fine-grained limestone which is dark blue when first quarried, weathering to a pale grey. When polished, the colour is a rich brown. The stone is very durable and has good resistance to salt damage. The stone is used in buildings and paving applications.	102
Stanley	Quarried at Upton Wold, Moreton-in-the-Marsh. There are a number of beds from which the stone is quarried. The stone is used in many building applications and is an Oolithic limestone. The strength depends on the bed from which it is quarried.	17.3–34.7
Swaledale Fossil Limestone	The quarry is near Scotch Corner in North Yorkshire. The limestone is a light-coloured stone and ranges in colour from beige to blue, containing many fossils. The stone polishes nicely. A very durable stone, resistant to acid rain. Used in buildings and paving applications.	95
Totternhoe	Quarried in Totternhoe near Barnstable, this limestone is greyish white in colour often with a green tinge. The stone is gritty in texture due to the presence of shells. The stone is used in ashlar and moulded work but has limited resistance to weathering. In practice, the performance of the stone relates to the way the stone has been extracted,	29.8

(*continued overleaf*)

Table 3.1 (*continued*)

Stone	Description	Crushing resistance (MPa)
	seasoned and laid. In some circumstances the stone is known to acquire remarkable toughness after weathering.	
Wroxton	Quarried at Alkerton near Banbury, Oxfordshire, this stone is a greenish blue or brown in colour. Traditionally the stone is used in random walling. It has good resistance to salt and frost.	36
Magnesian Limestones		
Anston	Quarried near Sheffield and is a fine granular crystalline aggregate with cavities. Dispersed in the stone are black carbon particles. A rich cream colour. Used in the Houses of Parliament, Westminster and other historic buildings.	
Bolsover	Quarried in Bolsover, Derbyshire. Has a bright yellow brown colour and is a good durable building stone; also used for paving.	
Cadeby Brown	The quarry is located in Cadeby near Doncaster. The stone is a light-coloured stone and the lighter stone tends to be softer. The stone has been used as a building stone and is considered to be durable. More recently, the stone has been used as a road aggregate. There is also a Cadeby White and a Cadeby Light Gold limestone quarried from the same location.	48–103 variable (Cadeby Brown)
Mansfield Woodhouse	Obtained from Mansfield in Nottinghamshire. Yellow in colour; used for decorative internal work including carvings and mouldings.	
Huddlestone	Quarried at Shurburn in Yorkshire. Cream in colour and used for general building work.	
Roche Abbey	Quarried near Bawtry, Yorkshire. Light cream in colour. Suitable for general building work but can discolour.	
Park Nook	Quarried near South Milford, Yorkshire. Cream in colour and used for general building work.	
Westwood Ground Limestone	The mine is near the village of Westwood, Wiltshire. The stone is an Oolithic limestone, coarse grained and buff coloured.	18.9

Table 3.1 *(continued)*

Stone	Description	Crushing resistance (MPa)
Sandstones		
Appleton	Quarried in Halifax, West Yorkshire the stone was formally known as Greenmoor Rock or Grenoside. The stone colour can range from fawn to a darker mottled brown. The larger-sized stone is used for paving and the smaller sizes can be used for roofing slates and dressing stone.	143.2
Baxtonlaw	Quarried in Hunstandworth, Weardale, which is owned by Dunhouse Quarry Works. The stone is a pale cream/yellow millstone grit. This is a very durable stone used in building and paving, and is not susceptible to acid rain or air pollution.	123
Bearl	Quarried at Bearl Quarry near Stocksfield, Northumberland. The stone is white/buff in colour and is a durable building and paving stone; used in many cities and towns across the UK. The stone has good resistance to acid rain and air pollution.	59.2
Birchover Gritstone	Quarried at Birchover Quarry near Stanton-in-Peak, Derbyshire. A millstone, being pink to buff in colour. This is considered a very durable building and paving stone and has been used in many cities across the UK. The stone is durable and not affected by acid rain.	73.6
Blaxter Sandstone	The stone is quarried near Otterburn, Northumberland. It is pale yellow/buff in colour. The stone is durable and not affected by acid rain. The stone is used in many buildings across the UK, particularly in the north of the country.	55.1
Callow Hill	The stone is quarried near Monmouth and is a red/brown colour. This is viewed as a durable building and paving stone and has been used extensively in the UK. The stone has a variable resistance to acids but is still considered durable.	153.4
Cat Castle Buff	The stone is quarried in Barnard Castle, County Durham and is a creamy buff-coloured stone. The stone was used for the construction of railway bridges north of York. The stone is durable and not affected by acid rain.	94–115.5 (Cat Castle Grey 53–65)
Copp Crag	The stone is quarried at Byrness and is a yellow brown stone used in buildings and paving. The stone is considered to be very durable.	89.4
Corncockle	Quarried at Corncockle near Lockerbie, the stone is pale red/brown in colour and used as a durable building and paving stone.	72.5

(continued overleaf)

Table 3.1 (*continued*)

Stone	Description	Crushing resistance (MPa)
Corsehill	Quarried in Annan, Dumfriesshire. Dark red and bright pink in colour. Good weathering stone suitable for carvings, dressing and ashlar work.	67.6
Cove Red	Quarried at Cove Quarry, Kirkpatrick Fleming near Annan in Dumfries and Galloway. The stone is red/brown in colour and used in buildings. This is a durable stone not affected by acid rain but is susceptible to salt damage.	116.11
Craigleith	Quarried near Edinburgh and is whitish grey in colour. Very hard and durable and is good for ashlar or general building.	
Cromwell	Quarried in Cromwell Quarry, Halifax. The stone is a buff to grey durable sandstone having good resistance to acid rain. The stone is used in buildings and other applications.	180.1
Crossland Hill	Quarried at Wellfield Quarry near Huddersfield. The stone is a millstone grit, buff in colour and used in buildings. The stone is very durable and not affected by acid rain.	132
Crossley	Quarried at the Thumpas Quarry, Halifax. The stone is buff coloured and is a very durable building stone.	183.6
Doddington	Quarried in Doddington, Northumberland, the stone is a speckled light to deep purple/pink in colour with occasional rust colour. The stone is a very durable building and paving stone and is used in many cites and towns.	51.3–58.2
Dukes Gritstone	Quarried in Whatstandwell, Derbyshire, the stone is a millstone grit, pink/lilac in colour with iron black and buff markings. The quarry has been in existence for a long time and the stone has been used in buildings since the 19th century. The stone is durable and not affected by acid rain.	74.9
Dunhouse Buff	Quarried at Dunhouse Quarry, County Durham the stone is a millstone grit, buff coloured, which is durable and used in building and paving applications. There is also a Dunhouse Sandstone Grey from the same quarry, but extracted from a different face.	84.1 (Dunhouse Sandstone Grey 137.8)
Flash	Quarried at Bolehill Quarry near Matlock, Derbyshire, the stone is a millstone grit coloured cream with some veining. The stone is used in buildings, very durable and is not affected by acid rain.	64.2
Fletcher Bank	Quarried in Southowram, Halifax. The stone is a buff/grey millstone grit. It is used in buildings and is very durable and not affected by acid rain.	113.6

Table 3.1 (*continued*)

Stone	Description	Crushing resistance (MPa)
Silex	Quarried at Hipperholme in Halifax, Yorkshire. Even tinted, light coloured, very durable and hard. The stone has a non-slippery nature and is good for steps and landings.	166.4
Hailes	The stones from these quarries in Edinburgh, in the neighbourhood of Cragleith, are of three tints: white, pink and bluish grey. The white rock is the strongest and the most compact of the three. The blue rock has the most marked laminations and the blue colouration is due to carbonaceous matter. Stone from this quarry has been used in Edinburgh for over two centuries and is excellent weathering stone for general building. It can be obtained in large blocks and therefore is good for landings and steps.	
Bramley Fell	Originally quarried near Leeds, but the name also denotes the coarse millstone grits from Yorkshire at quarries such as Horsforth. It is light brown in colour, very durable and good for general building purposes, paving and steps.	
Dean Forest	Quarried in Gloucestershire. There are two kinds: grey and red. The grey varies in colour from grey to shades of blue. The stone is hard and suitable for general building work. The red varies from a light to a deep rich red colour and the lighter red stones are very hard.	
Hall Dale	Quarried at Hall Dale Quarry near Darley Dale, Derbyshire, the stone is a millstone grit and is yellow/brown in colour with the coarser-grained stone being more yellow. The stone is a durable building and paving stone. St Georges Hall in Liverpool was constructed using this stone.	102.2
Heddon	Quarried near Newcastle. Light brown in colour and is a durable stone for general building.	
Hillhouse Edge	Quarried at Holmfirth, Huddersfield. The stone is a millstone grit York Stone, which is a very durable and strong stone used in buildings and other applications.	124.7
Kenton	Quarried near Newcastle. Light brown in colour and is good for general building work and also fine and carved work.	
Locharbriggs	Quarried at Locharbriggs near Dumfries, the stone is dull red to pink in colour, being medium grained. Used for paving and buildings, the stone is durable and not susceptible to damage by acid rain. It has good frost resistance.	47.3

(*continued overleaf*)

Table 3.1 *(continued)*

Stone	Description	Crushing resistance (MPa)
Grinshill	Quarried in Clive near Shrewsbury, Shropshire. The stone is cream/buff in colour and is a fine-grained, durable sandstone used for facing and general building work. The stone is not considered good in locations with high salt concentrations.	38
Myddle Red Sandstone	Quarried at Clive near Shrewsbury, Shropshire. The stone is red/brown in colour and is very durable. The stone is used in paving and buildings.	21
Peakmoor Sandstone	Quarried at Stanton Moor near Matlock, Derbyshire. The stone varies from light brown to pink and is a fine to medium-grained millstone grit. The stone is durable and not affected by acid rain. Used in buildings and other applications.	72.5
Plumpton Red Lazonby	Quarried at Lazonby Fell Quarry near Penrith. The stone is fine to medium sandstone, pale red or dark pink in colour. Used in buildings and paving, the stone is very durable and not affected by acid rain.	93–118
Red St Bees	Quarried at Salton Bay, Cumbria, the stone is a fine-grained dull red sandstone. The stone is used in buildings and is less durable than some other sandstones.	78.6–100.1
Rockingstone	Quarried near Bolster Moor, Huddersfield. The stone is a millstone grit and is a very durable stone not susceptible to frost action.	85–119.9
Scotgate Ash	Quarried at Patley Bridge in Yorkshire. Greyish yellow in colour and is a fine-grained, hard, laminated sandstone suitable for staircases and pavements.	
Howley Park	Quarried at Morley, Yorkshire. The sandstone is light brown in colour. A fine-grained, durable but not hard stone used for dressings, stairs and landings.	
Robin Hood	Quarried near Wakefield in Yorkshire. Greenish grey in colour. Durable and suitable for landings and stairs.	
Prudham	Quarried at Fourstones, Northumberland. Light cream colour and is excellent for dressing, moulded and ashlar work. Largely used in the north of England and Scotland.	
Pennant	Quarried in the coal measures in Glamorganshire and Gloucestershire. The colour is dark grey, green or blue and it is a hard-wearing, strong stone. The stone can be worked to a fine surface and weathers well.	

Table 3.1 *(continued)*

Stone	Description	Crushing resistance (MPa)
Scout Moor	Quarried in Southowram, Halifax. The stone is a blue/green, fine-grained millstone grit. The stone is used in buildings and paving and is durable but has limited resistance to acid rain.	186
Shire	Quarried at Wingerworth near Glossop, Derbyshire the stone is a medium-grained millstone, grit coloured buff to grey. The stone is a durable stone not affected by acid rain and is used in buildings and paving.	77
Spynie	Quarried at Elgin, Moray. The stone is a yellow/buff sandstone. The stone is used for railway bridges in northern Scotland and is a durable stone.	62.1
Stainton	Quarried in County Durham the stone is a fine grained buff sand stone with a brown speckle. The stone is a very durable stone used in buildings and paving. The stone is used across many towns and cities in the United Kingdom and has good frost resistance.	48–55.3
Stanton Moor	Quarried near Stanton in the Peak near Matlock, Derbyshire the stone is a buff/pink colour and a fine to medium grained mill stone grit. The stone is a very durable stone not affected by acid rain and is used for building and paving.	79
Stoke Hall	Quarried in Grindleford, Derbyshire the stone is a fine to medium buff coloured mill stone grit. The stone is very durable with resistance to acid rain. Used in buildings and paving applications the stone is used in many towns and cities across the United Kingdom.	103
Stoneraise red sandstone	The quarry is near Penrith and the stone is a fine to medium grained pale red sandstone with a speckle due to the quartz content. A very durable stone not affected by acid rain. The stone is used in building and paving applications.	75.6
Streatlam buff sandstone	Quarried at Moresby, Whitehaven. The sandstone is a pale yellow, fine to medium grained sandstone. The stone is durable and has good resistance to acid rain. The stone is used for cladding, paving, flooring and load bearing masonry.	78.6
Tenyard Hard York Sandstone	Quarried in Keighley, West Yorkshire. The stone is a pale yellow/brown fine grained sandstone. The stone is durable having good resistance to acid rain and good frost resistance being used in buildings and paving.	160.9
Torrington	Quarried near Barnstable in Devon. The stone is a dark grey very fine grained sandstone. A very durable stone with good resistance to acid rain, high frost resistance and resistance to salts. Used in buildings and paving the stone can be used as load bearing masonry.	248.8

(continued overleaf)

Table 3.1 (*continued*)

Stone	Description	Crushing resistance (MPa)
Waddington	Quarried near Clitheroe Lancashire. The stone is buff to grey in colour and is a fine to coarse grained mill stone grit. This is a very durable stone with good resistance to acid rain. The stone is used in buildings and paving.	90.4
Wattscliffe Lilac Gritstone	Quarried near Elton in Derbyshire. The stone is a fine to medium-grained millstone grit which is lilac to pink in colour. This is a very durable stone not affected by acid rain. The stone is used in buildings, as load-bearing masonry and cladding.	79.1
Woodkirk Sandstone	The quarry is near Morley, Leeds. The stone is a fine-grained stone, grey/buff to light brown in colour. There are two stone types: Woodkirk Brown and Woodkirk Type M. The stone is very durable but has limited resistance to acid rain. Used in building and paving applications.	98 (Woodkirk Brown) 101.5 (Woodkirk Type M)
Irish Sandstones		
Mountcharles	Quarried at Mountcharles, County Donegal. Warm cream in colour and bleaches white after a few years' exposure. One of the finest sandstones available. The appearance and weathering qualities are arguably the best for city conditions. Many of the largest and finest buildings in Dublin, London, Belfast and Londonderry have been built of this stone.	Not tested
Shamrock Stone	Quarried at Doonagore, Liscannor, County Clare. The stone is a hard, clean, close-grained millstone grit which is blue/grey in colour. Good for kerbs, landings, stairs and flagstones. Exceedingly durable and will not cause slipping. Not easily dressed.	
Ballway	Quarried at Ballycastle, County Antrim. Fine-grained sandstone, easily worked and finishes well. Light grey in colour and used for general work.	
New Milling	Quarried at Dungannon, County Tyrone. Light cream in colour. Used for general building and dressing.	
Magnesian Sandstones		
Red Mansfield	Quarried near Mansfield, Nottinghamshire. Reddish brown in colour, Oolithic in structure. Very durable and suitable for general building, carving and mouldings.	

Table 3.1 (*continued*)

Stone	Description	Crushing resistance (MPa)
White Mansfield	Quarried near Mansfield, Nottinghamshire in the same district as Red Mansfield. Whitish brown in colour. Very similar to the red variety but not so durable.	49.6
Granites		
Grey Aberdeen	Quarried in Aberdeenshire. Takes a high polish and is suitable for columns and ornamental work but is largely used for kerbs and sets.	
Aberdeen Corennine Pink	Quarried in Aberdeen. It is a close-grained pink colour and is largely used for constructional and decorative work.	
Aberdeen Peterhead Red	Quarried in Aberdeen and is of a coarse red variety. It can be made into large polished columns and is used for constructional and decorative purposes.	
Aberdeen Rubinslaw	Quarried at Rubinslaw near Aberdeen. Fine grained, grey in colour and can be obtained in large blocks. The stone takes a high polish, is extremely durable and ideal for monuments and decorative purposes. It contains black and white micas with felspars and quartz.	
Cornish Grey	Quarried in various parts of Cornwall. This is largely used for engineering works such as bridges and similar constructions.	
Guernsey Granite	Quarried in Guernsey. This varies from reddish brown to grey/blue in colour and is mainly used for paving.	
Castlewellan	Quarried at Castlewellan, County Down, Ireland. This stone is a mottled grey colour. It is a fine, beautiful stone and is reputed to be one of the best in the isles. It has been used for the Eddystone lighthouse and the base and pedestal of the Albert Memorial.	
Dalkey	Quarried at Arklow, County Wicklow, Ireland. This is blue/grey in colour and a good building stone.	
Newry	Quarried in Newry, County Down, Ireland. Grey in colour, very durable and suitable for general building purposes.	
Kingstown	Quarried in Kingstown, County Dublin, Ireland. This is grey in colour and a very hard stone, and is therefore generally used as paving sets.	

(*continued overleaf*)

Table 3.1 (*continued*)

Stone	Description	Crushing resistance (MPa)
English Marbles		
Green Purbeck	Quarried at Quarr Farm, Corfe Castle, Dorset. The stone is a green marble which is very dense and takes a polish. The stone is used for architectural details such as paving and columns.	145.7
Hopton Wood	Fossil limestone quarried at Hopton in Derbyshire. Takes a good polish and is considered to be a good marble. Used for decorative features such as columns, arches, ashlar walls and staircases. Grey to brown in colour with gradations.	
Purbeck Marble	Quarried at Purbeck in Dorset and is mottled grey in colour but takes a fine polish. The rock has an abundance of paludina shells and is much used for decorative work in churches, such as slender columns. Used in Westminster Abbey.	98.2
Mento	Quarried in Galway, Ireland. The marble is black in colour.	
Letternaphy	Quarried in Clifden, County Galway. The stone is green in colour, also known as Connemara Marble or Irish Green. Two kinds are available: the first is made up of bands, striped different shades of green and interlaced with white streaks; the second is very mottled.	
Kilkenny	Quarried in Kilkenny, Ireland. Black in colour and when dressed shows figures of shells.	
Ulveston Marble-Fawn and Mottled	Quarried near Baycliffe, Ulveston, Cumbria. There are several beds of stone and the colour varies from an oatmeal or dark cream colour known as "Fawn" to a light mottled brown known as "Mottled". It is not a marble in the geological sense but the stone is very dense and takes a polish, which means that it is called a marble. The stone has a good resistance to weathering and salt damage.	144 (Fawn) 100 (Mottled)
Churchtown	Quarried at Churchtown, Ireland. Red in colour and takes a good polish.	

Originally stone used in buildings was lime washed, a practice which originated from Medieval times and afforded the stonework some protection against decay. Figure 3.22 is a photograph showing lime wash used on the walls at Pembroke Castle.

There are general rules of construction which prevent the decay of stonework as follows:

- Vertical joints should be avoided as they are a source of weakness.
- Heavier stones should be positioned on the outside and point inwards, with smaller stones filling the inner wall.
- If the wall is free standing then the wall should taper inwards towards the top, as for a drystone wall, to avoid the migration downwards of stones on the inner wall due to the ingress of moisture.
- The wall should contain through stones which span the wall.
- Stone should not be laid in more than three vertical courses a day.
- Lay the stone with bedding planes horizontally or at right angles to the applied load.

Laying stone with the bedding vertical will cause the bedding planes to buckle and can cause delamination of the stone face. This is often seen around windows and openings; where the stone is too small to be used in the correct manner, it is turned through 90° to fit the proposed application. Figure 3.23 is a photograph showing stone laid in the wrong orientation, resulting in such a failure.

This should not be confused with contour scaling, which is where the sandstone will have a thin layer of clay running at right angles to the sandstone beds. The stone can be laid correctly, but the action of water on the clay causes the clay to fracture and delaminates the face of the stone.

As discussed earlier, cement renders are not suitable on stone walls and unfortunately the most common repair of stonework that has delaminated or suffered contour scaling is to render the wall, which only exacerbates the problem.

The problems associated with using cement mortar and render on stone walls is not confined to the structural damage of sanding and dusting, which results from moisture being trapped in the stone. Water is a good conductor of heat, and trapped moisture in the

Figure 3.22: Photograph of lime-washed walls at Pembroke Castle.

Figure 3.23: Photograph showing delamination of stone face due to vertical bedding.

wall therefore transmits heat generated internally to the external surface, making the wall less thermally efficient.

Most of the strength of a wall is from the interlocking stone and the friction that exists between the masonry units. Hence, in Victorian structures the joints were kept to a minimum thickness. Early mortar was mud or clay, and the main reason for the mortar was to keep out the rain, allow movement and prevent seeds and vegetation propagating in the wall. Thus the stones in the wall should remain in place without the need for mortar. Walls were constructed to taper with increasing height. This gave the wall its structural stability, with the resultant force from the horizontal and vertical loads travelling downwards through the middle third of the base. Such a taper can be demonstrated by looking at Victorian railway walls, where mortars were very weak and consisted of only lime mixed with water. Adding cement to mortars has made the joints more "glue" like, and the need for the taper in such structures has now been removed.

We will discuss mortars later in this book. Ideally, when working with stone, lime pointing should be used to retain the breathable characteristics of the wall. However, if a cement mortar is preferred (which makes the wall non-breathable), a good mix for pointing is one part lime, six to seven parts coarse sharp sand and one part cement. Using lime ensures that the mortar does not dry out before it hardens. If too much cement is used the mortar will crack, allowing the ingress of moisture and causing further cracking when it freezes. Many joints fail due to not being deep enough, having the wrong mortar mix, not being mixed well or not being well packed. Fresh pointing should be kept moist whilst it is setting to prevent cracking, particularly in dry weather.

When pointing stonework or laying stone the mortar should not be stronger than the surrounding stone, and should be flush with the face of the stonework. It should be finished with a soft brush as this closes any crevasses left by the trowel and prevents the ingress of moisture.

If re-pointing stone walls the mortar should be raked out to a depth twice the thickness of the joint. The joint should be wetted prior to the application of the new pointing.

In summary, stonework should be laid so that the mortared joints do not allow the passage of moisture – to avoid freezing causing cracks and blown stone and mortar. Interestingly, when brickwork fails in this manner it is usually confined to a single brick, whereas in stonework the problem usually involves a larger area of stonework.

In some parts of the country, for example Surrey, Bargate stone is traditionally used and small pieces of the stone are pressed into the joints to provide a decorative finish. This is known as garneting.

The use of iron in stonework is not a good idea, and with the introduction of metal windows and railings problems arose with the iron corroding. The resulting expansion of the metal splits the accommodating stone, causing structural damage to the masonry.

Modern timber frame construction

There are several types of modern timber frame building – for example a facing brick or rendered block outer skin with an internal timber frame. This usually comprises a timber frame with 50 × 125–150 mm timber studs at 400 mm centres with noggins placed at 600 mm centres. Insulation is placed between the studs and the plasterboard, with a plaster skim on the internal face. A cavity exists between the timber frame and the outer skin, and wall ties are employed to attach the studs to the external skin.

In this type of construction the windows are usually set further back than in traditional cavity construction, to ensure they are sited in the timber frame. Also, the roof space usually reveals the timber frame on the gable wall. The timber frame is the load-bearing part of the construction and carries the floors and trussed roof structure.

Another type of timber frame construction is a timber frame with walls having either a render or cladded finish. These usually comprise a timber frame with 50 × 125–150 mm timber studs at 400 mm centres with noggins placed at 600 mm centres. Attached to the external face is an exterior-quality plywood sheet, against which a breathable membrane is applied to prevent moisture penetration. Treated timber battens of 25 mm × 38 mm at 600 mm centres are attached vertically to form a drained cavity and support either the stainless steel lath to which the render is applied or the timber cladding. Internally, insulation is applied between the frame with a plasterboard and skim finish.

A typical timber frame construction can be seen in Figure 3.24.

Solid brick construction

Bricks have been used in construction for many years, and early bricks were of mud as evidenced in the Middle East as early as 7500 BC. The development of fired brickwork followed, and one of the earliest examples was noted in 2900 BC in the Indus Valley in the Middle East, but it is understood that fired brick had been used since 5000 BC. The Romans used bricks in the UK as early as 43 AD until they left in the 5th century. They also developed hydraulic lime mortar by adding volcanic ash to lime, which produced a mortar that hardened with the addition of water. This made masonry much stronger and led to much greater structures being possible. Domes, arches and barrel vaults were achieved using this latest technology, and the first slipform walls were also developed at this time.

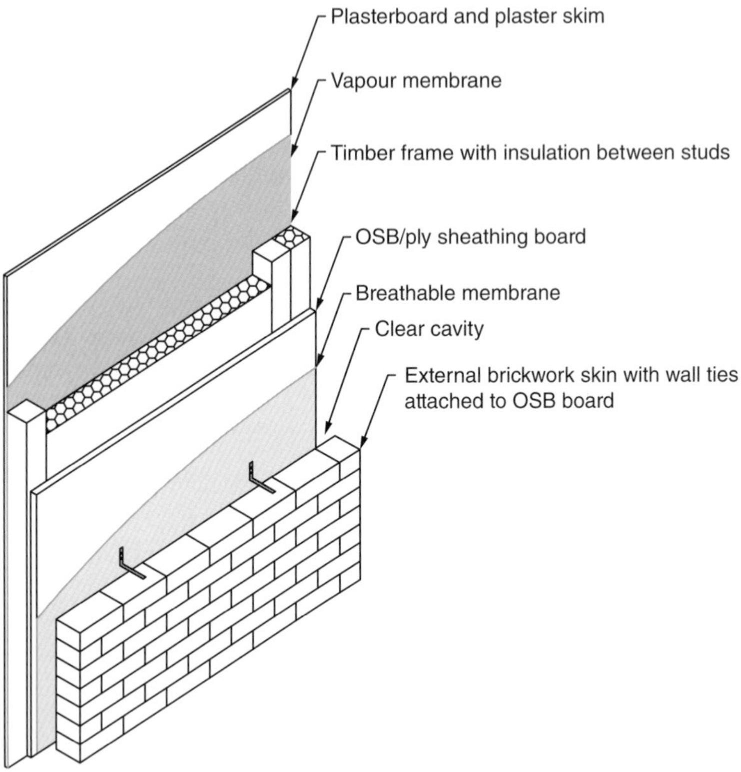

Figure 3.24: A modern timber frame construction.

Until recently, brickwork was employed for small to medium-sized buildings, but following research undertaken in the 1950s by the Swiss Federal Institute of Technology and the Building Research Establishment, tall structures were made possible.

Solid brick walls have no cavity and are generally 225 mm thick; after 1920, these had largely been replaced by cavity walls. Header bricks tie the two courses together and the strength of the brickwork is dependent on the number of headers and the type of brick and mortar used. In larger houses, such as Georgian houses, particularly where they span over more than one floor, walls can be 325 mm thick or even wider. Not only does this increase the compressive strength of the wall, it also spreads the load over a wider area at foundation level. Generally the wall will thicken at ground level and rather than being placed on a concrete foundation may be laid directly onto the subsoil.

Different bonds can be used in solid wall construction, and some examples can be seen in Figure 3.25.

Cavity construction

Cavity construction can comprise an internal skin of brick or block and an external skin of brick or rendered block.

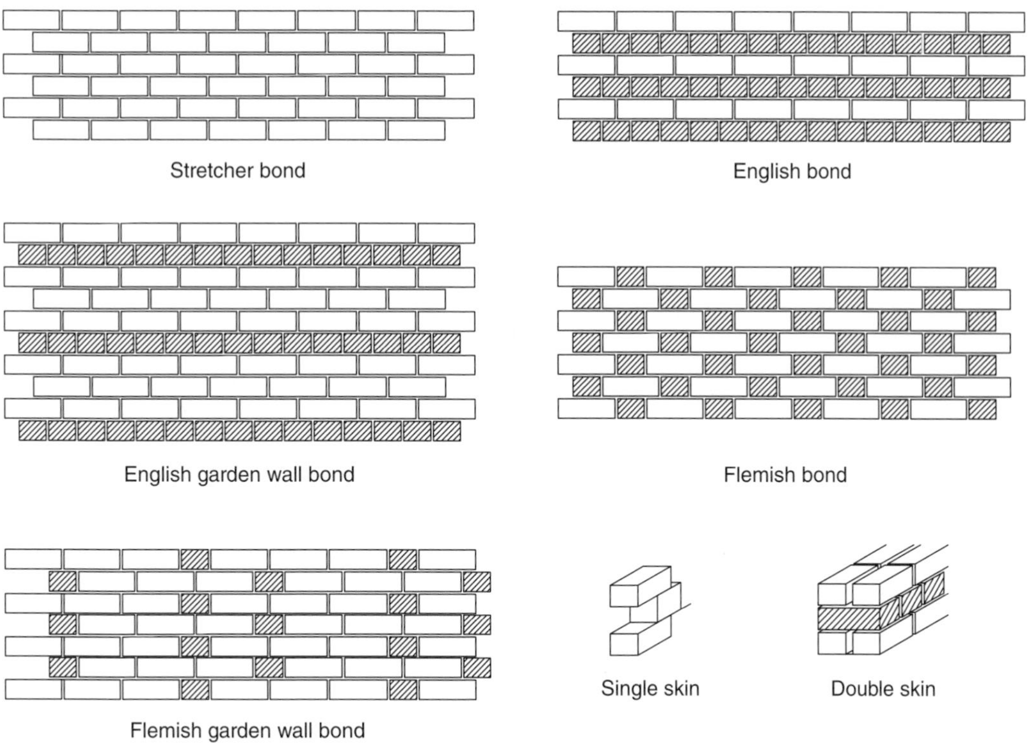

Figure 3.25: Different brick bonds.

After about 1920 cavity walls became widespread in the UK and across Europe, when it was realised that such a provision gave additional insulating properties. Earlier examples of cavity walls exist from the Victorian period, and early cavity walls were bonded using brick ties, thus creating a bridge for damp. In the 1930s, blockwork became more popular as the internal skin.

However, during the 1920s–1930s snapped headers were used to give the appearance of Flemish bond brickwork, which is a solid wall-type construction. Close examination is required to confirm the type of wall construction in these circumstances. Other bonds involved two stretcher bond bricks with a header between, forming a void as seen in Figure 3.26.

This bond was called the rat trap, Chinese bond or rowlock bond, and is a combination of solid and cavity construction.

Between 1930 and 1940, many lime mortars were replaced using cement mortars – although lime could be used to improve workability. Mortars in the early 1900s contained industrial waste such as ash, fine aggregate or brick dust. The pointing at this time was usually very fine (no greater than 8–10 mm) and was finished slightly recessed or flush.

During the 1960s–1970s cavities began to be filled with cavity wall insulation and since the 1980s the amount of insulation has increased. In the 1920s aerated concrete blocks were invented in Sweden and since the 1960s they have been improved and used extensively in the construction industry to provide further insulating properties to the

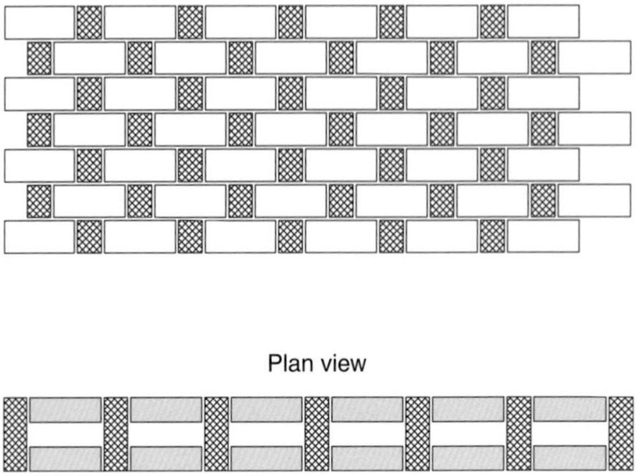

Figure 3.26: Snapped header rat trap bond.

internal skin of the cavity wall. By the late 1980s lightweight blocks were more common than dense blocks.

Typically these walls are between 250 and 300 mm thick, and the load-bearing element is the internal skin of the wall which carries the floors and roof.

Steel construction

As early as the 9th century China was using cast iron as a building material. During the 18th century cast iron production and technology improved to increase production and provide a cost-effective product. In 1779 the Iron Bridge in Shropshire was constructed using cast iron. Following this the Industrial Revolution saw steel and iron technologies advance, and large-scale structures became possible.

The Industrial Revolution also saw the introduction of cast iron into buildings, and in the 18th century the House of Commons introduced cast iron columns to support galleries.

Until 1840 the main construction material was cast iron, but during the early 1800s cast iron began to be replaced by wrought iron and many railway bridges were constructed using wrought iron. After 1850 riveted wrought iron girders became more prominent and cast iron beams were used less often, except in ornamental structures.

Wrought iron beams were not used after 1890, and mild steel sections became more readily available with the publication of the *Dorman Long Section Handbook* in 1887.

At the end of the 1800s wrought iron was no longer available and steel structures such as the Forth Bridge, constructed in 1890, showed what could be achieved using steel. Further development in the use of steel came in the 1900s through the use of welding, and this advanced steel fabrication. By the 1960s the use of rivets had ceased, to be replaced by welding.

Rolled steel was used for the first time in a furniture emporium in County Durham in 1900 and in 1906 the Ritz Hotel in London was the first substantial steel frame building in the UK.

There are advantages and disadvantages in the use of these materials, as seen below.

Wrought iron

Wrought iron is good in tension but weak in compression. This material shrinks as it cools and although this can be a disadvantage, it also has the advantage of forming tight connections. Wrought iron can be welded by heating two pieces together and beating them whilst hot. Wrought iron resists shocks and is good for gates and barriers, but is prone to rust and corrosion. Furthermore, gallic acid attacks wrought iron and for this reason it should not be used alongside oak. If exposed to heat wrought iron bends and provides some warning before it fails.

Cast iron

Cast iron is good in compression but poor in tension; it also has poor resistance to shocks and sudden impacts and is brittle in nature. Cast iron cannot be welded, twisted or riveted. Material failure occurs without warning and, if overloaded or exposed to sudden impact, the material is prone to sudden failure. Although the material will sustain heat, it will fail without warning in fire.

Steel

Steel has a high elasticity and can be welded together to form strong fabrications. The failure mechanism of steel is that it will generally buckle or become plastic (stretch) prior to failure, and does not fail without warning. Steel tables are available from steel manufacturers and the properties of various structural sections are available to undertake design.

Steel frame housing

Steel frame houses were constructed in the UK following World War II and comprise a steel frame, which is the load-bearing part of the structure, with an outer skin which can be brick or rendered block construction or profiled steel cladding. The steel frame is usually coated with bitumen or a proprietary moisture-resistant material to prevent damp passing to the steel frame from the outer skin. The internal wall typically comprises quilt insulation with a timber sub-structure and plasterboard finish. Two examples of a steel frame construction can be seen in Figures 3.27 and 3.28. The roof space normally comprises steel trusses, and as such this type of construction is easily identifiable.

The main problem associated with this type of construction is to ensure the steel frame does not suffer corrosion. The most vulnerable areas are at the base of the steel frame and under windows in rooms that are prone to condensation, such as kitchens or bathrooms. For this reason most mortgage lenders insist on structural surveys of this type of property to ensure the steel has not suffered any deterioration. This is achieved by removing bricks at a lower level and exposing the steel frame, so that an examination of its condition can be undertaken. A photograph of the steel frame exposed by the removal of the outer skin of brickwork can be seen in Figure 3.29. In this example the steelwork appears to be in very good condition with no corrosion.

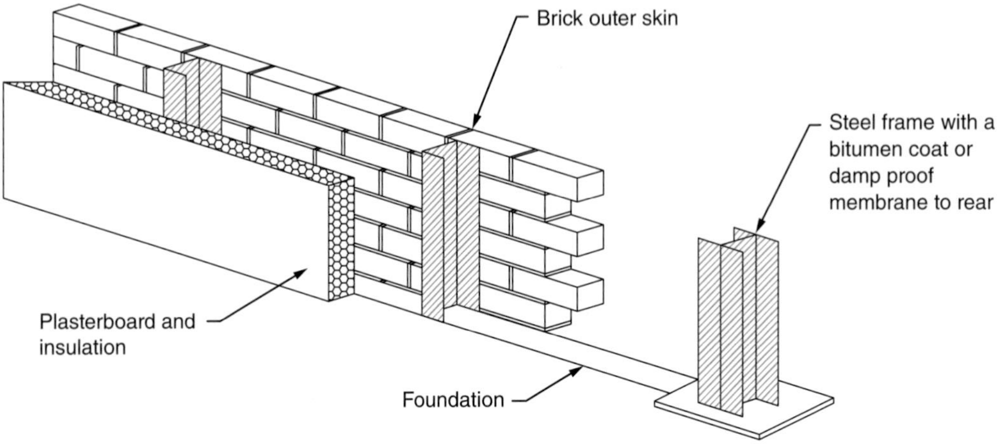

Figure 3.27: Example of a steel frame building.

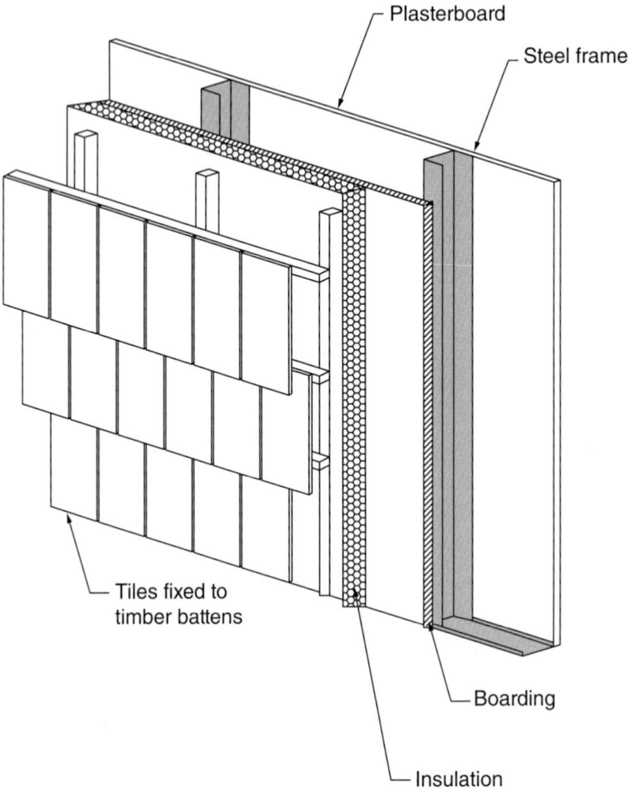

Figure 3.28: Another example of a steel frame building.

Figure 3.29: Photograph showing exposed frame of a steel frame house.

Commercial steel portal frames

Portal frame buildings can be used for supermarkets, warehouses, agriculture and other applications. The design of these structures is usually undertaken using structural design software, whereby a structural model is created and analysed using various wind-loading scenarios. The analysis will involve not only the individual structural steel members but also the connections and overall stability of the building.

The dead or permanent loading on such structures is relatively light compared with the imposed/variable and wind loads. The imposed/variable and dead/permanent loads are vertical but the wind loading is horizontal. The dead loads are relatively small and usually involve cladding and insulation. The side rails to which the cladding is attached transfer the permanent/dead loads and wind loads to the columns.

Wind loading is very important and can vary depending on a building's location, the topology of the site, the size of the building, exposure and its relationship to surrounding buildings. Wind loadings are generally carried to the foundation using bracing. The horizontal wind loads are usually carried by the columns, which experience shear and bending. Diagonal bracing in the roof and sides is used to ensure the horizontal loads are distributed effectively and efficiently to the columns. These are usually located spanning from the gable wall at eaves level and along the side walls to the first column from the gable.

The shape of the building is a key consideration, as one can imagine. If we consider a narrow building in relation to its length, then the building could be susceptible to overturning should high wind loads be encountered. In such circumstances the overturning will have to be overcome by the building's dead weight or the possibility of increasing the weight of the foundation. Consequently, a building of this nature may have to be constructed using a large pad foundation to overcome the resulting wind loading.

In general, this type of structure is constructed using large widths to overcome such problems. For spans up to 15 m, steel frame structures are primarily susceptible to deflection and it is this quality that is the usual concern to designers. Greater spans will place a larger emphasis on the wind loading.

For cost efficiency it is usually considered that the shorter the distance a load has to be carried to the foundation, the more cost-effective the building is. However, the usage of the building and other considerations – such as the placement of door openings – may inhibit the use of columns or structural members in key locations, thus inducing larger loads and consequently larger section sizes in other parts of the building.

Considering the above, a short span across the gable of say up to 15 m would normally have a frame spacing of between 3 and 5 m. For longer spans of 45–60 m the spacing of the frames is 10–12 m. This is based on experience and is for guidance only to emphasise the points above. As one can see, the shorter span of 15 m will be narrower in width and require more frequent load transfer to the foundation to prevent overturning. The larger building is not so susceptible to overturning due to the ratio of its length to span, and therefore the frames and transfer of load to the foundation can be less frequent.

The columns will experience the largest bending moment at the eaves or knee, where they connect to the rafters, and therefore the depth is increased to resist the moment by the use of a haunch. This also provides increased depth for the bolted joint connection between the column and rafter.

The bottom of the columns can be fixed or pinned. A fixed connection usually comprises anchor plates fixed into a concrete base. A pinned connection has a single pin or bolt from the column to the foundation block and there is no moment at the connection. However, if the column has two or more bolts and/or has haunches at its base, the connection is considered fixed. The connection type will affect the strength of the structure, as we saw in Chapter 2.

Precast concrete construction

Precast concrete construction occurred predominantly after World War II to provide cost-effective and easily constructed social housing that was in demand following the destruction of much of the housing stock.

Primarily, these types of house are of a modular form and comprise a concrete frame sited on a concrete slab. A typical section of an Airey construction is seen in Figure 3.30.

The concrete frame is in-filled with precast concrete panels. In 1980 the discovery of corroded reinforcement in the columns of Airey-type houses led to the discovery of similar problems all over the country. This has resulted in difficulties mortgaging this type of property. The government introduced a scheme of assistance under the Housing Defects Act 1984, which offered assistance for owners of prefabricated reinforced concrete houses designed before 1960 that were sold to the public before the defects were generally known.

The Building Research Establishment undertook research on the problems and defects found in Airey houses and published a report in 1981. Further reports in 1982 led to the Building Research Establishment expressing concern over 17 common types of prefabricated concrete house.

The investigation concentrated on the condition of the structure, mainly the reinforced concrete load-bearing components. The main problem concerns moisture penetration, which affects the embedded steel reinforcement causing corrosion and rusting of

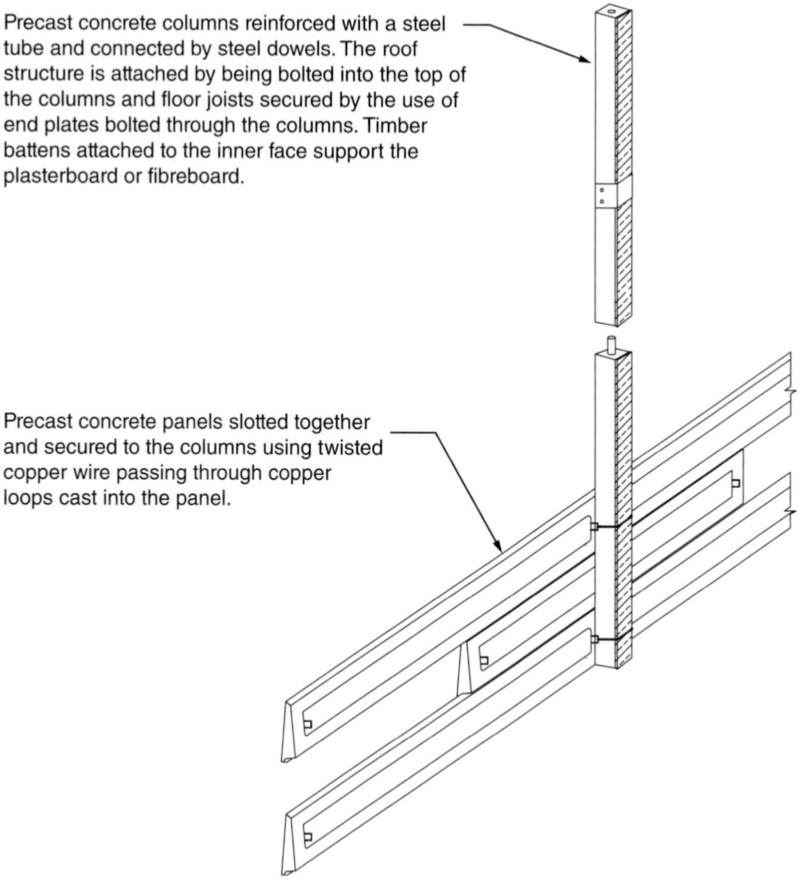

Precast concrete columns reinforced with a steel tube and connected by steel dowels. The roof structure is attached by being bolted into the top of the columns and floor joists secured by the use of end plates bolted through the columns. Timber battens attached to the inner face support the plasterboard or fibreboard.
Precast concrete panels slotted together and secured to the columns using twisted copper wire passing through copper loops cast into the panel.

Figure 3.30: A typical precast concrete Airey construction.

the steel. When the steel corrodes and rusts it expands up to seven times its original volume, spalling the concrete and consequently compromising the structural integrity of the concrete. The time scale varies considerably, and this may take as long as 30 years.

These properties can be repaired by specialist companies and in 1985 the National House Building Council (NHBC) approved arrangements for a warranty scheme for the repair of such homes. A subsidiary company was later established known as PRC Homes Ltd. As a designer or builder, a system of repairs is submitted to the NHBC for approval and upon receipt they offer a 10-year guarantee. Any repairs to this type of property have to be supported by a PRC (precast reinforced concrete) certificate and without this it is not possible to obtain a mortgage. In 1996 PRC Homes Ltd ceased to exist, but any repairs still have to be undertaken using a recognised licensed repairer.

Under the Housing Defects Act 1984 the following types of prefabricated concrete house were designated as defective: Airey, Boot, Butterley, Cornish Unit, Dorran, Dyke, Gregory, Hawkley SGS, Myton, Newland, Orlit, Parkinson, Reema, Schindler, Stent, Stonecrete, Terran, Underdown, Unity, Waller, Wates, Wessex, Winget and Woolaway. Some of the more common constructions are detailed below, with the main problems affecting their construction.

Airey

The Airey construction comprises reinforced concrete columns 104 mm × 57 mm in cross-section, which are reinforced with steel tubing. Precast concrete cladding panels are secured to the columns using copper wire to form the external weather-proof skin of the building.

The internal walls are blockwork to the ground floor and to the first floor plasterboard, which is fixed to timber batons running along the length of the concrete columns.

The main defect is the corrosion of the reinforcing tubes in the columns, causing cracking and spalling of the concrete.

Cornish Unit

The ground floor comprises precast concrete columns, with a joint at the base to a concrete plinth. Two unreinforced concrete panels are slotted into grooves in the columns to form the inner and outer skin of the external walls.

At the top of the columns is a precast concrete cornice which transfers the load from the roof and floor to the columns. The first floor and roof are constructed in timber and form a mansard roof, which is regarded as an independent timber structure supported on the columns below.

The main problem with Cornish dwellings is corrosion of the reinforcement in the columns, resulting in longitudinal cracking along the columns. A typical Cornish dwelling can be seen in Figure 1.10 of Chapter 1. The photograph shows an original construction and on the right-hand side of the photograph there is a Cornish dwelling which has been repaired under the Housing Defects Act 1984.

Reema

These types of property are constructed using hollow precast concrete panels which are lightly reinforced and accommodate a ring beam and columns at their edges. The columns and ring beam contain more substantial reinforcement. The units comprise a precast concrete floor anchored to the first floor ring beam and the roof is of timber construction.

The main problems are as a result of corrosion of the reinforcement in some panels. This can affect the window and door surrounds and some floor beams may be affected. The deterioration is largely as a result of carbonation or high chloride content.

Terran

These are constructed with storey-height reinforced concrete panels some 16 inches (405 mm) wide. The panels have an aggregate finish and are supported on a timber base kerb. This kerb is placed over a concrete slab foundation. Panels are bolted together and a timber baton on the flanges of the concrete allows the plasterboard to be attached internally. The wall plate is of timber construction supporting a low-pitch roof covered with preformed sheeting.

The main problems are corrosion of the reinforcement in the panels.

Unity

The building is constructed using precast reinforced concrete columns at approximately 3 foot centres (915 mm). The external wall is faced with precast concrete slabs, typically with an exposed aggregate finish. In earlier examples the inner skin is also precast concrete panels and the internal and external panels are tied using a copper tie across the cavity. However, in some cases the internal skin can be plasterboard. In later construction the external cladding is tied directly to the columns using copper ties and the inner wall is of block construction.

The roof and first floor comprise pressed steel beams bolted to the columns.

The main problem is deterioration of the columns, which is particularly prevalent where the inner skin is of plasterboard. Where the inner and outer skins are concrete, these can offer a means of alternative support to prevent collapse.

Less serious is the displacement of the external panels, and this can lead to water penetration.

Parkinson

Parkinson construction is identified by codes, and all types are of different layout but the same construction. The properties are formed of a series of prefabricated reinforced concrete beams and columns which interlock and bolt together at their junction. Ring beams 50 mm thick run in pairs, forming a cavity between the inner and outer skin. The inner and outer beams support concrete block walls, thus forming an inner and outer skin.

The main concern is corrosion of the reinforced concrete components, particularly around the beam/column connections, which can result in failure of the blockwork panels and partial collapse. This is only where considerable corrosion has occurred and is unlikely to lead to progressive collapse of the structure. The defective joint would be evident long before the collapse occurred.

Wates

The Wates construction comprises precast reinforced concrete panels which are load bearing. A precast concrete ring beam locates and holds the panels at first-floor and eaves level. The joints between the panels are filled with cement mortar. The foot of the panel is directly bolted to a slab foundation.

At first floor timber joists are attached to a wall plate bolted to the ring beam. The roof is of timber construction, with timber trusses bolted to the eaves ring beam.

The main problem is deterioration within the panels and at the joints between the panels. In some cases deterioration can occur at the ring beams at eaves and first-floor level.

Woolaway

The Woolaway construction comprises a lightweight, storey-height precast concrete column spaced at 762 mm centres. At the base of the column is a plinth which locates the columns. This acts as a ring beam and further ring beams are located at first-floor and eaves level. There are no fixings between the columns and ring beams. The columns and ring beams are cross-shaped in section and the half-storey-height panels are bolted to the flanges of the column and beams.

The roof is of timber construction, and gable walls and party walls are in-filled with blockwork. The external lining is covered in a render and the internal skin usually has a softboard lining.

The main problem with this type of construction is water penetration, which seriously corrodes the reinforcement, particularly to columns and ring beams. The lightweight concrete components contain small air pockets, which mean that corrosion of the reinforcement can remain undetected until serious deterioration has occurred. All concrete components in the external walls of Woolaway-type houses are prone to serious deterioration of the reinforcement, but deterioration of the ring beam can cause serious lateral instability.

Repairs

In some cases these types of structure have been repaired by cladding the concrete with an external brick skin. This offers two benefits, firstly protecting the wall and secondly allowing an opportunity for increased insulation in the cavity. In some cases the load is transferred from the concrete frame to the external wall, and care has to be taken if undertaking alterations on such properties to fully understand the load paths. It is advisable to seek the advice of a suitably qualified person, not only for the structural parameters but also to ensure the PRC certificate is issued.

Chapter 4 Steel

Steel properties

The properties of steel sections are available from Corus Construction & Industrial in a publication entitled "Structural sections to BS4: Part 1: 1993 and BSEN 10056 1999". This provides the dimensions of the steel sections, ratios for local buckling, second moment of area, radius of gyration, elastic modulus, plastic modulus, buckling parameter, torsional index, warping constant, torsional constant, the area of the section and the mass per metre. These parameters are used in the design of steel members.

Steel is usually graded as one of the following: S275, S355 or S460. The S means structural and the numbers relate to the yield strength (N/mm^2). The most common grade of steel used in buildings is S275.

Some of the most common section types can be seen in Figure 4.1.

We have already considered end restraints and the effect these have on the behaviour and strength of structural members in Chapter 2, and this is developed further in Chapter 8. The behaviour of steel sections is no exception, and when designing steel members consideration has to be given to the lateral restraint of the flange and the end supports.

Lateral torsional buckling

If we consider a steel beam with no restraint to the flange between the supports, the top flange can experience sideways lateral movement relative to the flange below. A plastic ruler held on its edge can be used to demonstrate this. If the ruler is held over a table as a cantilever and loaded on the top edge, the ruler will experience a sideways movement which is the lateral torsional buckling. In this case, as in reality, it is unlikely that any load placed on a beam will be located exactly centrally and as such there is an eccentric loading of the beam causing twisting.

Such restraint can be afforded by a concrete floor laid and fixed over the top of the beam, and this prevents rotation of the beam. Another example is where other beams connect into the side of the subject beam and afford varying degrees of lateral restraint depending on numbers and spacing. Without this, if there is no restraint to the flange lateral torsional buckling, a resulting twisting or skewing of the beam can be experienced. During the design of beams this action is taken into account by increasing the effective length of the beam.

The effect of end restraints on a beam

In addition to this, the end restraints also affect the strength of a beam and this is dependent on the type of connection. If no lateral restraint exists at its end, then for strength calculation purposes the beam's effective length is considered to be greater than

Structural Design of Buildings, First Edition. Paul Smith.

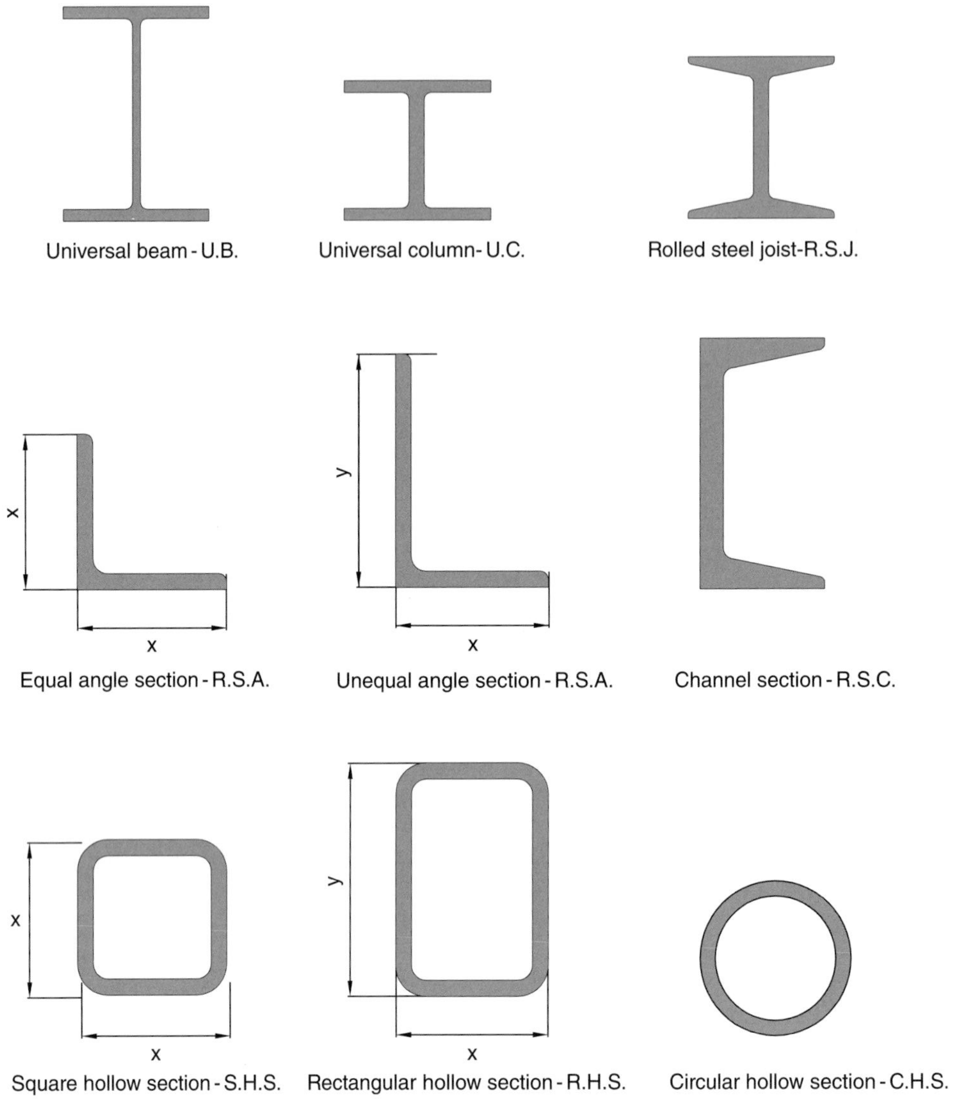

Figure 4.1: Rolled steel sections.

its actual length. If there are restraints at the end of the beam, or it is stiffened along its length, then the effective length can be reduced and the length can be considered to be only 0.7 of its actual length under the right conditions, thus reducing the size of the beam required. The same principle also applies to columns and the effective length for a column that is restrained in direction at both ends (i.e., is fixed) is 0.7 of its actual length between the supports. A column that is pinned at both ends is not considered to be restrained in direction, making its effective length equal to its actual length.

To this end the design standards provide intermediate and end restraint conditions depending on the type of connection. For example, a beam member resting on a support will have much less strength than a beam where the compression flange is laterally restrained by being bolted to a column.

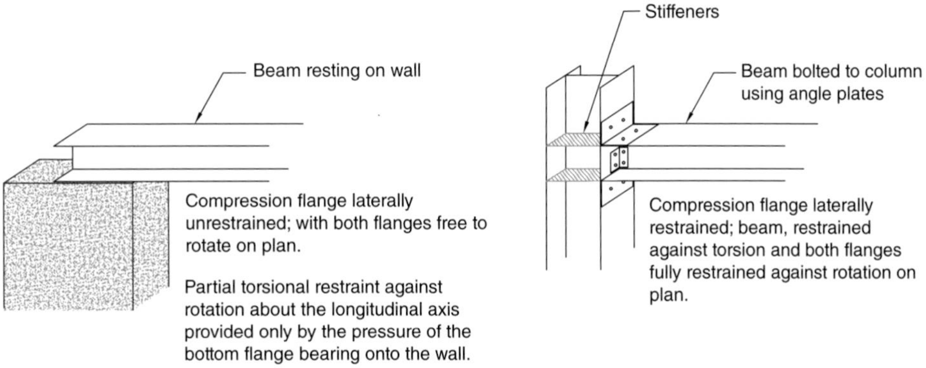

Figure 4.2: Diagrams showing conditions of end restraints for structural calculation examples.

Let us consider a beam that is 4.0 m long and loaded with a dead load of 5 kN/m and an imposed load of 4 kN/m. We assume there is no intermediate lateral restraint (i.e., no restraint between the supports). The beam rests on a pier and under the design standards this is considered to have restraints at the supports as shown in Figure 4.2. The bottom flanges are said to be supported with no positive connection.

The following calculations have been undertaken using structural engineering software, and we gratefully acknowledge Tekla (UK) Ltd for their approval in the use of this software.

STEEL BEAM ANALYSIS & DESIGN (EN1993-1-1:2005).

In accordance with EN1993-1-1:2005 incorporating corrigenda February 2006 and April 2009 and the UK national annex

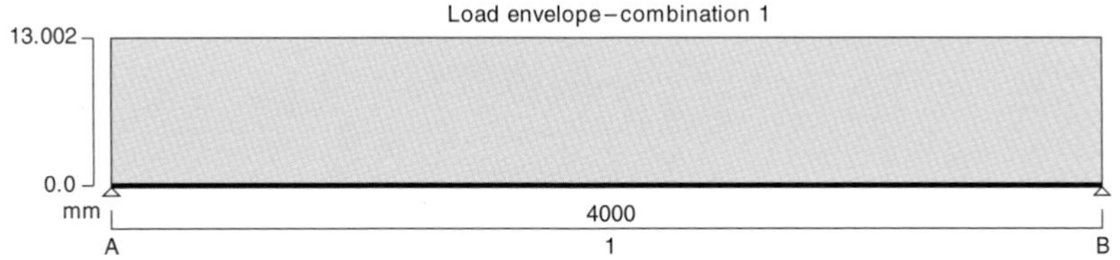

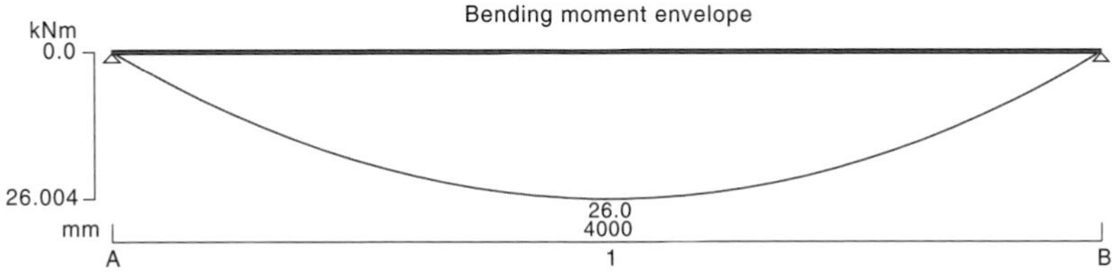

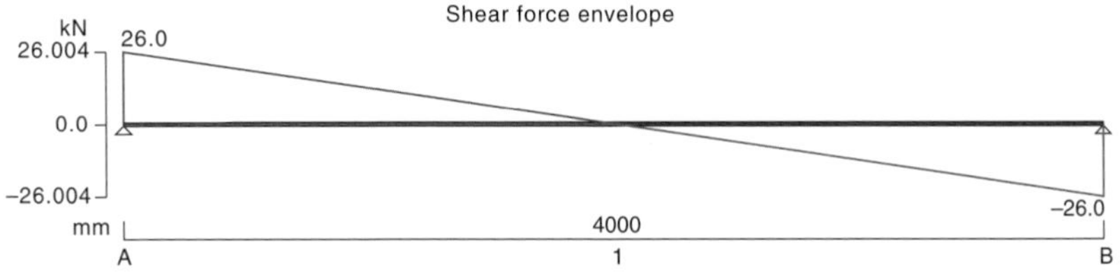

Support conditions

Support A	Vertically restrained Rotationally free
Support B	Vertically restrained Rotationally free

Applied loading

Beam loads	Permanent full UDL 5 kN/m Variable full UDL 4 kN/m Permanent self-weight of beam × 1

Load combinations

Load combination 1	Support A	Permanent × 1.35 Variable × 1.50
	Span 1	Permanent × 1.35 Variable × 1.50
	Support B	Permanent × 1.35 Variable × 1.50

Analysis results

Maximum moment;	M_{max} = **26** kNm;	M_{min} = **0** kNm
Maximum shear;	V_{max} = **26** kN;	V_{min} = **−26** kN
Deflection;	δ_{max} = **4.7** mm;	δ_{min} = **0** mm
Maximum reaction at support A;	R_{A_max} = **26** kN;	R_{A_min} = **26** kN
Unfactored permanent load reaction at support A;	$R_{A\_Permanent}$ = **10.4** kN	
Unfactored variable load reaction at support A;	$R_{A\_Variable}$ = **8** kN	
Maximum reaction at support B;	R_{B_max} = **26** kN;	R_{B_min} = **26** kN
Unfactored permanent load reaction at support B;	$R_{B\_Permanent}$ = **10.4** kN	
Unfactored variable load reaction at support B;	$R_{B\_Variable}$ = **8** kN	

Section details

Section type;	**UKB 178×102×19 (Corus Advance)**
Steel grade;	**S235**

EN10025-2:2004 – Hot rolled products of structural steels

Nominal thickness of element;	$t = \max(t_f, t_w)$ = **7.9** mm
Nominal yield strength;	f_y = **235** N/mm^2
Nominal ultimate tensile strength;	f_u = **360** N/mm^2
Modulus of elasticity;	E = **210000** N/mm^2

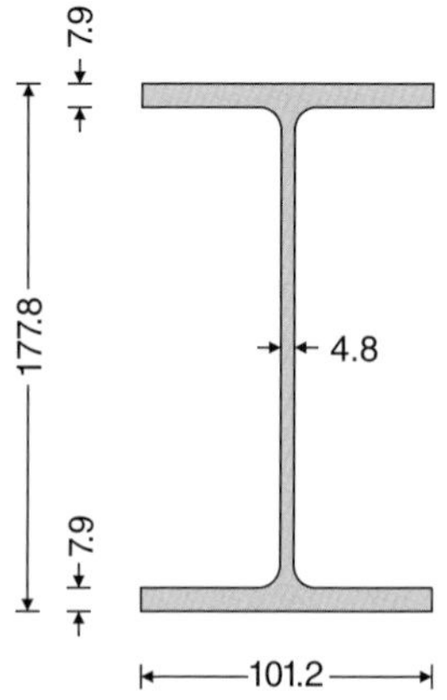

Partial factors – Section 6.1

Resistance of cross-sections;	γ_{M0} = **1.00**
Resistance of members to instability;	γ_{M1} = **1.00**
Resistance of tensile members to fracture;	γ_{M2} = **1.10**

Lateral restraint

Span 1 has lateral restraint at supports only

Effective length factors

Effective length factor in major axis; $K_y = \mathbf{1.000}$

Effective length factor in minor axis; $K_z = \mathbf{1.000}$

Effective length factor for torsion; $K_{LT.A} = \mathbf{1.200}$; + 2 × h

$K_{LT.B} = \mathbf{1.200}$; + 2 × h

Classification of cross-sections – Section 5.5

$\varepsilon = \sqrt{[235\ \text{N/mm}^2/f_y]} = \mathbf{1.00}$

Internal compression parts subject to bending – Table 5.2 (sheet 1 of 3)

Width of section; $c = d = \mathbf{146.8}$ mm

$c/t_w = 30.6 \times \varepsilon <= 72 \times \varepsilon$; Class 1

Outstand flanges – Table 5.2 (sheet 2 of 3)

Width of section; $c = (b - t_w - 2 \times r)/2 = \mathbf{40.6}$ mm

$c/t_f = 5.1 \times \varepsilon <= 9 \times \varepsilon$; Class 1

Section is class 1

Check shear – Section 6.2.6

Height of web; $h_w = h - 2 \times t_f = \mathbf{162}$ mm

Shear area factor; $\eta = \mathbf{1.000}$

$h_w/t_w < 72 \times \varepsilon/\eta$

Shear buckling resistance can be ignored

Design shear force; $V_{Ed} = \max(\text{abs}(V_{max}), \text{abs}(V_{min})) = \mathbf{26}$ kN

Shear area – cl 6.2.6(3); $A_v = \max(A - 2 \times b \times t_f + (t_w + 2 \times r) \times t_f, \eta \times h_w \times t_w) = \mathbf{985}$ mm^2

Design shear resistance – cl 6.2.6(2); $V_{c,Rd} = V_{pl,Rd} = A_v \times (f_y/\sqrt{[3]})/\gamma_{M0} = \mathbf{133.7}$kN

PASS – Design shear resistance exceeds design shear force

Check bending moment major (y–y) axis – Section 6.2.5

Design bending moment; $M_{Ed} = \max(\text{abs}(M_{s1\_max}), \text{abs}(M_{s1\_min})) = \mathbf{26}$ kNm

Design bending resistance moment – eq 6.13; $M_{c,Rd} = M_{pl,Rd} = W_{pl.y} \times f_y/\gamma_{M0} = \mathbf{40.2}$ kNm

Slenderness ratio for lateral torsional buckling

Correction factor – Table 6.6; $k_c = \mathbf{0.94}$

$C_1 = 1/k_c^2 = \mathbf{1.132}$

Curvature factor; $g = \sqrt{[1 - (I_z/I_y)]} = \mathbf{0.948}$

Poisson ratio; $\nu = \mathbf{0.3}$

Shear modulus; $G = E / [2 \times (1 + \nu)] = \mathbf{80769}$ N/mm^2

Unrestrained length; $L = 1.2 \times L_{s1} + 2 \times h = \mathbf{5156}$ mm

Elastic critical buckling moment;

$M_{cr} = C_1 \times \pi^2 \times E \times I_z/(L^2 \times g) \times \sqrt{[I_w/I_z + L^2 \times G \times I_t/(\pi^2 \times E \times I_z)]} = \mathbf{25.6}$kNm

Slenderness ratio for lateral torsional buckling; $\bar{\lambda}_{LT} = \sqrt{(W_{pl.y} \times f_y/M_{cr})} = \mathbf{1.253}$

Limiting slenderness ratio; $\bar{\lambda}_{LT,0} = \mathbf{0.4}$

$\bar{\lambda}_{LT} > \bar{\lambda}_{LT,0}$ – ***Lateral torsional buckling cannot be ignored***

Design resistance for buckling – Section 6.3.2.1

Buckling curve – Table 6.5; b

Imperfection factor – Table 6.3; $\alpha_{LT} = \mathbf{0.34}$

Correction factor for rolled sections; $\beta = \mathbf{0.75}$

LTB reduction determination factor; $\phi_{LT} = 0.5 \times \left[1 + \alpha_{LT} \times (\bar{\lambda}_{LT} - \bar{\lambda}_{LT,0}) + \beta \times \bar{\lambda}_{LT^2}\right] = \mathbf{1.234}$

LTB reduction factor – eq 6.57; $\chi_{LT} = \min\left(1/\left[\phi_{LT} + \sqrt{\left(\phi_{LT^2} - \beta \times \bar{\lambda}_{LT^2}\right)}\right], 1, 1/\bar{\lambda}_{LT^2}\right) = \mathbf{0.549}$

Modification factor; $f = \min\left(1 - 0.5 \times (1 - k_c) \times \left[1 - 2 \times (\bar{\lambda}_{LT} - 0.8)^2\right], 1\right) = \mathbf{0.982}$

Modified LTB reduction factor – eq 6.58; $\chi_{LT,mod} = \min(\chi_{LT}/f, 1) = \mathbf{0.559}$

Design buckling resistance moment – eq 6.55; $M_{b,Rd} = \chi_{LT,mod} \times W_{pl.y} \times f_y/\gamma_{M1} = \mathbf{22.5}$ kNm

FAIL – Design bending moment exceeds design buckling resistance moment

Check vertical deflection – **Section 7.2.1**
Consider deflection due to variable loads

Limiting deflection; $\delta_{lim} = L_{s1}/360 = \mathbf{11.1}$ mm
Maximum deflection span 1; $\delta = \max(\text{abs}(\delta_{max}), \text{abs}(\delta_{min})) = \mathbf{4.682}$ mm

PASS – Maximum deflection does not exceed deflection limit

We can see that the buckling resistance moment is exceeded and the beam fails.

In our next example let us consider the same beam but this time at the other extreme, whereby the beam is connected to a column at both ends and the type of connection ensures that the restraint is much stronger. However, we will still assume no intermediate lateral restraint.

The restraint condition at the ends of the beam is fully restrained against rotation on plan. See Figure 4.2.

The following calculations have been undertaken using structural engineering software, and we gratefully acknowledge Tekla (UK) Ltd for their approval in the use of this software.

STEEL BEAM ANALYSIS & DESIGN (EN1993-1-1:2005).

In accordance with EN1993-1-1:2005 incorporating corrigenda February 2006 and April 2009 and the UK national annex

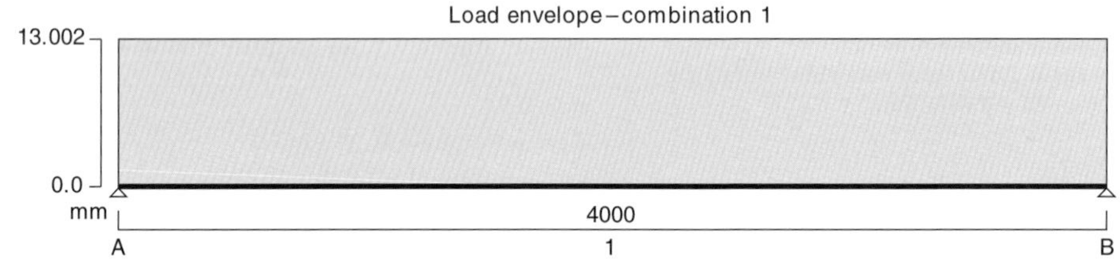

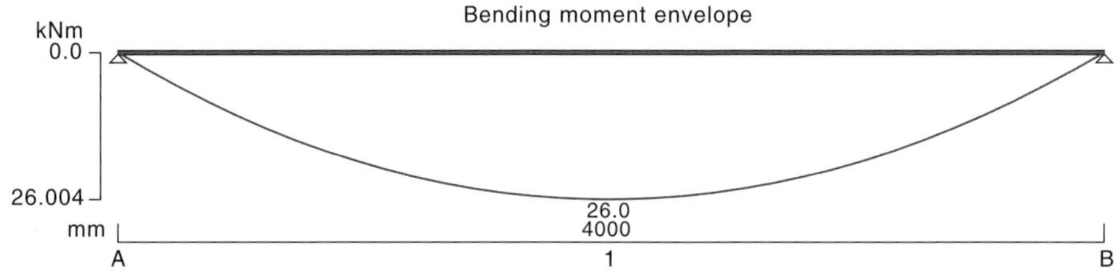

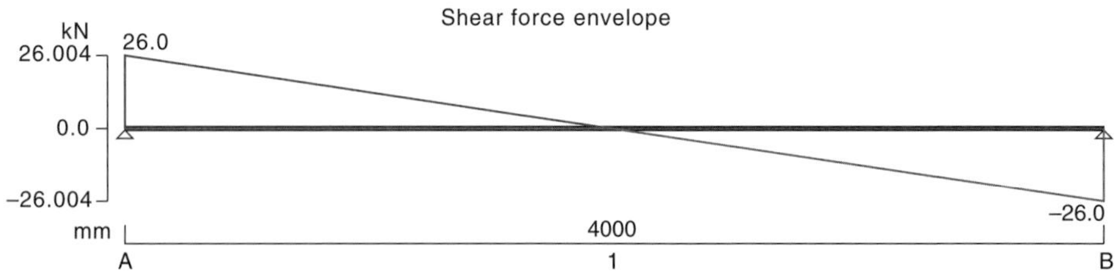

Support conditions

Support A	Vertically restrained Rotationally free	
Support B	Vertically restrained Rotationally free	

Applied loading

Beam loads	Permanent full UDL 5 kN/m Variable full UDL 4 kN/m Permanent self-weight of beam × 1	

Load combinations

Load combination 1	Support A	Permanent × 1.35 Variable × 1.50
	Span 1	Permanent × 1.35 Variable × 1.50
	Support B	Permanent × 1.35 Variable × 1.50

Analysis results

Maximum moment;	M_{max} = **26** kNm;	M_{min} = **0** kNm
Maximum shear;	V_{max} = **26** kN;	V_{min} = **−26** kN
Deflection;	δ_{max} = **4.7** mm;	δ_{min} = **0** mm
Maximum reaction at support A;	R_{A_max} = **26** kN;	R_{A_min} = **26** kN
Unfactored permanent load reaction at support A;	$R_{A\_Permanent}$ = **10.4** kN	
Unfactored variable load reaction at support A;	$R_{A\_Variable}$ = **8** kN	
Maximum reaction at support B;	R_{B_max} = **26** kN;	R_{B_min} = **26** kN
Unfactored permanent load reaction at support B;	$R_{B\_Permanent}$ = **10.4** kN	
Unfactored variable load reaction at support B;	$R_{B\_Variable}$ = **8** kN	

Section details

Section type;	**UKB 178×102×19 (Corus Advance)**
Steel grade;	**S235**

EN10025-2:2004 – Hot rolled products of structural steels

Nominal thickness of element;	$t = \max(t_f, t_w)$ = **7.9** mm
Nominal yield strength;	f_y = **235** N/mm^2
Nominal ultimate tensile strength;	f_u = **360** N/mm^2
Modulus of elasticity;	E = **210000** N/mm^2

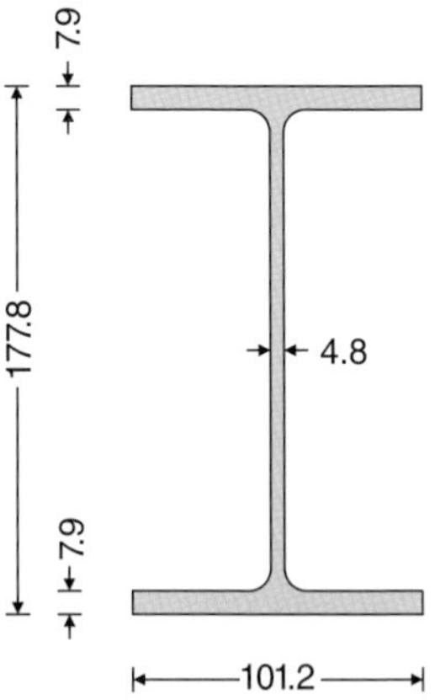

Partial factors – Section 6.1

Resistance of cross-sections; γ_{M0} = **1.00**

Resistance of members to instability; γ_{M1} = **1.00**

Resistance of tensile members to fracture; γ_{M2} = **1.10**

Lateral restraint

Span 1 has lateral restraint at supports only

Effective length factors

Effective length factor in major axis; K_y = **1.000**

Effective length factor in minor axis; K_z = **1.000**

Effective length factor for torsion; $K_{LT.A}$ = **0.700**;
$K_{LT.B}$ = **0.700**;

Classification of cross-sections – Section 5.5

$\varepsilon = \sqrt{[235\ \text{N/mm}^2/f_y]} = \mathbf{1.00}$

Internal compression parts subject to bending – Table 5.2 (sheet 1 of 3)

Width of section; c = d = **146.8** mm

$c/t_w = 30.6 \times \varepsilon <= 72 \times \varepsilon$; Class 1

Outstand flanges – Table 5.2 (sheet 2 of 3)

Width of section; $c = (b - t_w - 2 \times r)/2$ = **40.6** mm

$c/t_f = 5.1 \times \varepsilon <= 9 \times \varepsilon$; Class 1

Section is class 1

Check shear – Section 6.2.6

Height of web; $h_w = h - 2 \times t_f$ = **162** mm

Shear area factor; η = **1.000**

$h_w/t_w < 72 \times \varepsilon/\eta$

Shear buckling resistance can be ignored

Design shear force; $V_{Ed} = \max(\text{abs}(V_{max}), \text{abs}(V_{min}))$ = **26** kN

Shear area – cl 6.2.6(3); $A_v = \max(A - 2 \times b \times t_f + (t_w + 2 \times r) \times t_f, \eta \times h_w \times t_w)$ = **985** mm^2

Design shear resistance – cl 6.2.6(2); $V_{c,Rd} = V_{pl,Rd} = A_v \times (f_y/\sqrt{[3]})/\gamma_{M0}$ = **133.7**kN

PASS – Design shear resistance exceeds design shear force

Check bending moment major (y–y) axis – Section 6.2.5

Design bending moment; $M_{Ed} = \max(\text{abs}(M_{s1\_max}), \text{abs}(M_{s1\_min}))$ = **26** kNm

Design bending resistance moment – eq 6.13; $M_{c,Rd} = M_{pl,Rd} = W_{pl.y} \times f_y/\gamma_{M0}$ = **40.2** kNm

Slenderness ratio for lateral torsional buckling

Correction factor – Table 6.6; k_c = **0.94**

$C_1 = 1/k_c^2$ = **1.132**

Curvature factor; $g = \sqrt{[1 - (I_z/I_y)]}$ = **0.948**

Poisson ratio; ν = **0.3**

Shear modulus; $G = E/[2 \times (1 + \nu)]$ = **80769** N/mm^2

Unrestrained length; $L = 0.7 \times L_{s1}$ = **2800** mm

Elastic critical buckling moment; $M_{cr} = C_1 \times \pi^2 \times E \times I_z/(L^2 \times g) \times \sqrt{[I_w/I_z + L^2 \times G \times I_t/(\pi^2 \times E \times I_z)]}$ = **56.4** kNm

Slenderness ratio for lateral torsional buckling; $\bar{\lambda}_{LT} = \sqrt{(W_{pl.y} \times f_y/M_{cr})}$ = **0.845**

Limiting slenderness ratio; $\bar{\lambda}_{LT,0}$ = **0.4**

Library item: LTB slenderness ratio output

$\bar{\lambda}_{LT} > \bar{\lambda}_{LT,0}$ – ***Lateral torsional buckling cannot be ignored***

Design resistance for buckling – Section 6.3.2.1

Buckling curve – Table 6.5; b

Imperfection factor – Table 6.3; α_{LT} = **0.34**

Correction factor for rolled sections; β = **0.75**

LTB reduction determination factor; $\phi_{LT} = 0.5 \times \left[1 + \alpha_{LT} \times (\bar{\lambda}_{LT} - \bar{\lambda}_{LT,0}) + \beta \times \bar{\lambda}_{LT^2}\right] = \mathbf{0.843}$

LTB reduction factor – eq 6.57; $\chi_{LT} = \min\left(1/\left[\phi_{LT} + \sqrt{\left(\phi_{LT^2} - \beta \times \bar{\lambda}_{LT^2}\right)}\right], 1, 1/\bar{\lambda}_{LT^2}\right) = \mathbf{0.792}$

Modification factor; $f = \min\left(1 - 0.5 \times (1 - k_c) \times \left[1 - 2 \times (\bar{\lambda}_{LT} - 0.8)^2\right], 1\right) = \mathbf{0.970}$

Modified LTB reduction factor – eq 6.58; $\chi_{LT,mod} = \min(\chi_{LT}/f, 1) = \mathbf{0.816}$

Design buckling resistance moment – eq 6.55; $M_{b,Rd} = \chi_{LT,mod} \times W_{pl.y} \times f_y/\gamma_{M1} = \mathbf{32.9}$ kNm

PASS – Design buckling resistance moment exceeds design bending moment

Check vertical deflection – Section 7.2.1
Consider deflection due to variable loads

Limiting deflection; $\delta_{lim} = L_{s1}/360 = \mathbf{11.1}$ mm

Maximum deflection span 1; $\delta = \max(\text{abs}(\delta_{max}), \text{abs}(\delta_{min})) = \mathbf{4.682}$ mm

PASS – Maximum deflection does not exceed deflection limit

We can see in this example that the beam passes the design standards. Even though the same beam has been employed under the same loading conditions, the second beam is much stronger due to the end restraint conditions.

The effective length used for strength calculation purposes in the second analysis is 0.7lt (where lt is the length of the beam between the supports), whereas in the unrestrained example the effective length employed is 1.2lt + 2h (where h is the depth of the beam).

This can be demonstrated using a plastic ruler, which we consider will act in a similar manner to a column, and we apply a load to the top of the ruler – for example pushing the ruler with one's hand, it can be seen that it will move laterally. However, if the same ruler is clasped in a vice at one end and the same force applied to the top of the ruler, the lateral displacement is less and therefore the column is considered stronger and can sustain a much greater load. Further explanation can be found in Chapter 8, where we discuss the slenderness ratio and the Euler load.

Bending failure

We have previously seen how a beam subjected to a load will experience bending, and this induces longitudinal stresses across the beam. Figures 2.10 and 2.11 in Chapter 2 explain this theory. It can be seen that when all the section is in tension and compression above and below the neutral axis, if the load is increased the section will reach a point where it is considered to be fully plastic and the compression and tension stresses are said to have yielded. At this point plastic hinges form, the maximum moment has been reached and consequently the beam fails.

Local buckling

When a load is applied to a beam and the bending process commences, localised failure of the beam can occur due to bending before the plastic limit and full plastic moment is reached. This can occur in the flange or web of the section, but only where these parts experience compression and are too thin to sustain the load. This results in the section experiencing tearing or buckling.

Shear failure

A beam can fail in shear when the shear load is greater than the shear capacity of the section under consideration. It is the web of a steel beam that predominantly resists shear loads, and the web will experience both tension and compression stresses. The shear force is greatest at a support. The steel section yields at this point when the shear failure load is reached. Plastic hinges will also begin to form in the flanges of the steel during this process. If the web is too thin, buckling may occur in the web prior to the yield point and this appears as folds or ripples within the steel characterised by being at a 45° angle to the face of the web.

Web bearing and buckling

This type of failure occurs directly over a support or under a high concentrated load, where the load transfers through the flange to the web. Excessive loads can cause the web to crease or buckle as a consequence of the high stresses that the section is subjected to.

The different types of failure can be seen in Figure 4.3.

If full lateral restraint is provided, for example by a concrete floor which is capable of resisting a defined lateral load in the compression flange, then the buckling effects do not have to be checked. However, structural members not provided with full lateral restraint should be checked for buckling.

Deflection

Interestingly, deflection is not a mode of failure but is a serviceability consideration with respect to how the deflection affects finishes applied to the structure, for example plaster will crack under deflection. The limits placed on the deflections of structural members are based on the tolerances of the finishes to withstand the deflections under unfactored imposed loads only.

Fire and corrosion

Steel sections will require fire resistance depending on the usage and occupancy of the building in which they are employed, and this is outlined in the Building Regulations. Steel can be afforded such protection using plasterboard, where 12.5 mm will provide half-hour fire protection. Fire-retardant plasterboard is available, which is coloured pink, and a single sheet will offer one-hour fire protection.

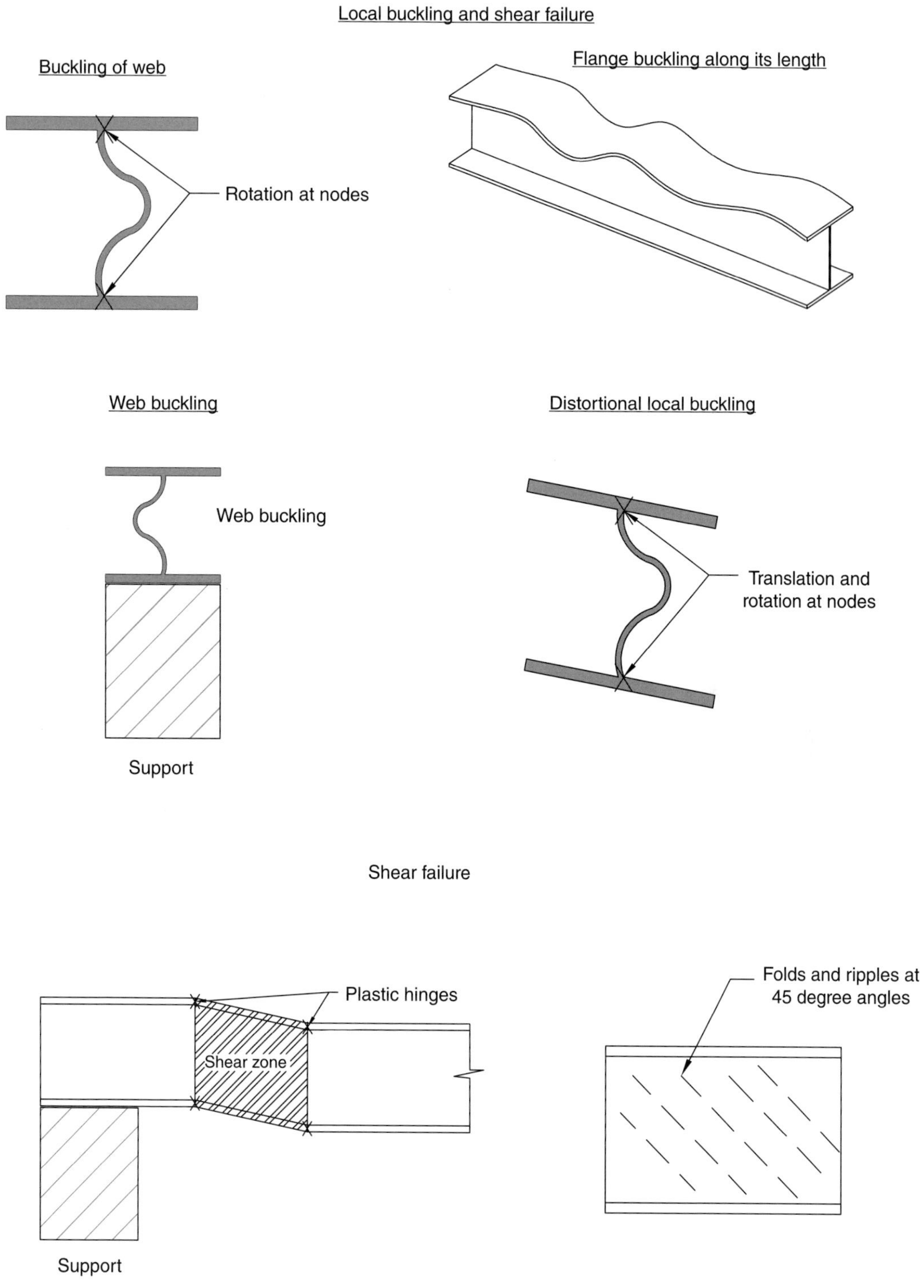

Figure 4.3: Five modes of failure of steel sections.

Other means of fire protection include encasing with concrete and fire-retardant sprays and intumescent coatings. At approximately 300°C steel begins to lose its strength, and at 600°C the steel becomes plastic.

Corrosion can also reduce the strength of steel considerably, and steel exposed to corrosive environments – including water – will require protective coatings.

Chapter 5 Concrete

The history of cement and concrete

The history of mortar and cement is covered in Chapter 8, which explains the development over time, but in the early 1800s cement became available and was known as Portland cement through its origin on the Island of Portland off the Devon coast near Weymouth.

In 1898 the Glenfinnan Viaduct was constructed on the West Highland Railway Line in Scotland, which is located between Malake and Fort William. This is a 21-arch viaduct and when it was completed, it was the largest concrete structure in Britain. Most people will recognise this structure as the viaduct that carries the Hogwarts Express in the Harry Potter movies. This demonstrates how far the use of concrete had developed during this century.

Cement

The most common cement is Ordinary Portland Cement, manufactured by heating limestone to form a clinker containing calcium silicate. The mixture is then ground to a fine powder and calcium sulphate (gypsum) is added to regulate the hydration. This is the chemical reaction which occurs when the powder is mixed with water. During the hydration volumetric changes occur in the concrete which can lead to linear contraction of up to 0.4 mm per metre. The amount of shrinkage is dependent on the volume of water used and thus the water:cement ratio. Another parameter affecting shrinkage is the rate of drying; in hot weather shrinkage tends to be greater, thus curing should be regulated to prevent excessive shrinkage.

A variety of other cements are available, such as the following.

Rapid-hardening cement

Rapid-hardening cement is finer than Ordinary Portland Cement and consequently the finer particles in the cement achieve strength and cure much more quickly.

Sulphate-resisting cement

This is usually darker in colour than Ordinary Portland Cement due to the greater quantity of tetracalcium aluminoferrite. When sulphates attack concrete they react with tricalcium aluminate constituent and weaken the concrete. In sulphate-resisting cement this constituent is reduced and substituted with tetracalcium aluminoferrite.

Structural Design of Buildings, First Edition. Paul Smith.

Sulphates are present underground and in seawater, and although under normal circumstances these types of cement are able to resist the onset of sulphate attack, this is dependent on the cement content and the concentration of sulphates.

Low-heat cement

This is used where large volumes of concrete are needed and the control of the evolution of the heat and strength is much slower to ensure cracking does not occur through the high temperatures encountered when using ordinary cement. However, the end strength is not affected. The heat released is much the same, but the rate is controlled.

White cement

White cement is produced using white china clay and limestone to reduce the coloured components of the mix and produce a white concrete for decorative purposes.

All these cements are produced in the same way and the proportions are altered depending upon their particular usage.

Hydrophobic cement

This cement is coated to resist hydration, and this allows the cement to be stored for long periods of time in moist atmospheres. During manufacture a coat of water-repellent compound such as oleic acid is added which is rubbed off by friction during use in the cement mixer, and therefore the cement behaves in the normal manner.

Water and workability – now known as consistence

The workability of concrete is the ease with which it can be worked or compacted. The workability can be tested using a slump test, which provides an indication of changes in the water and materials contained in the concrete. The test is conducted using a conical mould, which is 100 mm in diameter at the top and 200 mm at the bottom and 300 mm high. The mould is filled in four layers and each is tapped with a tamping rod 25 times. The top layer is tapped or rodded and the surplus concrete is struck off level with the mould. The mould is then inverted and removed, leaving the concrete which slumps. The amount of slump is measured against the side of the mould. EN 12350-2 sets out the standards for this test.

The consistency of the materials used and the water content will clearly affect the slump. Hence, increased volumes of water will increase the slump. Variations in water should be kept to a minimum, and the minimum amount of water necessary to provide workability should be used. Saturated aggregates will add to the volume of water used, and such variations need to be regulated.

The amount of water affects the strength of the concrete, and this is determined by the water:cement ratio. If a water:cement ratio of 0.4 is specified then the volume of water per 25 kg bag of cement should be $0.4 \times 25 = 10\,\text{kg}$, which is 10 litres (density of water 1000 kg/m^3). Note that this also includes any water or moisture in the aggregates. In many cases, when foundations are poured, the concrete lorries delivering the ready mix wash out and pour the concrete water residue into the foundation trench, thus weakening the ordered mix.

Clearly there is always a temptation to increase the water content to make the concrete more workable, and thus make the work easier, but the consequences of doing so adversely affect the strength.

The term "consistency" is used instead of "workability" in the new standards BS EN 206-1 and BS 8500. The consistency is graded into classes identified using the letter S. For example, consistency class S2 has a limit on the slump of 50–90 mm, with a maximum deviation on range limit 30–110 mm.

Failure of concrete

Concrete, unlike steel, is not an elastic material and the stress–strain curve is more curved. See Figure 5.1 for a graph of stress versus strain for concrete. For stresses up to half the maximum, the curve is more or less a straight line. Beyond this point, the curvature becomes increasingly pronounced as the strain increases and the maximum stress point is reached. Beyond this point the strain can be increased, but is accompanied by a decline in the stress until the strain reaches its ultimate value and the concrete crushes.

There are two ways in which a reinforced concrete section can fail, referred to as type 1 and type 2.

Type 1	Due to failure of the steel in tension.
Type 2	Due to failure of the concrete in compression.

Type 1 failures are very rare due to the large strains which can occur in the steel after it has yielded and before it reaches its tensile strength. See Figure 5.2, which shows the stress–strain relationship for this type of failure.

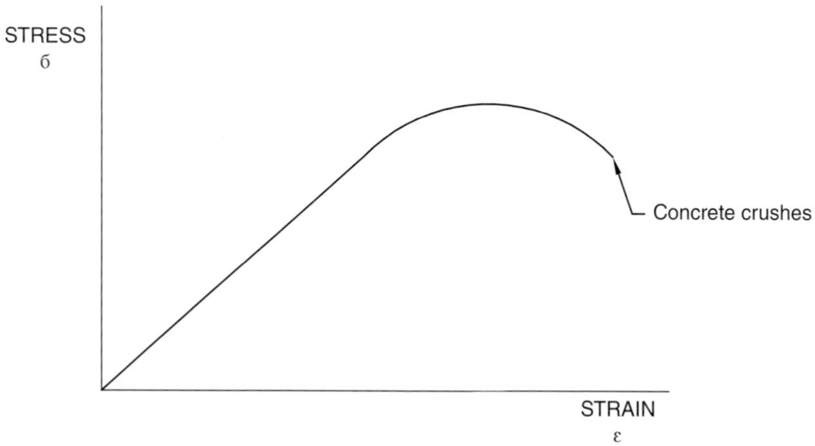

Figure 5.1: The stress–strain relationship for concrete.

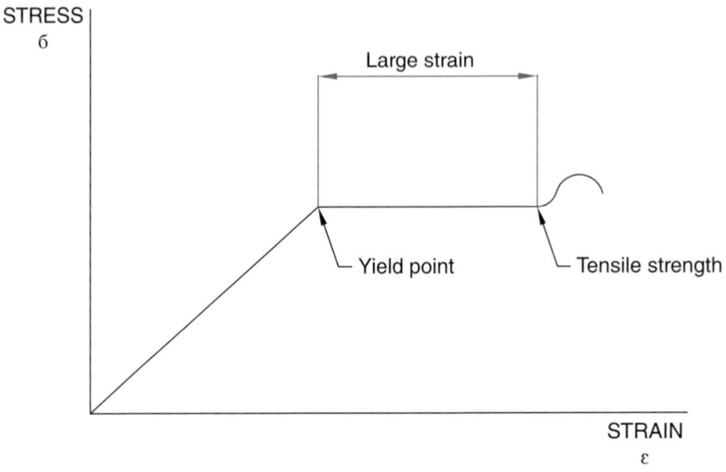

Figure 5.2: The stress–strain relationship for concrete where the steel yields before the concrete.

Since the design standards specify a minimum amount of reinforcement or steel, the failure described as type 1 can be neglected for all practical purposes.

Meanwhile, however, type 2 failures can be classified into two groups.

Primary tension failure

This occurs when relatively low percentages of steel compared to concrete are used. Remember that concrete is not good at resisting tensile forces, and this is the role of the reinforcement. As a load and resulting bending moment is applied to a beam, it will have a tension and a compression zone. The demarcation between these zones is the neutral axis. Within these zones is a compressive force and a tension force, separated by a distance known as the lever arm. Thus, the moment applied to the beam is resisted and should be less than the compression force multiplied by the lever arm – which is equal to the tension forces multiplied by the lever arm. This results in a moment of resistance to the applied moment.

Now, when the steel yields it can no longer absorb or resist any further increase in the tensile forces and moments applied, and therefore for the resistance moment to increase the lever arm has to increase. For this to happen clearly the neutral axis has to move and more of the section becomes in tension and less of the concrete in compression. Eventually so much of the concrete is in the tension zone that it finds it difficult to resist, and the section begins to fail. Essentially the steel yields whilst the concrete is still stressed elastically. Further increase in bending moment cannot produce an increase in the steel stress, so the required increase in the moment of resistance of the section is produced by an increase in the lever arm. Consequently, this mode of failure is characterised by a gradual rise in the neutral axis with increase in load until the depth of concrete in the compression zone is so reduced that the concrete crushes.

This type of failure is characterised by large visible cracks and large deflections which provide ample warning of impending collapse.

This type of failure is considered desirable, and the design codes ensure that all sections are designed to meet this requirement. These sections are referred to as "under-reinforced sections". This is not to be confused with the section not containing enough reinforcement.

Primary compression failure

This occurs when the steel percentage is high and the concrete crushes whilst the steel is still stressed elastically. There is very little noticeable cracking or deflection, and the section may fail explosively without any visual warning. Such sections are called "over-reinforced sections". Owing to this behaviour their use is forbidden under the design codes, because an over-reinforced section may lead to an explosive failure without warning if the load is sufficient.

Strength of concrete

Concrete is formed through the addition of water to the constituent parts, which allows hydration. The concrete will increase in strength over time and this can continue for a number of years. The strength of the concrete will depend on the original mix proportions and the environment in which it is placed. When designing structural members in concrete the strength provided is normally the strength gained after 28 days, although a 70- or 90-day strength can be specified depending on its application.

Within the first few months concrete will generally reach 90% of its strength, but will continue to strengthen depending upon the environment and the type of concrete used. If moisture is present and the right environment prevails, in theory the hydration process will continue and the concrete can continue to gain marginal strength infinitum.

Concrete is designed by referring to a compressive strength, and this is determined by compressive tests as outlined in BS EN 12390-1 (2000) for cubes. Under BS EN 12390-1 the concrete is formed into 100 mm cubes and tested using a compression test after 28 days.

Under Eurocode 2 the compressive strength of concrete is determined by concrete strength classes relating to a 5% cylinder strength fck or the cube strength fckcube. The concrete is prepared by making cylinders using 100 mm diameter, 200 mm deep moulds and then tested by compression. This procedure is outlined in EN 206-1. The cubes will not all fail at the same compressive strength, and a normal distribution curve results. Design standards dictate that the characteristic strength of concrete shall be taken as the value at which not more than 5% of the test results fail.

Hence, for a concrete of strength 35 N/mm^2, this means that the concrete has characteristic compressive cube strength of 35 N/mm^2. In some circumstances this can be written as C28/35, where the 28 is the characteristic cylinder strength followed by the 35 which is the characteristic cube strength (both in N/mm^2).

To specify the concrete the designer can use a designated mix which specifies the strength class of the concrete, for example RC 28/35. This means that the concrete is to be used for reinforcement and normally a maximum aggregate size would also be specified.

Concrete mix designs

Concretes were originally designed and specified in accordance with BS 5328:1 Section 5, however this standard was withdrawn in 2003 and replaced with BS EN 206-1 "Concrete specification performance production and conformity" and the BS complementary standard BS 8500 published in 2002. Part 2 of this standard is intended for producers and specifies the basic requirements for concrete relating to the types of concrete listed in BS 8500-1; it also specifies the requirements relating to delivery and testing.

Within the standard, alterations to the names of mixes were introduced and "designated mixes" are now known as "designated concrete" and "standard mixes" as "standardised prescribed concretes".

Proprietary concrete

This type of concrete allows the supplier to produce a concrete for a particular application without declaring the ingredients or composition of the concrete. However, the supplier is required to provide some assurance that the concrete will meet the expected requirements. It is usually used in large contracts, where substantial amounts of the same concrete are required – for example for driveways.

Designed concrete

This type of specification requires the strength to be specified and this may be a designed concrete, where the specification states the limitations on its use. For example, use in an industrial floor. The person specifying the concrete should include the following:

Compressive strength
Water:cement ratio
Minimum cement content
Type of cement
Maximum aggregate size
Consistence class or slump

It is particularly important to use designed concrete where special cements are required, or where concretes are exposed to seawater or chlorine.

Designated mix – now known as "designated concrete"

Designated mixes are those where a designation is assigned to the concrete specification and the strength and other factors – such as workability – have been decided and the concrete mix must meet these criteria. These mixes are usually delivered as ready-mixed concrete. The mixes are prepared under strict quality control conditions.

These concretes usually have suffices as follows: GEN, which is a general-purpose concrete; RC, which is a concrete for reinforcement; PAV, which is for paved areas; and FND, for concretes in corrosive environments such as sulphate environments. The new standards changed the way in which concrete is specified, and under the new standards concrete is typically specified as follows:

- *Concrete designation.* For example, concrete designation produced in accordance with EN 206-1 and BS 8500-2.
- *Designation of the concrete.* For example, GEN1 or RC35.
- *Maximum aggregate size.* For example, if this is greater than 20 mm then the aggregate size will need to be specified.

- *Consistence class*. This is the slump graded from S1 to S4. For example, a slump of 50–90 mm will have a consistence class 2.

Standard mixes – now known as "standardised prescribed concretes"

Prescribed mixes are specified by fixed-weight proportions of their constituents. The grades associated with these mix proportions are the minimum characteristic strength expected to be achieved and are similar to lower-strength designated concretes. Mix proportions and materials are provided in the standards to achieve the required grades. These are used with the suffix ST and have codes ST1 to ST5.

Prescribed

This type of concrete is where the person specifying the concrete prescribes the exact composition and constituents of the concrete. A simple example is four parts gravel, one part sand and one part cement. The strength is not defined.

Creep

When a concrete section is subjected to a sustained load over time a deformation takes place, and this process is known as creep. The deformation can result in cracking, which will then allow damage to finishes such as plaster. The creep can be reduced through the aggregate and the mix design. This is a phenomenon that is not confined to just concrete, but also affects other materials.

Environment

The environment in which concrete is employed is considered in Eurocode 2 Table 4.1, which is based on EN 206-1. Such considerations include the durability and description of the environment. The environment is provided with a class designation ranging from no risk of corrosion or attack through a range of possibilities including corrosion by carbonation, chlorine attack, seawater, freezer thaw attack and chemical attack. The class designation determines the cover to the reinforcement and the grade of concrete to be used. For the UK the cover and grade are provided in BS 8500-1 as a National Annex to the Eurocode.

Air-entrained concrete

Air entrainers produce tiny bubbles in the concrete mixture to reduce the damage caused by the freeze–thaw process and the action of de-icing salts. The bubbles are small and cannot be seen with the naked eye, but have the advantage of increasing the workability of the concrete. However, although the durability of the mix is improved, the use of such products reduces the strength of the concrete. An air entrainer such as washing-up liquid is sometimes used by builders to the side of shuttering to ensure that when the shuttering is struck, the concrete does not spall.

Accelerators and retarders

These are added to increase or reduce the hydration process and consequently speed up or slow down the setting. This may be important depending on the environment and site conditions under which the concrete is being poured. Accelerators are made from chlorides or nitrate compounds. Retarders comprise sugars, glucose or citric acid.

Plasticizers

Plasticizers increase the workability of a mortar or concrete and can be used instead of water to achieve the same effect. The advantage of a plasticizer rather than increasing the water content is that the water:cement ratio is maintained and the concrete does not lose any strength.

Fly ash, silica flume and ground granulated blast furnace slag

These materials are added to concrete to replace the cement content in the mix. The production of concrete through the use of cement adds a considerable burden to the environment, and the use of recycled products such as these can reduce greenhouse emissions. These products have hydraulic properties, which improve the concrete and reduce the amount of cement required.

Anti-corrosion

Corrosion is a particular problem when using reinforced concrete, and de-icing salts can attack the reinforcement. The reinforcement can be treated using corrosion-resistant reinforcement such as low carbon/chromium reinforcement, which is usually marked by the steel mill on the reinforcement bar. Other measures include epoxy-coated reinforcement, which is identified because it is a light green in colour. Stainless steel or galvanised reinforcement can also be employed, and this is usually grey in colour. We have already discussed the various types of concrete and how the correct choice can limit chemical attack and anti-corrosion. We have also discussed the use of cover to the reinforcement, depending on the environment in which the concrete is to be used.

Chapter 6 Timber

Grading of timber

Timber is graded in accordance with BS 5268 and the strength is determined by considering a number of parameters including moisture content, knots, wane, slope of the grain and splits. There are two methods by which to assess the grade of the timber. The first is a visual inspection undertaken by a suitably qualified person in compliance with BS 4978. The second is machine grading undertaken by measuring the resistance of the timber member to flexing, which is a measure of its strength. BS EN 14081 and BS EN 519:1995 outline the requirements for machine grading.

Strength is grouped into classes, namely C14–C50 for softwoods and D30–D70 for hardwoods. Trusses are graded TR26 and are softwood. The grade of softwood timber usually available is either C16 or C24. BS 5268 also includes a TR20 strength grade. A typical grading stamp can be seen in Figure 6.1.

Moisture

The moisture content is important in both softwoods and hardwoods, and the design codes reduce the strength and stiffness of the timber depending on the exposure of the timber during its life as a structural member and the initial moisture content. Moisture will cause splitting and twisting of the timber as it dries, and the strength of oak is greatly affected depending on whether it is specified as kiln dried or green. Timber should always be purchased when possible as dry graded, and the average water content for timber should not exceed 20%, with no single piece exceeding 24%.

Under BS 5268 clause 2.6.2 and Eurocode 5 clause 2.3.1.3 the timber will be allocated a service class depending on the exposure. Service class 1 assumes a temperature of 20°C and a relative humidity exceeding 65% for a relatively short period. Typically the average moisture content will be 12%. This usually applies to timbers in heated environments.

Service class 2 assumes a temperature of 20°C and a relative humidity of the surrounding air not exceeding 85% for a relatively short period. Typically the average moisture content will be 15–18%. This typically applies to timbers in covered buildings.

Finally, service class 3 applies to fully exposed timbers and those used in external applications. Typically the moisture content will be in excess of 20%.

If we consider the beams below with the same parameters but one in service class 1 and the other in service class 3, we can see that this can make the difference between the beam passing and failing.

Structural Design of Buildings, First Edition. Paul Smith.

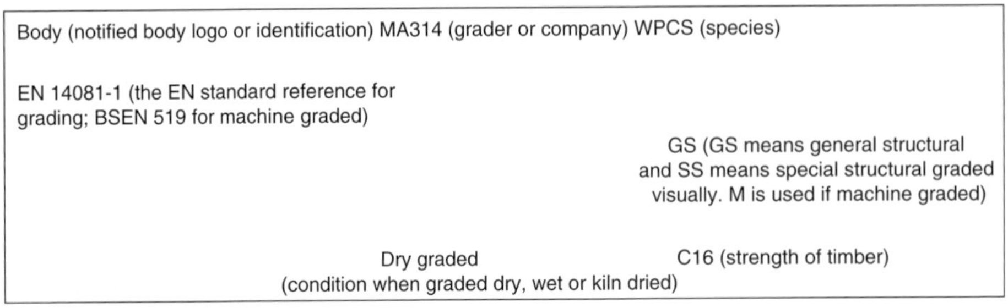

Figure 6.1: Typical grading stamp for timber.

Consider a beam 2.5 m long with a dead load of 2.0 kN/m and an imposed load of 1.0 kN/m. The beam has a bearing of 100 mm and is a Grade C16 timber, 75 mm × 195 mm. First, let us consider the beam in a service class 1 environment.

The following calculations have been undertaken using structural engineering software and we gratefully acknowledge Tekla (UK) Ltd for their approval in the use of this software.

TIMBER BEAM ANALYSIS & DESIGN TO BS 5268-2: 2002.

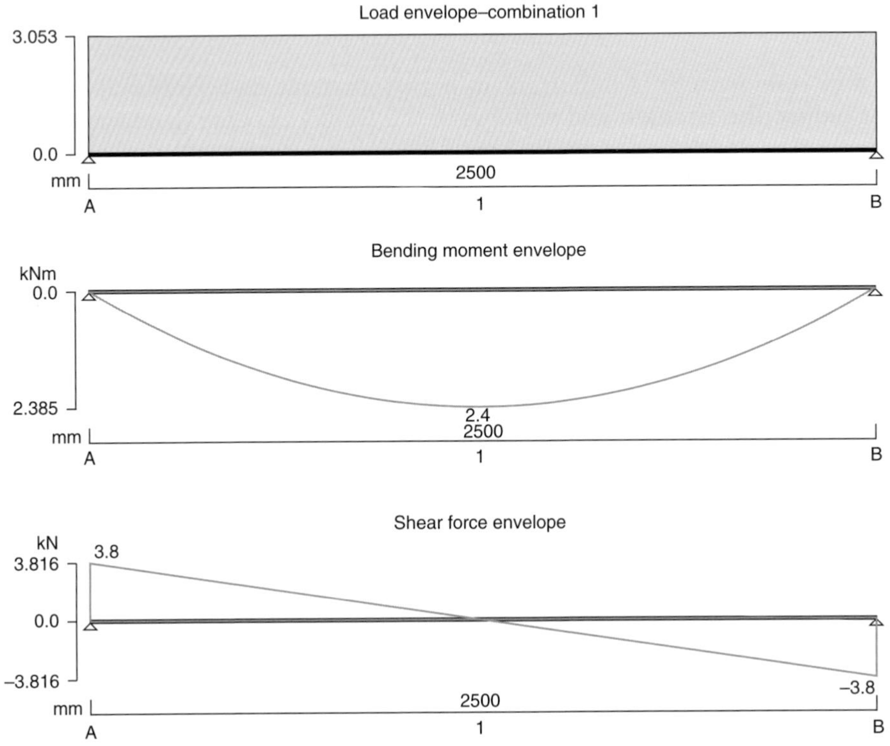

Applied loading
Beam loads

Dead full UDL 2.000 kN/m
Imposed full UDL 1.000 kN/m
Dead self-weight of beam × 1

Load combinations

Load combination 1	Support A	Dead × 1.00
		Imposed × 1.00
	Span 1	Dead × 1.00
		Imposed × 1.00
	Support B	Dead × 1.00
		Imposed × 1.00

Analysis results

Maximum moment;	M_{max} = **2.385** kNm;	M_{min} = **0.000** kNm
Design moment;	M = max(abs(M_{max}),abs(M_{min})) = **2.385** kNm	
Maximum shear;	F_{max} = **3.816** kN;	F_{min} = **−3.816** kN
Design shear;	F = max(abs(F_{max}),abs(F_{min})) = **3.816** kN	
Total load on beam;	W_{tot} = **7.633** kN	
Reactions at support A;	R_{A_max} = **3.816** kN;	R_{A_min} = **3.816** kN
Unfactored dead load reaction at support A;	R_{A_Dead} = **2.566** kN	
Unfactored imposed load reaction at support A;	$R_{A\_Imposed}$ = **1.250** kN	
Reactions at support B;	R_{B_max} = **3.816** kN;	R_{B_min} = **3.816** kN
Unfactored dead load reaction at support B;	R_{B_Dead} = **2.566** kN	
Unfactored imposed load reaction at support B;	$R_{B\_Imposed}$ = **1.250** kN	

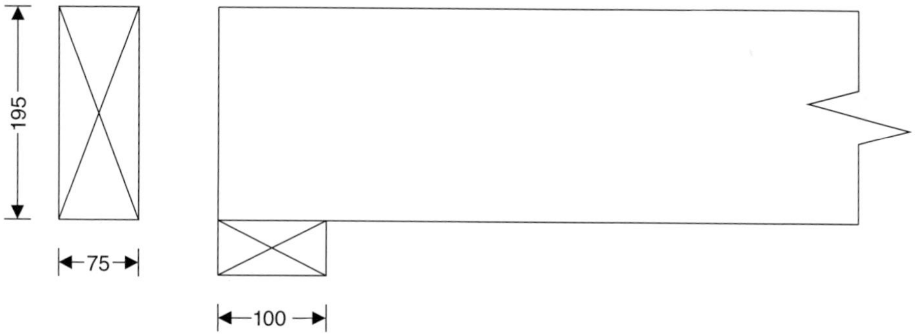

Timber section details

Breadth of sections;	b = **75** mm
Depth of sections;	h = **195** mm
Number of sections in member;	N = **1**
Overall breadth of member;	b_b = N × b = **75** mm
Timber strength class;	**C16**

Member details

Service class of timber;	**1**
Load duration;	**Long term**
Length of bearing;	L_b = **100** mm

Section properties

Cross-sectional area of member;	$A = N \times b \times h =$ **14625** mm^2
Section modulus;	$Z_x = N \times b \times h^2/6 =$ **475313** mm^3
	$Z_y = h \times (N \times b)^2/6 =$ **182813** mm^3
Second moment of area;	$I_x = N \times b \times h^3/12 =$ **46342969** mm^4
	$I_y = h \times (N \times b)^3/12 =$ **6855469** mm^4
Radius of gyration;	$i_x = \sqrt{(I_x/A)} =$ **56.3** mm
	$i_y = \sqrt{(I_y/A)} =$ **21.7** mm

Modification factors

Duration of loading – Table 17;	$K_3 =$ **1.00**
Bearing stress – Table 18;	$K_4 =$ **1.00**
Total depth of member – cl.2.10.6;	$K_7 = (300\ mm/h)^{0.11} =$ **1.05**
Load sharing – cl.2.9;	$K_8 =$ **1.00**

Lateral support – cl.2.10.8

Ends held in position and members held in line, as by purlins or tie rods at centres not more than 30 times the breadth of the member

Permissible depth-to-breadth ratio – Table 19;	**4.00**
Actual depth-to-breadth ratio;	$h/(N \times b) =$ **2.60**

PASS – Lateral support is adequate

Compression perpendicular to grain

Permissible bearing stress (no wane);	$\sigma_{c\_adm} = \sigma_{cp1} \times K_3 \times K_4 \times K_8 =$ **2.200** N/mm^2
Applied bearing stress;	$\sigma_{c\_a} = R_{A\_max}/(N \times b \times L_b) =$ **0.509** N/mm^2
	$\sigma_{c\_a}/\sigma_{c\_adm} =$ **0.231**

PASS – Applied compressive stress is less than permissible compressive stress at bearing

Bending parallel to grain

Permissible bending stress;	$\sigma_{m\_adm} = \sigma_m \times K_3 \times K_7 \times K_8 =$ **5.557** N/mm^2
Applied bending stress;	$\sigma_{m\_a} = M/Z_x =$ **5.018** N/mm^2
	$\sigma_{m\_a}/\sigma_{m\_adm} =$ **0.903**

PASS – Applied bending stress is less than permissible bending stress

Shear parallel to grain

Permissible shear stress;	$\tau_{adm} = \tau \times K_3 \times K_8 =$ **0.670** N/mm^2
Applied shear stress;	$\tau_a = 3 \times F/(2 \times A) =$ **0.391** N/mm^2
	$\tau_a/\tau_{adm} =$ **0.584**

PASS – Applied shear stress is less than permissible shear stress

Deflection

Modulus of elasticity for deflection;	$E = E_{min} =$ **5800** N/mm^2
Permissible deflection;	$\delta_{adm} = \min(14\ mm, 0.003 \times L_{s1}) =$ **7.500** mm
Bending deflection;	$\delta_{b\_s1} =$ **5.777** mm
Shear deflection;	$\delta_{v\_s1} =$ **0.540** mm
Total deflection;	$\delta_a = \delta_{b\_s1} + \delta_{v\_s1} =$ **6.317** mm
	$\delta_a/\delta_{adm} =$ **0.842**

PASS – Total deflection is less than permissible deflection

Using the same parameters and analysing the same beam but in service class 3 or exposed location, we have the results below.

These calculations have been undertaken using structural engineering software and we gratefully acknowledge Tekla (UK) Ltd for their approval in the use of this software.

TIMBER BEAM ANALYSIS & DESIGN TO BS 5268-2: 2002.

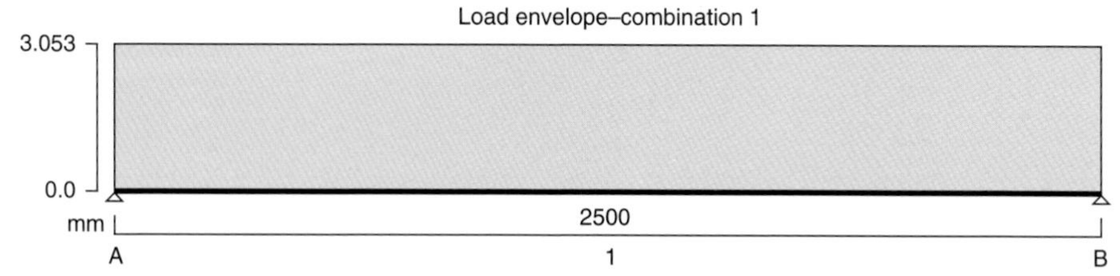

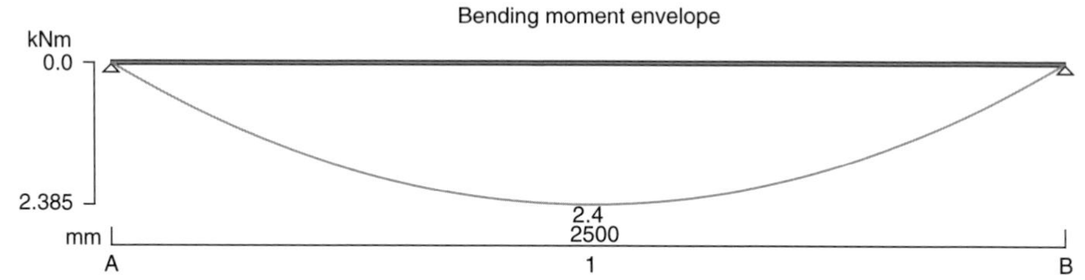

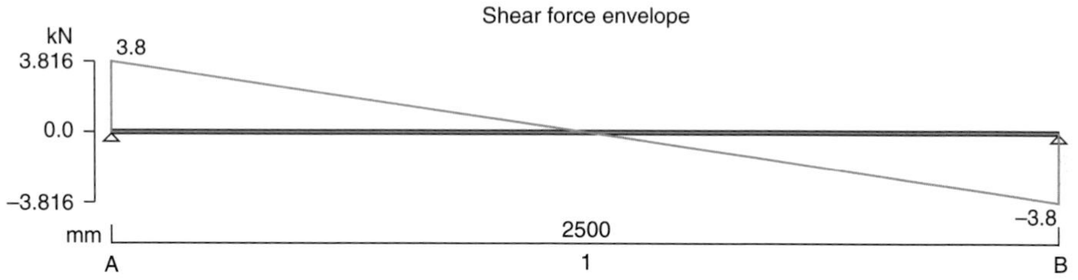

Applied loading
Beam loads

Dead full UDL 2.000 kN/m
Imposed full UDL 1.000 kN/m
Dead self-weight of beam × 1

Load combinations

Load combination 1	Support A	Dead × 1.00
		Imposed × 1.00
	Span 1	Dead × 1.00
		Imposed × 1.00
	Support B	Dead × 1.00
		Imposed × 1.00

Analysis results

Maximum moment;	M_{max} = **2.385** kNm;	M_{min} = **0.000** kNm
Design moment;	M = max(abs(M_{max}),abs(M_{min})) = **2.385** kNm	
Maximum shear;	F_{max} = **3.816** kN;	F_{min} = **−3.816** kN
Design shear;	F = max(abs(F_{max}),abs(F_{min})) = **3.816** kN	
Total load on beam;	W_{tot} = **7.633** kN	
Reactions at support A;	R_{A_max} = **3.816** kN;	R_{A_min} = **3.816** kN
Unfactored dead load reaction at support A;	R_{A_Dead} = **2.566** kN	
Unfactored imposed load reaction at support A;	$R_{A\_Imposed}$ = **1.250** kN	
Reactions at support B;	R_{B_max} = **3.816** kN;	R_{B_min} = **3.816** kN
Unfactored dead load reaction at support B;	R_{B_Dead} = **2.566** kN	
Unfactored imposed load reaction at support B;	$R_{B\_Imposed}$ = **1.250** kN	

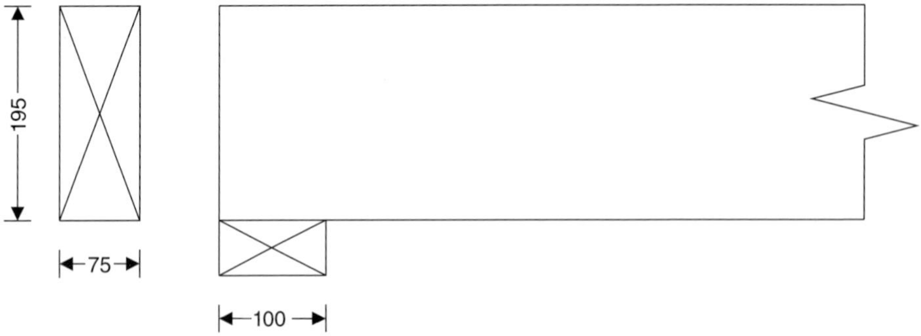

Timber section details

Breadth of sections;	b = **75** mm
Depth of sections;	h = **195** mm
Number of sections in member;	N = **1**
Overall breadth of member;	b_b = N × b = **75** mm
Timber strength class;	**C16**

Member details

Service class of timber;	**3**
Load duration;	**Long term**
Length of bearing;	L_b = **100** mm

Section properties

Cross-sectional area of member;	A = N × b × h = **14625** mm^2
Section modulus;	$Z_x = N \times b \times h^2/6$ = **475313** mm^3
	$Z_y = h \times (N \times b)^2/6$ = **182813** mm^3
Second moment of area;	$I_x = N \times b \times h^3/12$ = **46342969** mm^4
	$I_y = h \times (N \times b)^3/12$ = **6855469** mm^4
Radius of gyration;	$i_x = \sqrt{(I_x/A)}$ = **56.3** mm
	$i_y = \sqrt{(I_y/A)}$ = **21.7** mm

Modification factors

Service class for bending – Table 16;	K_{2m} = **0.80**
Service class for tension – Table 16;	K_{2t} = **0.80**
Service class for compression – Table 16;	K_{2c} = **0.60**
Service class for shear – Table 16;	K_{2s} = **0.90**
Service class for modulus of elasticity – Table 16;	K_{2e} = **0.80**
Duration of loading – Table 17;	K_3 = **1.00**
Bearing stress – Table 18;	K_4 = **1.00**
Total depth of member – cl.2.10.6;	$K_7 = (300\ \text{mm}/h)^{0.11}$ = **1.05**
Load sharing – cl.2.9;	K_8 = **1.00**

Lateral support – cl.2.10.8

Ends held in position and members held in line, as by purlins or tie rods at centres not more than 30 times the breadth of the member

Permissible depth-to-breadth ratio – Table 19;	**4.00**
Actual depth-to-breadth ratio;	h/(N × b) = **2.60**

PASS – Lateral support is adequate

Compression perpendicular to grain

Permissible bearing stress (no wane);	$\sigma_{c\_adm} = \sigma_{cp1} \times K_{2c} \times K_3 \times K_4 \times K_8$ = **1.320** N/mm^2
Applied bearing stress;	$\sigma_{c\_a} = R_{A\_max}/(N \times b \times L_b)$ = **0.509** N/mm^2
	$\sigma_{c\_a}/\sigma_{c\_adm}$ = **0.385**

PASS – Applied compressive stress is less than permissible compressive stress at bearing

Bending parallel to grain

Permissible bending stress; $\sigma_{m\_adm} = \sigma_m \times K_{2m} \times K_3 \times K_7 \times K_8 = \mathbf{4.446}\ N/mm^2$

Applied bending stress; $\sigma_{m\_a} = M/Z_x = \mathbf{5.018}\ N/mm^2$

$\sigma_{m\_a}/\sigma_{m\_adm} = \mathbf{1.129}$

FAIL – Applied bending stress exceeds permissible bending stress

Shear parallel to grain

Permissible shear stress; $\tau_{adm} = \tau \times K_{2s} \times K_3 \times K_8 = \mathbf{0.603}\ N/mm^2$

Applied shear stress; $\tau_a = 3 \times F/(2 \times A) = \mathbf{0.391}\ N/mm^2$

$\tau_a/\tau_{adm} = \mathbf{0.649}$

PASS – Applied shear stress is less than permissible shear stress

Deflection

Modulus of elasticity for deflection; $E = E_{min} \times K_{2e} = \mathbf{4640}\ N/mm^2$

Permissible deflection; $\delta_{adm} = \min(14\ mm, 0.003 \times L_{s1}) = \mathbf{7.500}\ mm$

Bending deflection; $\delta_{b\_s1} = \mathbf{7.222}\ mm$

Shear deflection; $\delta_{v\_s1} = \mathbf{0.675}\ mm$

Total deflection; $\delta_a = \delta_{b\_s1} + \delta_{v\_s1} = \mathbf{7.896}\ mm$

$\delta_a/\delta_{adm} = \mathbf{1.053}$

FAIL – Total deflection exceeds permissible deflection

We can see from the above examples that the same timber beam exposed to a different environment of humidity and moisture can result in the difference between the beam passing the design criteria or failing in deflection and permissible bending stress.

There are two methods by which the moisture content of timber can be reduced, and these are seen below.

Air-dried timber

This is where timber is stacked in open stores or sheds with air spaces between the stacked timbers.

Kiln-dried timber

This is where timber is dried in heated environments with ventilation. This is expensive compared with air-dried timber and requires specialist equipment which is used in a controlled environment. This reduces the moisture content much more quickly.

The water content of timber essentially softens the cell walls in the timber, resulting in a stretching of the fibres which makes them easier to break. This reduces the stiffness and strength of the timber – notably the bending and compressive stresses.

A freshly felled tree will have a water content in excess of 50% and as it dries the volume changes, reducing the size of the timber. Interestingly, up to a water content of 20% the thickness and width of the timber can increase by as much as 0.25% for a 1% increase in water content.

Dimensions of timber

It can be shown that the strength of timber increases directly as a result of an increase in breadth, inversely as a result of an increase in length and as a square of depth. Therefore, increasing the depth of a timber member will increase its strength much more than increasing its width. However, there are limitations to consider. For example, it is not practical to have a beam supporting floorboards that are only 12 mm wide (since the beam or joist will not be wide enough to sustain the impact of a nail or screw without splitting). Thus, most joists are a minimum 38 mm thick. This is shown in more detail in the bending and deflection sections below.

Shear

The maximum shear parallel to the grain is 1.5 times the average value of the horizontal shear stress at the level of the neutral axis:

$$\Gamma_{a,11} = 1.5P/A \qquad (6.1)$$

where

$\Gamma_{a,11}$ = maximum applied horizontal shear stress

P = vertical shear force

A = cross-sectional area

This value must not exceed the permissible shear stress Γ_{adm}.

As can be seen, as the cross-sectional area increases the maximum applied horizontal shear stress reduces – thus making the timber section stronger.

Bending

To satisfy the design criteria the applied bending stress must not exceed the permissible bending stress. The applied bending moment is given by the expression

$$\sigma_{m,a,11} = M/Z \qquad (6.2)$$

where

$\sigma_{m,a,11}$ = maximum applied bending stress

M = applied moment

Z = elastic or sectional modulus ($bd^2/6$)

b = breadth

d = depth

As can be seen, since Z is a function of the breadth and depth, as the depth increases the square of this number reduces the maximum applied bending stress for any applied moment. Thus, increasing the depth of the member increases the moment of resistance more so than an increase in breadth.

Deflection

Finally, let us consider the deflection, which is limited under BS 5268 by the span of the beam to 0.003 × span or 14 mm for floor joists.

The equation for the actual deflection is a sum of the deflection caused by bending deflection and the deflection caused by shear deflection. Hence:

$$\Delta_{total} = \Delta_{bending\ deflection} + \Delta_{shear\ deflection} \tag{6.3}$$

The type of loading will determine the formula to use for deflection, and these can be seen in Figure 2.9 in Chapter 2. Essentially, if we consider the deflection due to bending we can see that the formula is a function of the second moment of area ($bd^3/12$) for a rectangular section. Thus, if the depth is increased one can see that the cube of this number increases the second moment of area considerably, and more so than the width. This has the effect of reducing the deflection due to bending.

Hence, an increase in the breadth of a section of timber will increase its strength but not as much as an increase in its depth.

Chapter 7 Foundations

Purpose of foundations

The purpose of a foundation is to transfer loads to the ground without any subsidence or settlement, and thus ensure the building remains stable. The foundation will exert a bearing pressure on the soil and the soil will have a bearing capacity which it can sustain.

The history of foundations

In 1875 the Public Health Act made it incumbent on local authorities to improve the structural stability of houses, with the particular focus to prevent fires. In 1878 the Building Act gave details on wall and foundation thicknesses. Bylaws were soon passed by local government and following this date foundations were required to be of concrete, 225 mm deep using Portland cement or hydraulic lime. The mix proportions proposed at the time were in the region of 1:1:4 or 1:1:5 cement:sand:stone. The later Building Act of 1894 made proposals to increase the thickness of external and party walls and interestingly, brick foundations were proposed rather than the earlier concrete type. Figure 7.1 explains the foundation and wall thicknesses taken from the London Building Act 1894. This publication preferred brick foundations to concrete.

This model appears to have remained unchanged for a number of years, until the London Building Act 1930 which made the requirement that the width of the brick footing should be twice the thickness of the wall. Concrete could be employed but this was not a necessity.

In 1930 bylaws in London introduced concrete foundations and the width of the foundation was three times the thickness of the wall. The projection of the foundation from the wall and the thickness of the foundation were not to be less than the thickness of the wall.

In 1939 bylaws were introduced which required a brick or concrete foundation. The total width of the foundation was not less than twice the thickness of the wall. The depth of the foundation was taken as 1.33 times the projection of the foundation from the wall, and not less than 225 mm.

In 1965 the National Building Regulations standardised the foundation requirements throughout England and Wales, with the exception of the London Boroughs, where the London Acts outlined above prevailed. In 1984 the Building Regulations were consolidated under one Act and resulted in the Building Regulations 1985. The requirements for foundations are seen in Part A of the Building Regulations and provide foundation dimensions for buildings on various types of subsoil. Essentially, the thickness of the foundation must not be less than the projection of the foundation from the wall, and not less than 150 mm.

Structural Design of Buildings, First Edition. Paul Smith.

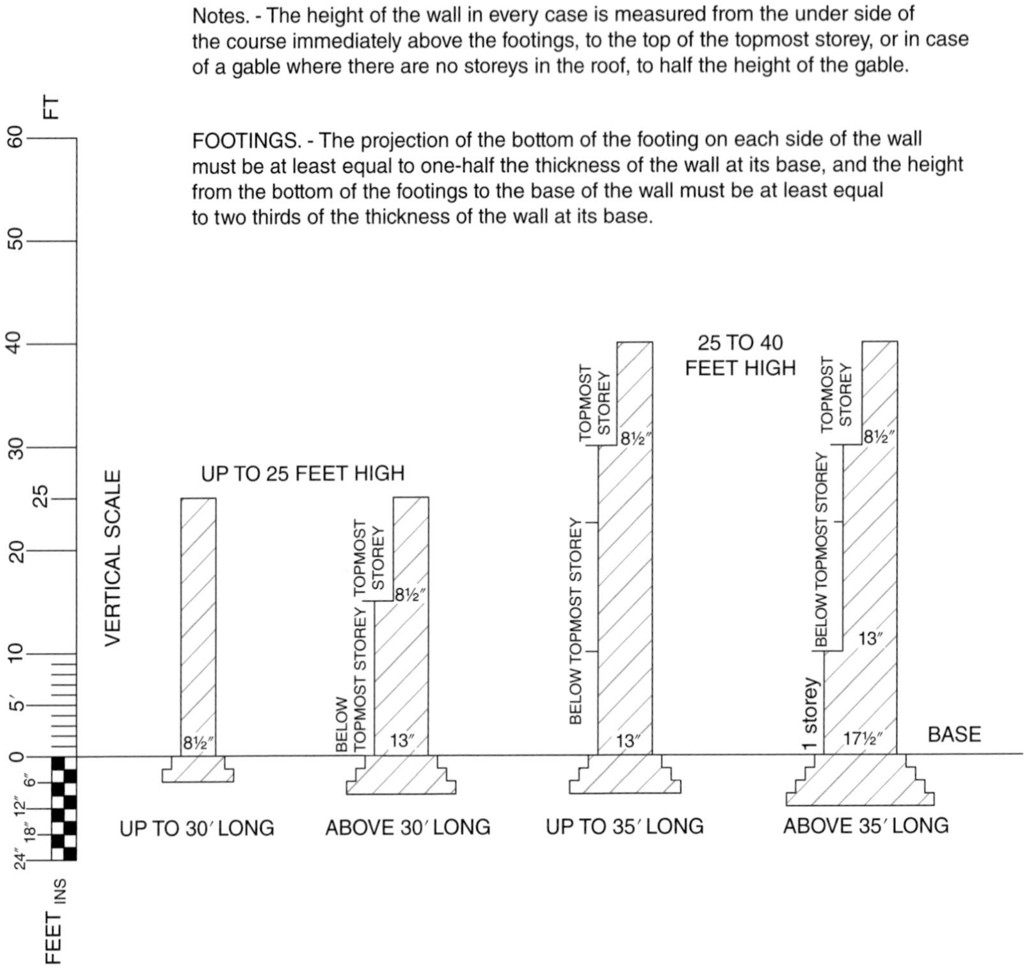
Notes. - The height of the wall in every case is measured from the under side of the course immediately above the footings, to the top of the topmost storey, or in case of a gable where there are no storeys in the roof, to half the height of the gable.
FOOTINGS. - The projection of the bottom of the footing on each side of the wall must be at least equal to one-half the thickness of the wall at its base, and the height from the bottom of the footings to the base of the wall must be at least equal to two thirds of the thickness of the wall at its base.
FT
60
50
40
30
25
20
10
5′
0
6″
12″
18″
24″
FEET INS
VERTICAL SCALE
UP TO 25 FEET HIGH
25 TO 40 FEET HIGH
TOPMOST STOREY
BELOW TOPMOST STOREY
1 storey
8½″
13″
17½″
BASE
UP TO 30′ LONG
ABOVE 30′ LONG
UP TO 35′ LONG
ABOVE 35′ LONG

Figure 7.1: Thickness of foundations, party walls and external walls taken from London Building Act 1894.

Table 10 in Part A1/2 of the Building Regulations provides the minimum width for foundations based on the type of soil and the load exerted on the foundation. It also provides a simple field test to identify the type and strength of soil based on the pliability and excavation of the soil. For example, a clay that can be "indented slightly by thumb" is regarded as a stiff clay.

Different materials will have different bearing capacities, thus for gravel or sand it can be seen that the minimum width of foundation is less than for a weaker, more silty soil.

Building Regulation requirements

Section 2E of Part A1/2 of the Building Regulations focuses on foundations and states that "*The building shall be constructed so that the combined dead, imposed and wind loads are sustained and transmitted by it to the ground a) safely; and b) without causing such deflection or deformation of any part of the building, or such movement of the ground, as will impair the stability of any part of another building.*"

"The building shall be constructed so that ground movement caused by: a) swelling, shrinkage or freezing of the subsoil; or b) land-slip or subsidence (other than subsidence arising from shrinkage, in so far as the risk can be reasonably foreseen), will not impair the stability of any part of the building."

Foundations should be located centrally under all external and internal walls and taken to a depth below the influence of drains and/or surrounding trees, and taken to natural undisturbed ground of adequate ground-bearing capacity. Later in this chapter we will see why this is important, so as not to introduce an eccentric load to the foundation.

The Regulations also prescribe the concrete mixes suitable for foundations – either a Grade ST2 or Gen 1 concrete mix complying with BS 8500-2 should be used. Alternatively, in proportion of 50 kg Portland cement to not more than 200 kg (0.1 m^3) fine aggregate and 400 kg (0.2 m^3) coarse aggregate.

The minimum thickness of a foundation should be 150 mm, or the thicknesses as outlined in Table 10 of Part A1/2 of the Building Regulations, whichever is the greater. The Regulations base the dimensions on strip or trench-fill foundations, although other types are available (see below) which require specialist design.

Stepped foundation

When strip or trench-fill foundations have to be stepped, there are strict criteria under the Building Regulations that need to be complied with so as to prevent shear failure in the concrete. *"Foundations stepped on elevation should overlap by twice the height of the step, by the thickness of the foundation, or 300 mm, whichever is the greater.*

For trench-fill foundations the overlap should be twice the height of the step or 1 m, whichever is greater; steps in foundation should not be of greater height than the thickness of the foundation." Taken from Building Regulations Part A; see Figure 7.2.

Foundations should unite at each change in level.

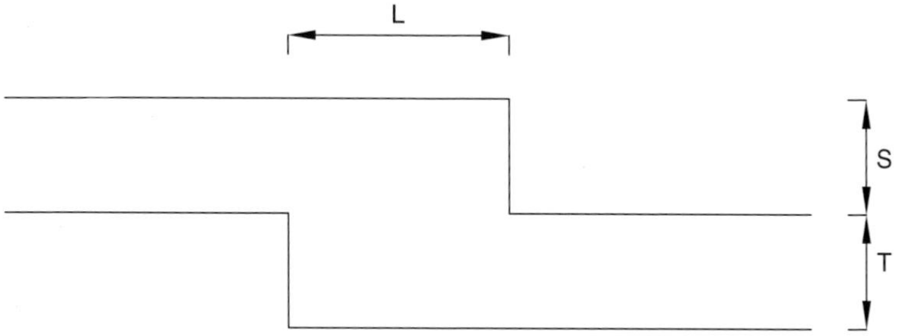

Minimum overlap L = twice the height of the step, or thickness of foundation or 300 mm, whichever is greater.

S should not be greater than T.

(For trench-fill foundations, minimum overlap L = twice height of step, or 1 m whichever is greater).

Figure 7.2: Diagram of stepped foundation.

Types of foundation

Strip foundation

Early strip foundations were constructed of brick or stone, widened at the base of the wall to distribute the load of the wall over a wider area. This was a technique used in Victorian and Edwardian properties. The foundations were typically constructed at a shallow depth, possibly no more than 450–500 mm below ground level. The wall would thicken below the ground surface, corbelling the stone or brickwork to a width that would ensure the load could be sustained by the receiving soil. Figure 7.3 is a photograph of a Georgian property with a stepped brick foundation.

Strip foundations are used to sustain line loads and are shallow. Since the 1900s concrete has been used and typically a foundation will be a minimum of 150 mm deep (usually 225 mm) and 150 mm wider on each side of the wall. Shallow foundations are constructed on suitable ground where the material is strong enough to support the foundation loads. An example of such a design is provided below.

The following calculations have been undertaken using structural engineering software and we gratefully acknowledge Tekla (UK) Ltd for their approval in the use of this software.

Figure 7.3: Photograph showing a stepped brick foundation in a Georgian property.

STRIP FOOTING ANALYSIS AND DESIGN (BS 8110-1:1997).

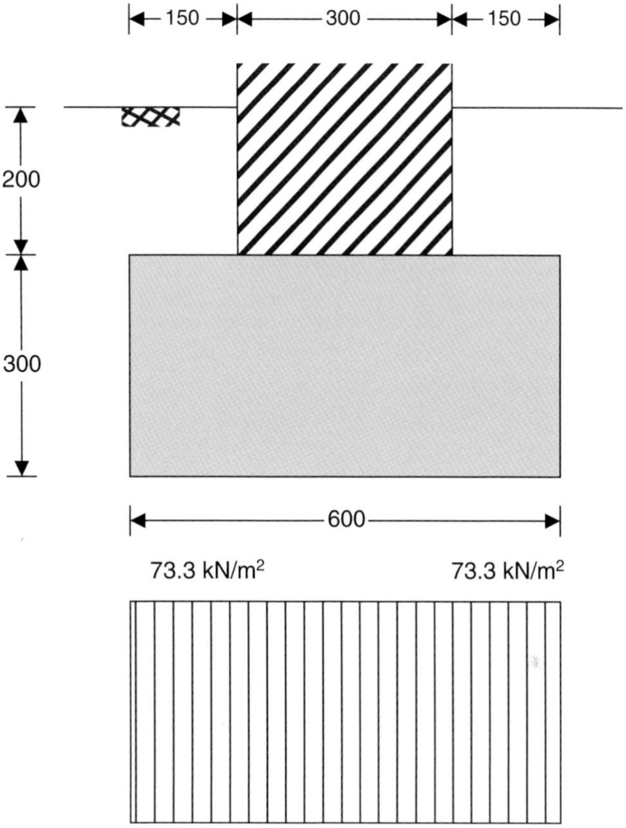

Strip footing details

Width of strip footing;	B = **600** mm
Depth of strip footing;	h = **300** mm
Depth of soil over strip footing;	h_{soil} = **200** mm
Density of concrete;	ρ_{conc} = **23.6** kN/m^3

Load details

Load width;	b = **300** mm
Load eccentricity;	e_P = **0** mm

Soil details

Density of soil;	ρ_{soil} = **20.0** kN/m^3
Design shear strength;	ϕ' = **25.0** deg
Design base friction;	δ = **19.3** deg
Allowable bearing pressure;	$P_{bearing}$ = **150** kN/m^2

Axial loading on strip footing

Dead axial load;	P_G = **25.0** kN/m
Imposed axial load;	P_Q = **10.0** kN/m
Wind axial load;	P_W = **0.0** kN/m
Total axial load;	P = **35.0** kN/m

Foundation loads	
Dead surcharge load;	F_{Gsur} = **2.400** kN/m^2
Imposed surcharge load;	F_{Qsur} = **1.500** kN/m^2
Strip footing self-weight;	$F_{swt} = h \times \rho_{conc}$ = **7.080** kN/m^2
Soil self-weight;	$F_{soil} = h_{soil} \times \rho_{soil}$ = **4.000** kN/m^2
Total foundation load;	$F = B \times (F_{Gsur} + F_{Qsur} + F_{swt} + F_{soil})$ = **9.0** kN/m
Calculate base reaction	
Total base reaction;	$T = F + P$ = **44.0** kN/m
Eccentricity of base reaction in x;	$e_T = (P \times e_P + M + H \times h)/T$ = **0** mm
	Base reaction acts within middle third of base
Calculate base pressures	
	$q_1 = (T/B) \times (1 - 6 \times e_T/B)$ = **73.313** kN/m^2
	$q_2 = (T/B) \times (1 + 6 \times e_T/B)$ = **73.313** kN/m^2
Minimum base pressure;	$q_{min} = \min(q_1, q_2)$ = **73.313** kN/m^2
Maximum base pressure;	$q_{max} = \max(q_1, q_2)$ = **73.313** kN/m^2
	PASS – Maximum base pressure is less than allowable bearing pressure
Material details	
Characteristic strength of concrete;	f_{cu} = **15** N/mm^2
Calculate base lengths	
Left-hand length;	$B_L = B/2 + e_P$ = **300** mm
Right-hand length;	$B_R = B/2 - e_P$ = **300** mm
Calculate rate of change of base pressure	
Length of base reaction;	$B_x = B$ = **600** mm
Rate of change of base pressure;	$C_x = (q_1 - q_2)/B_x$ = **0.000** kN/m^2/m
Calculate minimum depth of unreinforced strip footing	
Average pressure to left of strip footing;	$q_L = q_1 - C_x \times (B_L - b/2)/2$ = **73.313** kN/m^2
Minimum depth to left of strip footing;	$h_{Lmin} = (B_L\text{-}b/2)\times\max(0.15\times[(q_L/1\ \text{kN/m}^2)^2/(f_{cu}/1\ \text{N/mm}^2)]^{1/4},1)$ = **150** mm
Average pressure to right of strip footing;	$q_R = q_2 + C_x \times (B_R - b/2)/2$ = **73.313** kN/m^2
Minimum depth to right of strip footing;	$h_{Rmin} = (B_R\text{-}b/2)\times\max(0.15\times[(q_R/1\text{kN/m}^2)^2/(f_{cu}/1\text{N/mm}^2)]^{1/4},1)$ = **150** mm
Minimum depth of unreinforced strip footing;	$h_{min} = \max(h_{Lmin}, h_{Rmin}, 300\ \text{mm})$ = **300** mm
	PASS – Unreinforced strip footing depth is greater than minimum

The depth from ground level to the bottom of the foundation is important to ensure that the foundation does not experience movements in the soil due to climatic changes such as heat, frost and moisture changes. Except where strip foundations are founded on rock, they should have a minimum depth of 450 mm to their underside to avoid the action of frost – except in clay soils where this is increased to 750 mm to avoid the influence of vegetation. However, these are minimum values and will normally need to be increased to transfer the loads to suitable subsoil. We will cover this in more detail later in this chapter.

Trench-fill foundation

Trench-fill foundations are trenches with a minimum width of 450 mm. When excavated the sides are trimmed and prepared and then mass concrete is poured into the trench, levelled at the top to receive the wall construction. This type of construction can be quicker than strip foundations, since there is a labour saving in the construction of the sub-surface walls.

Concrete mixes for strip and trench-fill foundations are generally ST2 or Gen 1 mix. ST2 is a standard mix consisting of 4–20 mm stone, graded sand and water. BS 5328: 1 Section 4 provides the mix proportions: 50 kg ordinary Portland cement, 200 kg fine aggregate and 400 kg coarse aggregate. ST2 has a compressive strength of 10 N/mm^2.

Gen 1 is a designated mix. BS 5328:1 Section 5 specifies the mix. The mix strength is also 10 N/mm^2.

An example of the design of a trench-fill foundation can be seen below.

The following calculations have been undertaken using structural engineering software and we gratefully acknowledge Tekla (UK) Ltd for their approval in the use of this software.

MASS POUR FOOTING ANALYSIS AND DESIGN (BS 8110-1:1997).

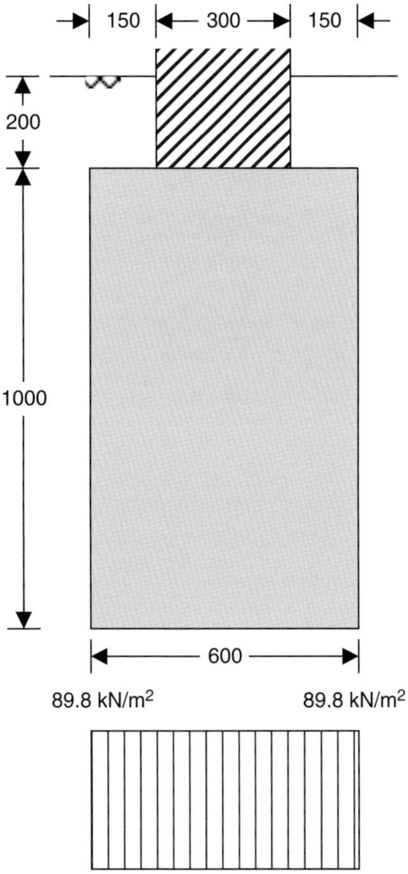

Strip footing details

Width of strip footing;	B = **600** mm
Depth of strip footing;	h = **1000** mm
Depth of soil over strip footing;	h_{soil} = **200** mm
Density of concrete;	ρ_{conc} = **23.6** kN/m^3

Load details
Load width; $b = \mathbf{300}$ mm
Load eccentricity; $e_P = \mathbf{0}$ mm

Soil details
Density of soil; $\rho_{soil} = \mathbf{20.0}$ kN/m^3
Design shear strength; $\phi' = \mathbf{25.0}$ deg
Design base friction; $\delta = \mathbf{19.3}$ deg
Allowable bearing pressure; $P_{bearing} = \mathbf{150}$ kN/m^2

Axial loading on strip footing
Dead axial load; $P_G = \mathbf{25.0}$ kN/m
Imposed axial load; $P_Q = \mathbf{10.0}$ kN/m
Wind axial load; $P_W = \mathbf{0.0}$ kN/m
Total axial load; $P = \mathbf{35.0}$ kN/m

Foundation loads
Dead surcharge load; $F_{Gsur} = \mathbf{2.400}$ kN/m^2
Imposed surcharge load; $F_{Qsur} = \mathbf{1.500}$ kN/m^2
Strip footing self-weight; $F_{swt} = h \times \rho_{conc} = \mathbf{23.600}$ kN/m^2
Soil self-weight; $F_{soil} = h_{soil} \times \rho_{soil} = \mathbf{4.000}$ kN/m^2
Total foundation load; $F = B \times (F_{Gsur} + F_{Qsur} + F_{swt} + F_{soil}) = \mathbf{18.9}$ kN/m

Calculate base reaction
Total base reaction; $T = F + P = \mathbf{53.9}$ kN/m
Eccentricity of base reaction in x; $e_T = (P \times e_P + M + H \times h)/T = \mathbf{0}$ mm

Base reaction acts within middle third of base

Calculate base pressures

$q_1 = (T/B) \times (1 - 6 \times e_T/B) = \mathbf{89.833}$ kN/m^2
$q_2 = (T/B) \times (1 + 6 \times e_T/B) = \mathbf{89.833}$ kN/m^2
Minimum base pressure; $q_{min} = \min(q_1, q_2) = \mathbf{89.833}$ kN/m^2
Maximum base pressure; $q_{max} = \max(q_1, q_2) = \mathbf{89.833}$ kN/m^2

PASS – Maximum base pressure is less than allowable bearing pressure

Material details
Characteristic strength of concrete; $f_{cu} = \mathbf{15}$ N/mm^2

Calculate base lengths
Left-hand length; $B_L = B/2 + e_P = \mathbf{300}$ mm
Right-hand length; $B_R = B/2 - e_P = \mathbf{300}$ mm

Calculate rate of change of base pressure
Length of base reaction; $B_x = B = \mathbf{600}$ mm
Rate of change of base pressure; $C_x = (q_1 - q_2)/B_x = \mathbf{0.000}$ kN/m^2/m

Calculate minimum depth of unreinforced strip footing
Average pressure to left of strip footing; $q_L = q_1 - C_x \times (B_L - b/2)/2 = \mathbf{89.833}$ kN/m^2
Minimum depth to left of strip footing; $h_{Lmin} = (B_L\text{-}b/2)\times\max(0.15\times[(q_L/1\text{ kN/m}^2)^2/(f_{cu}/1\text{ N/mm}^2)]^{1/4},1) = \mathbf{150}$ mm
Average pressure to right of strip footing; $q_R = q_2 + C_x \times (B_R - b/2)/2 = \mathbf{89.833}$ kN/m^2
Minimum depth to right of strip footing; $h_{Rmin} = (B_R\text{-}b/2)\times\max(0.15\times[(q_R/1\text{kN/m}^2)^2/(f_{cu}/1\text{N/mm}^2)]^{1/4},1) = \mathbf{150}$ mm
Minimum depth of unreinforced strip footing; $h_{min} = \max(h_{Lmin}, h_{Rmin}, 300\text{ mm}) = \mathbf{300}$ mm

PASS – Unreinforced strip footing depth is greater than minimum

Raft foundation

Raft foundations are usually employed on ground conditions which have weak bearing capacities or there is a likelihood of differential settlement due to changes in the soil or moisture. Raft foundations spread the load over the area of the slab, thus reducing the bearing pressure to the ground. This is achieved through reinforcement of the slab.

The extraction of moisture in clay soils can lead to volumetric changes in the soil, thus a raft foundation can be used to ensure any depressions or downward movement of the subsoil can be accommodated or the voids spanned by spreading the load of the supported structure over a wider area.

Ground heave is caused when clay soils rehydrate, causing the clay to swell and push on the underside of the foundation. In these circumstances a reinforced concrete slab can be employed which is suspended from a boot or edge beam, thus creating a void beneath the slab. The void is high enough to accommodate any movements in the ground.

Products are available which overcome the effects of heave, and can comprise a moulded hexagonal cellular polystyrene void former which is placed on the prepared ground. The reinforced concrete is poured over the product and this supports the concrete until it sets. The product is available in different grades and each grade is designed to support a safe working load for a period of approximately 16 hours. After this time the material disintegrates and the slab remains suspended; thus the ground beneath can rise or heave without adversely affecting the structural integrity of the slab.

Different types of raft foundation can be seen in Figures 7.4 and 7.5 below and will depend on the line load along the edges and the spans required.

Slab raft

This comprises a flat slab which can be reinforced at the top only or the top and bottom depending on the tension forces anticipated in the foundation and the particular conditions the slab has been designed for. An example can be seen in Figure 7.4.

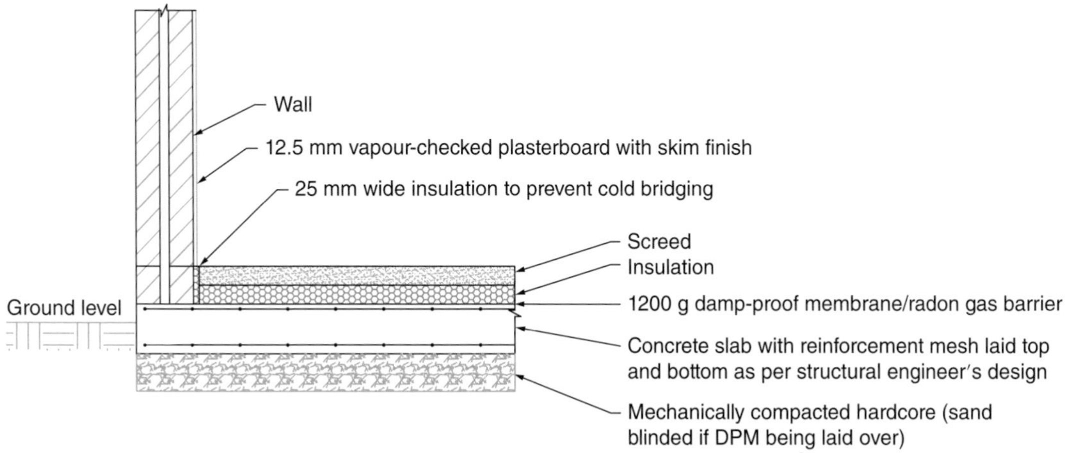

Figure 7.4: Diagram of a typical slab raft foundation.

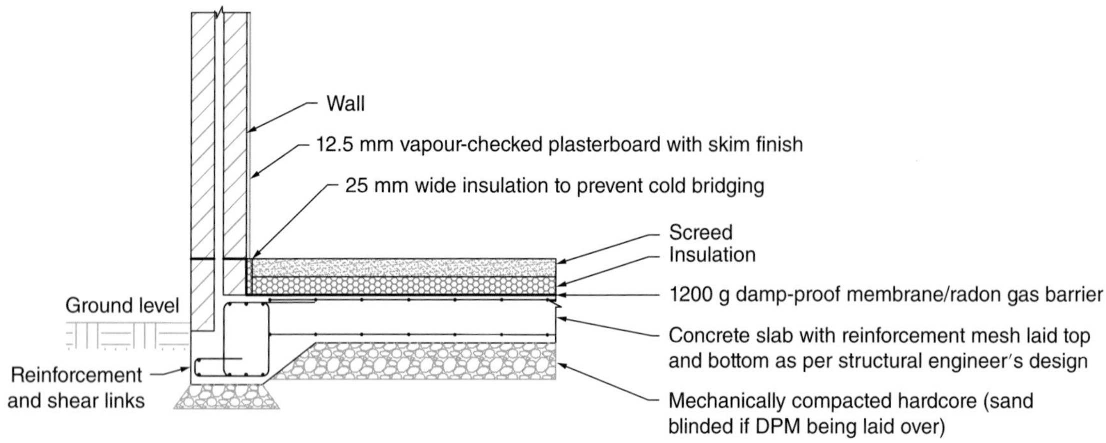

Figure 7.5: Diagram of a typical raft foundation with a tapered boot.

Boot raft

The boot design will allow a step to hide the outer skin of brickwork in the ground. Boot rafts can be tapered or square on the underside and usually require reinforcement shear links. Figure 7.5 shows a raft foundation with a tapered boot.

The strength of the concrete used will depend on exposure and loading, but generally an RC35 mix is suitable. This is a ready-mixed concrete specified in accordance with BS 5328:1 Section 5 and is used for reinforced concrete. The numbers relate to the strength of the mix and in this case the mix will have a compressive strength of 35 N/mm^2.

Piles

Piles are used where foundations have to reach a deep lower level of subsoil. Examples where piles might be used are:

- Made up ground.
- In areas such as salt plains which are unable to sustain the loads.
- Where high water tables are experienced.
- In soils subject to shrinkage and heave.

Piles have been used in construction for many years. The Romans used timber piles in the construction of the first bridge across the River Thames in 60 AD. The Royal Palace in Amsterdam is constructed using timber piles. In the UK, the most common timber used for timber piles is Douglas fir, although other timbers such as oak and elm are also suitable.

End pile

An end pile will be placed at a depth where there is a more stable stratum of soil and transfers the load to this soil stratum.

Friction pile

This pile transfers the load via friction from the side of the pile to the subsoil.

Combination

This type of pile will be driven into a lower, harder material but also relies on frictional resistance or enlarging the base to improve the bearing capacity.

There are different types of installation of piles, as seen below.

Replacement or bored piles

These types of pile are bored into the ground, replacing and removing the soil for the pile to be inserted. When the appropriate depth is achieved, the pile apparatus is then removed, a reinforcement bar placed and the hole filled with concrete. Clearly, in unstable soils which are likely to collapse, compressed air or bentonite slurry can be used to support the hole until the concrete is poured. The photograph in Figure 7.6 shows a replacement pile activity using a bored pile.

Figure 7.6: Photograph showing a bored pile foundation activity.

Displacement piles or driven piles

These types of pile are driven into the ground using a hammer and as the pile progresses it displaces the soil. Examples of this type of pile include vibrated piles, jacked piles and screw piles.

The piles can be of steel, timber, concrete or composite, with the type and material used in the pile dependent on the environment, the locality of adjacent buildings and the soil type. For example, vibration through the use of displacement piles may cause damage to adjacent buildings.

Steel piles can be prone to corrosion due to oxygen in the soil or the presence of moisture. Concrete piles can be cast in situ or in precast segments, but the use of concrete has to be given careful consideration in aggressive soils to ensure the piles do not suffer from sulphate attack or thaumasite attack.

Timber piles have been used for many years and are particularly good when immersed in water, since the alternatives of steel and concrete are prone to rust or corrosion. Timber piles that are used below the water level, and remain so, are resilient to biological deterioration and there are examples where structures have remained for long periods on timber pile foundations that are immersed under water. An example is the Royal Palace in Amsterdam. However, there are also examples where structures have been constructed on marshy land using timber piles and, following the lowering of the water table due to drainage and extraction, the piles have suffered decay. One such example is York Minster. In such cases, where the water table varies, it is possible to use timber piles below the water table and concrete piles above.

Modern techniques have led to the advancement of piles by using expanded polymers. Uretek PowerPiles™ are one such example, and these can be used to underpin or stabilise subsoils that contain voids. These work by drilling a small hole into the ground, approximately 36 mm in diameter. Then, a compressed 30 mm diameter shrink-wrapped prefabricated power pile is inserted into the hole. An extraction machine is connected to the power pile and a resin is injected as the pipe is withdrawn. The resin expands in the ground around the pile, compacting and displacing the weak ground. This stabilises the ground and as the resin hardens, increased support is offered to the structure above.

Bearing pressure

The bearing pressure is the pressure that is exerted by the foundation on the ground. This is determined by assessing the load to the foundation and dividing by the area of the foundation.

Bearing capacity

The bearing capacity of a soil is the pressure which it can sustain. Table 7.1 (based on Table 1 of BS 8004: 1986) provides presumed safe bearing capacities for rocks, cohesive soils and cohesionless soils. Further values are available in Tables 2 and 3 of BS 8004: 1986 for high-porosity chalk and Keuper Marl.

It is important to ensure that the bearing pressure is less than the bearing capacity of the soil, otherwise the building will fail. An example of such failure is the leaning tower

Table 7.1: Presumed safe bearing capacities for rocks, cohesive and cohesionless soils (based on Table 1 of BS 8004: 1986)

Category	Type of rock/soil	Presumed allowable bearing value (kN/m^2)[a]	Remarks
Rocks	Strong igneous and gneissic rocks in sound condition	10000	These values are based on the assumption that the foundations are taken down to unweathered rock.
	Strong limestone and strong sandstone	4000	
	Schists and slates	3000	
	Strong shales, strong mudstones and strong siltstones	2000	
Non-cohesive soils	Dense gravel, or dense sand and gravel	>600	Width of foundation not less than 1 m. Ground water level assumed to be a depth not less than below the base of the foundation.
	Medium dense gravel, or medium dense sand and gravel	<200 to 600	
	Loose gravel, or loose sand and gravel	<200	
	Compact sand	>300	
	Medium dense sand	100 to 300	
	Loose sand	<100, value depending on degree of looseness	
Cohesive soils	Very stiff boulder clays and hard clays	300 to 600	Susceptible to long-term consolidation settlement.
	Stiff clays	150 to 300	
	Firm clays	75 to 150	
	Soft clays and silts	<75	
	Very soft clays and silts	Not applicable	
Peat and organic soils		Not applicable	All these soils are highly compressible and foundations will be subject to considerable settlement over time if placed on them.
Made ground or fill		Not applicable	All made-up ground should be treated as suspect because of the likelihood of extreme variability.

[a]Permission to reproduce extracts derived from British Standards is granted by BSI.

of Pisa. Using the information above, we can show simplistic typical calculations of bearing pressure and bearing capacity.

Example

Let us propose a strip foundation with the top of the foundation 200 mm below ground level and located in a soil which has a density of 20 kN/m^3. The water table is outside the zone of influence of the foundation and the proposed foundation is

200 mm deep and 600 mm wide. If the loads from the building are dead load/m = 35 kN/m and imposed load/m = 9 kN/m we can calculate the total load to the foundation. See diagram in Figure 7.7.

Concrete load from foundation assuming density of concrete = 23.6 kN/m^3:

$$0.2\,\text{m deep} \times 0.6\,\text{m wide} \times 23.6 = 2.83\,\text{kN/m} \tag{7.1}$$

Surcharge of soil over foundation:

$$0.2\,\text{m deep} \times 0.6\,\text{m wide} \times 20\,\text{kN/m}^3 = 2.4\,\text{kN/m} \tag{7.2}$$

Hence, total load to bottom of foundation:

$$35 + 9 + 2.83 + 2.4 = 49.23\,\text{kN/m} \tag{7.3}$$

If this load is divided by the width of the footing (i.e., 600 mm/m length) then the bearing pressure becomes

$$49.23/(1 \times 0.6) = 82.1\,\text{kN/m}^2 \tag{7.4}$$

Assuming the bearing capacity of the soil is 120 kN/m^2, then 82.1 kN/m^2 < 120 kN/m^2 and the foundation is adequately designed for the subsoil. See Figure 7.8 for the bearing pressure diagram.

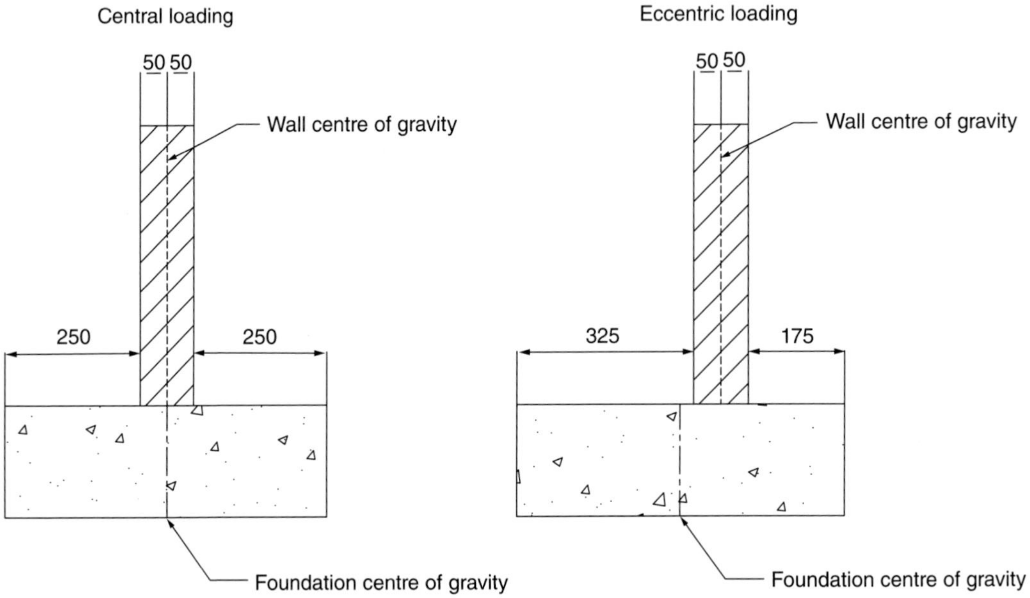

Figure 7.7: Showing central and eccentric loading to a foundation.

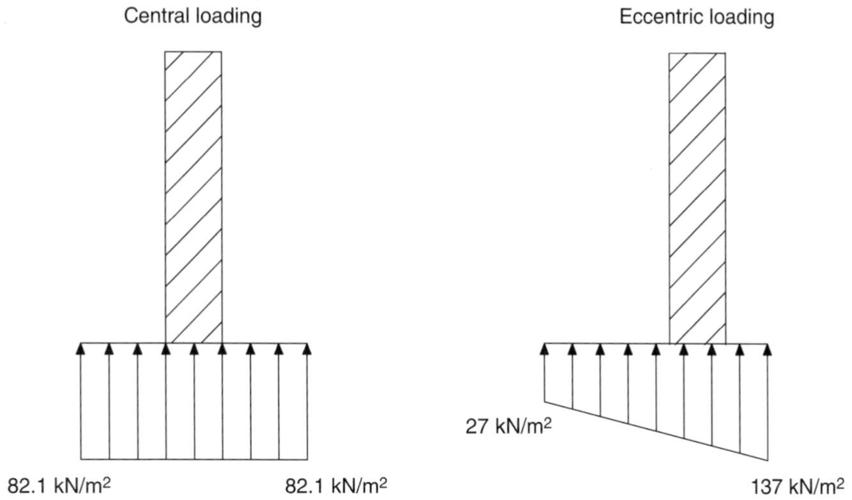

Figure 7.8: Showing the bearing pressure distribution for a centrally loaded and an eccentrically loaded foundation.

Eccentric loading on foundations

It is important that foundations are loaded centrally, since an eccentric loading on a foundation can make a difference. The eccentric load can increase the bearing pressure to one side of the foundation and in some cases the resultant force passing through the foundation may fall outside the middle third of the base.

Example

Let us consider the same foundation used in the example above and re-evaluate the bearing pressure if the wall is constructed 75 mm to one side of the foundation centre (i.e., the wall has an eccentric loading). See Figure 7.7.

The total base vertical reaction from above is 49.2 kN/m^2 = Rv

The base width B = 0.6 m

$$et = \text{eccentricity of base reaction in } x = (P \times ep/Rv) \quad (7.5)$$

where P = total axial load (dead and imposed load) = 44 kN/m
ep (eccentricity) = 75 mm

Hence, from equation (7.5), et = 44 × 0.75/49.2 = 67 mm

Calculating the base pressures

The formulae for the minimum and maximum bearing pressure are as follows, provided the resultant vertical force acts through the middle third of the foundation:

The eccentric load produces a moment.

The maximum bearing pressure is provided by direct pressure + pressure due to bending.

$$\frac{Rv}{B}\left(1-\frac{6et}{B}\right) = \text{minimum bearing pressure Q1} \quad (7.6)$$

$$\frac{Rv}{B}\left(1+\frac{6et}{B}\right) = \text{maximum bearing pressure Q1} \quad (7.7)$$

$$Q1 = (Rv/B)\times(1-6et/B) = (49.2/0.6)\times(1-6\times 0.067/0.6)$$
$$Q1 = (Rv/B)\times(1+6et/B) = (49.2/0.6)\times(1+6\times 0.067/0.6)$$

$$Q1 = 82\times 0.33 = 27\,\text{kN/m}^2$$
$$Q1 = 82\times 1.67 = 137\,\text{kN/m}^2$$

The bearing pressure diagram can be seen in Figure 7.8, and shows that the maximum bearing pressure is greater than the bearing capacity ($137\,\text{kN/m}^2 > 120\,\text{kN/m}^2$). Hence, a small shift of the wall on the foundation has resulted in the failure of the foundation.

Climatic and moisture changes

Climate and moisture changes in the soil can lead to volumetric changes. Clay soils are particularly prone to this, and the Building Research Establishment produces maps which identify the location of firm, shrinkable clays in the UK.

Clay is composed of flat layered minerals such as kaolinite, illite and montmorillonite, which are less than 0.002 mm in diameter. These small particles in the clay can hold water and moisture within their molecular structure. This means that when water is absorbed the layers are forced apart, hence a swelling or expansion occurs. Therefore, when exposed to moisture, clay will swell and expand causing heave.

In times of drought these types of soil will shrink, and this effect is exacerbated by the extraction of moisture by vegetation or trees. Trees have different capacities for extracting moisture from the ground, and the Institution of Structural Engineers, the Building Research Establishment and the NHBC all produce tables providing the recommended minimum distances for trees in close proximity to buildings. The NHBC guidelines concentrate on new builds; it may be different when assessing existing buildings.

The damage or potential for damage caused by trees is not an exact science, and consequently there are variations between the publications. These tables are based on the species of trees, the distance and maturity of the trees, and the location in the country – which is indicative of the type and nature of clay soil and the susceptibility to movement. To provide some context in relation to trees, if we consider a species such as willow, which is a high-water-demand species of tree, they can extract as much as 80 to 100 gallons of water from the ground each day.

Table 7.2 was provided by kind permission of J and I Richardson. The first dimension represents 75% of cases and is a fairly typical tree-to-damage distance for each tree. The figure in brackets is for rare/exceptional circumstances. It must be emphasised that these

Table 7.2: Minimum distance to buildings of various species of tree

Species	Distance (m)
Ash	11 (15+)
Beech	9 (15+)
Birch	9 (9+)
Cherry	6 (10+)
Cypress	5 (10+)
False acacia	10 (14+)
Fruit tree (includes ornamentals, hawthorns and sorbus)	6 (10+)
Lime	10 (15+)
Oak	13 (25+)
Plane	9 (15+)
Poplar	15 (22+)
Sycamore and maples	10 (15+)
Willow	12 (25+)

Reproduced by kind permission of J and I Richardson.

are only a rough guide and this is not an exact science, but is based on experience involving over 100 000 cases from laboratory notes and in the field. The book entitled *Tree Roots and Buildings* (Cutler and Richardson, Longman, 2nd edn, 1989) is an essential reference for published data in this area.

Section 2E4 of Part A1/2 of the Building Regulations provides for minimum depths of foundation to overcome the effects of frost and volume changes in clay soils. Currently this minimum depth is 750 mm. These are minimum depths and most Local Authorities will insist on a depth of 1.0 m to the underside of the foundation; if trees or vegetation are present, this may well be increased in areas of clay soils subject to volumetric changes.

Physical damage by trees

In addition to trees extracting water from the subsoil causing moisture movements, swelling and heave, tree roots can also cause physical damage to foundations by pushing against the structure and causing damage.

Underpinning

Underpinning is a method by which the foundation can be lowered to a soil depth that is capable of sustaining the bearing pressure exerted by the wall. Although the property may originally have been constructed on suitable ground strata, the environment may have changed. For example, a leaking drain may have reduced the bearing capacity of the soil directly underneath the foundation and firmer ground will be required to support the loads from the foundation. This would mean extending the foundation to an area

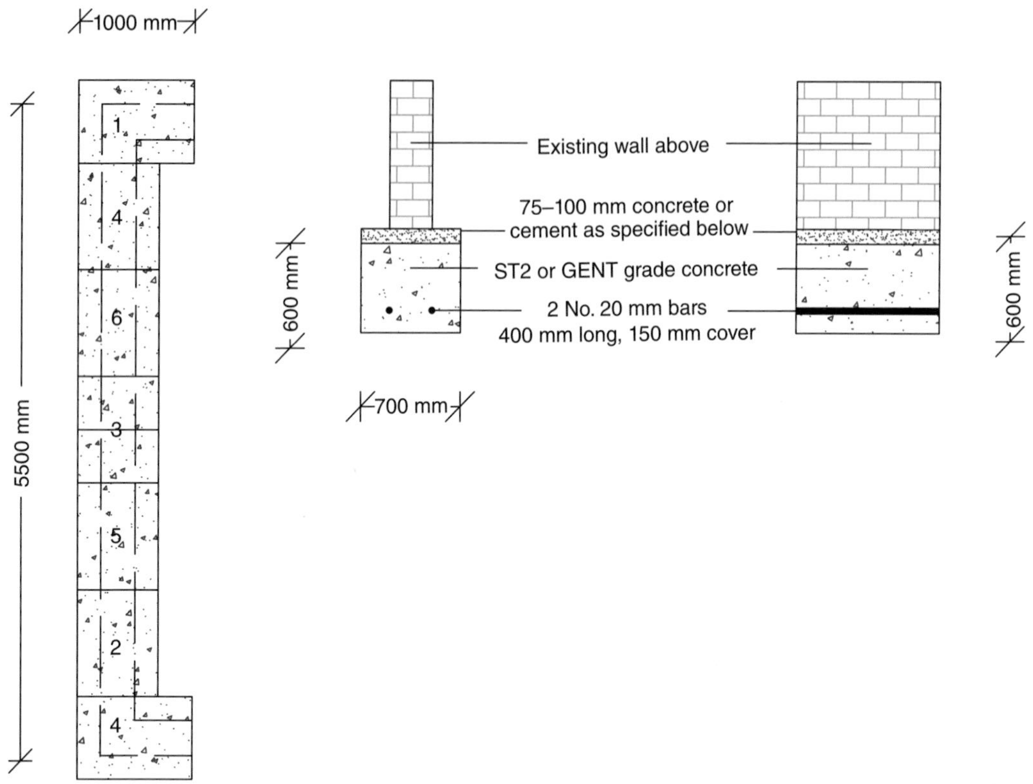

Figure 7.9: Diagram of underpinning.

below the influence of the offending drain. In other circumstances trees may have reached such a maturity that the moisture extraction from a shrinkable clay means that the foundation has to be lowered.

The diagram in Figure 7.9 shows a typical underpinning schedule and a cross-section of the arrangement. It is important to note the following:

- No more than 1.0 m should be excavated at any one time, assuming the masonry components of the wall are in a structurally good condition.
- Adjacent bays should not be excavated until one leg of the underpinning has been completed, pinned and secured sufficiently to support the wall above.
- Only bays with the same reference number should be excavated at the same time.
- No more than 20% of the wall should be left unsupported, and needles and props may be needed if the structural integrity of the wall above is poor.
- A gap should be left between the underside of the wall and the foundation to allow for a strong mix of concrete to be added 24 hours later or the gap should be dry packed with slate and a 1:3 mortar mix.
- A minimum of 24 hours should be left between each series of bays and before packing over the bays to the underside of the wall.

We strongly recommend that a structural engineer is consulted prior to undertaking any underpinning, as each project will require individual assessment.

Chapter 8 Walls

The strength of walls

In the design or construction of a wall we have to be mindful what elements are responsible for the strength of the wall and have some consideration for the likely loads to which it will be exposed. It is therefore important to understand the components of the wall that provide its strength.

Masonry unit

A masonry wall will normally be constructed using a concrete block, hollow block, aerated block, clay brick, concrete brick or stone. Each of these units has a compressive strength. Clay bricks are manufactured using clay which is moulded to a standard shape, normally 215 mm × 102.5 mm × 65 mm. The clay is then fired in a kiln to temperatures up to 1500°C.

The above measurements are for a metric brick, but imperial bricks had a range of dimensions and these are as follows:

225 mm × 107.5 mm × 67 mm
230 mm × 110 mm × 70 mm
230 mm × 110 mm × 73 mm
230 mm × 110 mm × 76 mm
230 mm × 110 mm × 80 mm

The earliest bricks were 51 mm deep and they have increased in depth over the years. Roman bricks varied in width and length, and could be up to 600 mm long.

Various faced finishes and textures can be provided to the brick using moulds, or sand added to the surface prior to firing the clay. In addition to this bricks can be frogged, cellular, hollow, perforated or solid, and Figure 8.1 shows some of the more common shapes.

Bricks can be further classified as facing, common or engineering. Facing bricks are used for aesthetic purposes. Common bricks are used for general construction. Engineering bricks are stronger bricks with low water absorption and are classified as Class A or B. This means they are more resistant to frost damage and for this reason they are generally used below the damp-proof course where the bricks will contain a larger volume of moisture. Engineering bricks can also be used where a larger compressive strength is required.

Structural Design of Buildings, First Edition. Paul Smith.

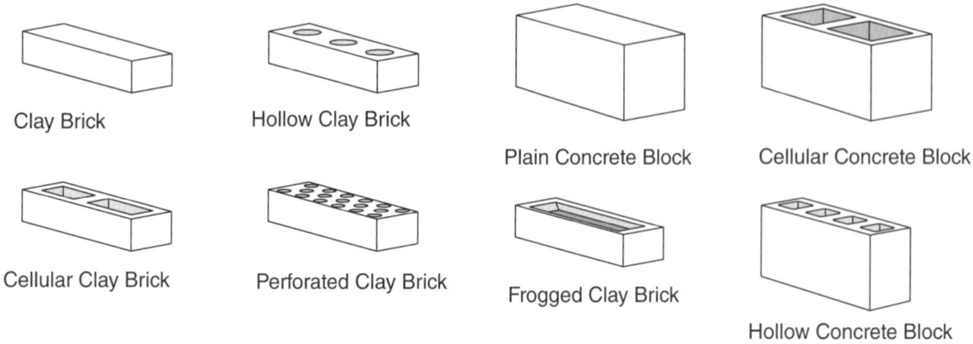

Figure 8.1: Shapes of bricks and blocks.

Table 8.1: Frost resistance and soluble salt content of brickwork based on British Standards

Designation	Frost resistance	Soluble salt content
FN	Resistant	Normal
FL	Resistant	Low
MN	Moderately resistant	Normal
ML	Moderately resistant	Low
ON	Not resistant	Normal
OL	Not resistant	Low

Frost resistance and soluble salts

The durability of brick depends on its resistance to frost, moisture and temperature changes, and the main damage in the UK is likely to arise from frost attack. For this reason bricks are classified depending on their resistance to frost damage. If the brick is likely to be exposed to freeze–thaw cycles then it will be required to be frost resistant and under BS 5628 will be designated frost resistant (F). If the brick is moderately frost resistant the designation will be (M), and finally where the brick is not likely to encounter freeze–thaw cycles and is not required to be frost resistant the designation will be (O).

In addition to this the soluble salt content is also important and this will be designated under BS 5628 as low (L) or normal (N).

So, a brick designated FL will have a high frost resistance with a low soluble salt content. See Table 8.1.

The more recent Eurocode BS EN 771 categorises the durability as F0, F1 and F2: for passive, moderate and severe exposure to frost, respectively. The soluble salt categories are classified as S0, S1 and S2, and reflect the percentage by mass of the amount of sodium, magnesium and potassium contained within the brick.

The higher the soluble salts content, the higher the risk of sulphate attack on mortar and clay bricks.

Concrete blocks

As the name suggests, concrete blocks are manufactured from concrete and the concrete can be aggregate or aerated. Aerated blocks are high in thermal qualities and are used extensively in house-building programmes throughout the UK – as the internal skin of a cavity wall. Blocks can be solid, hollow or cellular, and Figure 8.1 shows the more common shapes.

Aerated blocks typically have a compressive strength ranging from 2.8 to 3.5 N/mm^2, although higher strengths are available. Concrete blocks are usually 7 N/mm^2, but higher-strength blocks are available ranging to 35 N/mm^2.

It is difficult to ascertain the strength of a thermal block by sight, and some blocks have different scribbled patterns etched into the side to indicate their compressive strength. These depend on the manufacturer of the block.

Careful consideration should be given to the exposure conditions and strength when specifying bricks and blocks.

Mortar

Cement and mortar were originally believed to have been used by the Mesopotamians as early as the 4th century BC, then later in Egypt and by the Romans. The original mortars would have been weak and consisted of lime, sand and gravel.

Earlier mortars consisted of mud, and eventually lime mixed with water was introduced which added some bonding qualities to the wall, albeit this had little strength. The main purpose was to prevent air flow into the property, causing drafts.

In the 1st century lime-based mortars were developed by the Romans using the addition of water, lime, sand and hydraulic cement known as Pozzolana. This was originally a volcanic ash but today Pozzolanic materials include brick or tile dust, pulverised fuel ash (which is a waste product from power stations) and pumice (which is volcanic ash). These products are added to non-hydraulic limes and provide hydraulic properties such as allowing the lime to set without exposure to carbon dioxide. However, these additives actually cause further permeability of the mortar.

In the early 1800s cement became available and was known as Portland cement through its origin on the Island of Portland off the Devon coast near Weymouth.

There are two types of cement, hydraulic and non-hydraulic. Hydraulic means that the cement hardens due to hydration – a chemical reaction between the cement and the water. Non-hydraulic cement is where the cement does not harden with water – for example, lime hardens due to its reaction with carbon dioxide. Mortars made with hydraulic lime set more rapidly and are harder and less permeable.

Lime mortar NHL3.5, for example, should not be confused with non-hydraulic lime. This is a common mistake, but the initials NHL mean natural hydraulic lime. Non-hydraulic lime is lime putty and has very little strength, typically 0.3–0.5 N/mm^2. Hydraulic lime is stronger and is specified by a strength. The environment and usage will determine the strength of lime to be used. NHL3.5 means the mortar has a strength of 3.5 N/mm^2. See Table 8.2.

For cob construction or timber frame it is recommended to use lime putty or non-hydraulic lime.

Table 8.2: Lime mortars and their application

Application	Type of lime	Suggested mix ratio by volume	Notes
Pointing/building/ stone/brickwork	Fat lime mortar NHL2 or NHL3.5 Glaster®	Premixed or 3 sand:1 lime putty 1.5 sand*:1 hydraulic lime Premixed	The exact ratio will depend on the sand/aggregate used. The colour, texture and workability of the mortar is predominantly influenced by the selection of sand/aggregate. The softer the stone/brick the softer the mortar must be.
Flagstone bedding	NH3.5 or NHL5 Glaster®	2.5 sand*:1 hydraulic lime Premixed	For smaller tiles contact Tŷ-Mawr.
Pavings, copings, chimneys, parapets	NH5 or NHLZ Glaster®	2 sand*:1 hydraulic lime Premixed	Very exposed areas, high weathering applications.

*It is important to use a sharp, well-graded and well-washed sand.
Note that Glaster® is a lime putty blended with fine recycled glass aggregate.

Lime putty (non-hydraulic lime)

Lime is traditionally prepared by burning limestone. During this process the carbon dioxide is removed. The resultant material is known as quicklime or calcium oxide. If the quicklime is mixed with water a violent reaction ensues known as slaking, and a slurry results called slaked lime or calcium hydroxide. This is also known as lime putty and is kept in this form as slurry until it is used. When the lime putty is exposed to carbon dioxide in the air a set begins to take place and the lime putty begins to return to calcium carbonate. The longer lime putty is stored, the better it becomes because the lime particles continue to break down and any particles that have not undergone the violent reaction of slaking will do so.

Non-hydraulic lime is also available in powder form and can be purchased from building suppliers. This is added to water to produce lime mortar, which should be soaked for one or two days. It is not as good as lime putty and has a relatively short shelf life; the newer the lime, the better it will be.

Hydraulic lime

This lime is produced in much the same way as non-hydraulic lime but contains clay which is found in the limestone used in the process.

The strength of limes can be seen in Table 8.3.

Important rules in the use of lime mortars

- Choose the right type of lime and aggregate for your application.
- The mixture must be well mixed.

Table 8.3: Strength of lime mortars

Lime	Compressive strength after 28 days (N/mm^2)
Non-hydraulic lime putty	0.3–0.5
Hydraulic limes NHL2	1.3–2.0
NHL3.5	2.0–4.5
NHL5	5.0–10.0
Limecrete floors	4.0 increasing to 6.5 and 8.3 after 90 days

Table reproduced with kind permission of Tŷ-Mawr Lime Ltd. Copyright Tŷ-Mawr Lime Ltd.

- Always measure using a gauging box or bucket.
- Mortar must not be allowed to dry out too quickly. Dampen surrounding masonry.
- Hydraulic lime mortar must be used within two hours and then left to set in the wall.
- Do not use if the temperature is likely to fall to or below 5°C before carbonation has taken place (note that this can be weeks or months after application).
- Protect from frost, excessive sunlight and drying winds.
- Pointing should be kept moist for seven days. The carbonation can only complete in the presence of moisture.
- Depending on the hydraulic lime and the time of year, it should be possible to build at about the same rate as with Portland cement mortars but bear in mind that the mortar will continue to gain strength for up to a year – although an adequate set will be achieved in a matter of a week or two.

Reproduced by kind permission of Tŷ-Mawr Ltd. Copyright Tŷ-Mawr Ltd.

Cement

The advent of cement allowed for taller, larger structures since this new product offered better adhesion. Cement is prepared by burning chalk and clay and has the advantage of hardening under water. Cement mortars are much more impermeable than lime mortars and are much stronger. They also set more rapidly.

Using cement in a predominantly lime mortar is a great temptation to increase the rate of set. Such a mix may include one part cement to three parts lime and eight to twelve parts sand. The problem with this is that the cement weakens the mortar mix through a process called segregation.

All masonry moves to some degree; when lime mortar is used the masonry is more flexible and allows the egress of moisture from the beds. This makes a wall structure more breathable. The more cement is used, the stronger but less flexible the structure will be.

Characteristic strength of masonry

The characteristic strength of masonry is determined by knowing the strength of the masonry unit (i.e., the brick or block) and the prescribed mortar mix. Eurocode 6: Design of Masonry Structures provides details on how to determine the characteristic strength of masonry. An example can be seen below.

Design example

Let us propose a wall constructed using a group 1 clay masonry unit as taken from table 2.4 of the National Annex in E.C.6 giving a proposed K value of 0.5. By using table N.A.2 of E.C.6 it can be determined that if we have a mortar with a cement:sand ratio of 1:5, the mortar is classified as an M4 mortar which has a compressive strength fm of 4 N/mm^2.

Thus:

K = 0.5 is a constant. The following values can be obtained from Eurocode 6 for general-purpose mortar: $\alpha = 0.7$ and $\beta = 0.3$.

fb = the normalised mean compressive strength of the masonry units in the direction of the applied action.

$$fb = \text{conditioning factor} \times \text{shape factor} \times \text{mean compressive strength} \tag{8.1}$$

If using air-dried bricks, the conditioning factor = 1
Standard format size bricks have a shape factor = 0.85
Compressive strength of brick = 10 N/mm^2 say
Hence fb = 8.5
fm = 4 N/mm^2 compressive strength of the mortar
fk = characteristic strength of the masonry (N/mm^2)

$$fk = kfb^{\alpha}fm^{\beta} \tag{8.2}$$

Taken from clause 3.6.1.2 of Eurocode 6: Design of Structural Masonry.

In the example above:

$$fk = 0.5 \times 8.5^{0.7} \times 4^{0.3} = 3.38\,\text{N/mm}^2$$

This is the combined strength of the masonry and mortar.

This value is used to determine the strength of the wall not the compressive value of the masonry unit. Therefore, it can be seen that if two walls of the same brick and manufacturing control are constructed but using different mortar strengths, this alters the characteristic strength of the masonry and consequently a weaker mortar will lead to a weaker strength of wall.

Slenderness ratio

The slenderness is dependent on three parameters:

- The height of the wall or column.
- The restraint conditions at the ends of the wall or column.
- The cross-sectional area of the wall or column.

The slenderness ratio is the relationship between the effective height and the effective thickness, and the formula is as follows:

$$\text{Slenderness ratio} = \text{effective height}/\text{effective thickness} \tag{8.3}$$

If we consider walls or columns, then structures that are short and stocky will be more able to sustain larger loads than slender walls. Short and stocky walls generally fail in crushing, whereas slender walls are more prone to fail in bending.

In cathedrals, Masons used columns to support the roof structures and ceilings and the slenderness ratio can be seen to be less than 10 in most cases. This ensures the columns would fail in crushing rather than bending, for which the stone would have less resistance. The crushing strength of stone is very high, and the low value for the slenderness ratio ensured the strength of these structures. Typically the columns would sustain loads in excess of 100 tonnes/m^2. This is low in comparison with the loads that could be sustained by the stone material. Thus, the high compressive resistance and low slenderness ratio meant that the structures were arguably over-designed but certainly stable. This information would have been passed from generation to generation, by father to son as master masons. These masons understood stability and the need for low centres of gravity to sustain the stability of the huge structures.

Flexural stiffness and the second moment of area

To demonstrate the impact of the Euler load on wall stability and its relationship to the slenderness ratio, we first have to consider flexural stiffness. We shall demonstrate this with the example of a beam.

If we consider a beam which is loaded with a specified load, the beam will bend and this deflection will be dependent on the second moment of area (I) of the beam and the stiffness of the beam represented by E (Young's modulus). The flexural stiffness of the beam is a product of EI.

The second moment of area is a moment. If we consider a force, the moment due to that force is defined as the force multiplied by the perpendicular distance from the fulcrum or axis. Hence, similarly, if we consider a moment of area this is defined as the area times the distance from the axis. Thus, the axis is some distance y away from the centroid of the object (the point at which the area of the object is concentrated) and the first moment of area is given by

$$I = Ay \tag{8.4}$$

Let us consider a rectangle with breadth b and depth d and the axis along the edge of the rectangle as shown in Figure 8.2.

The distance from the centre of the area to the edge, y, is d/2.

$$\text{Hence, the first moment for a rectangle is bdxd}/2 \text{ or } bd^2/2 \tag{8.5}$$

If the first moment of area is the area times the distance once, then the second moment of area is the area times the distance twice.

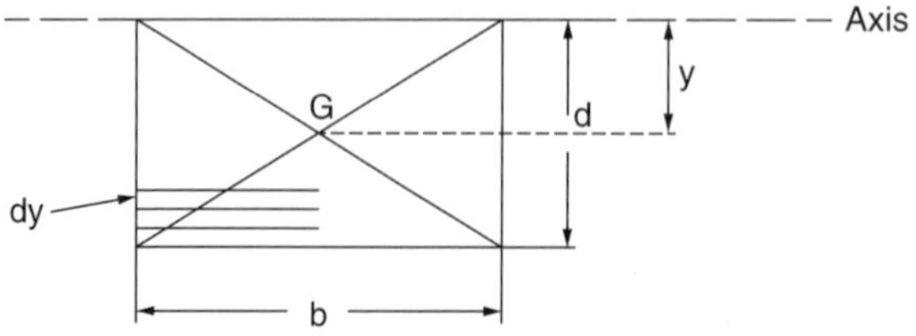

Figure 8.2: Second moment of area derivation diagram.

Thus let us consider the same shape and regard this as comprised of strips of width b and breadth dy.

$$\text{The first moment was } y \times \text{area} = y.bdy \tag{8.6}$$

Thus the second moment becomes y^2 times area.

$$\text{This is equal to } y^2 \times bdy \tag{8.7}$$

To determine the whole area we have to consider all the strips and consequently we integrate the formula as follows:

$$I = \int by^2 dy \tag{8.8}$$

The limits are d/2 and –d/2.

$$\text{Thus, the equation becomes } I = \int_{d/2}^{d/2} by^2 dy \tag{8.9}$$

$$\text{This gives us } I = \frac{bd^3/8}{3} + \frac{bd^3/8}{3} \tag{8.10}$$

$$\text{Thus } I = bd^3/24 + bd^3/24 = bd^3/12 \tag{8.11}$$

Note that the second moment of area is not to be confused with the second moment of inertia, which has the same symbol I.

Euler load

If we take a perfectly elastic and slender strut pinned at each end, the strut will buckle elastically at a load value known as the Euler load. This is provided by the formula as follows:

$$Pe = \pi^2 EI/L^2$$

as seen previously in Chapter 2 (equation (2.31)).

Pe = Euler load
E = Young's modulus
I = second moment of area
L = length

The type of end fixings and conditions will make a difference to the load that a particular member can sustain. If we consider the same length of similar material loaded in the same manner, the effective length can be shown to change depending on the end conditions. This can be shown mathematically, but is beyond the scope of this book. The Euler load (i.e., the load at which the member buckles) is shown as follows and is a function of its effective length:

$$\text{Fixed end}\ \ Pe = \pi^2 EI/(L/2)^2$$

as seen previously in Chapter 2 (equation (2.32)).

$$\text{One end pinned and one end fixed}\ Pe = \pi^2 EI/(0.7L)^2 \tag{8.12}$$

$$\text{Cantilever of one end fixed and one end free}\ Pe = \pi^2 EI/(2L)^2 \tag{8.13}$$

If we use a value of 3000 mm as the length of the member, and apply equations (2.31), (2.32), (8.12) and (8.13) the results as seen in Table 8.4 can be obtained. Assuming a value of EI that is constant at 677064909.6, the second moment of area and Young's modulus will be the same for the same member. We can see that the fixed end strut can sustain a load four times greater than the pinned end.

As discussed previously, the restraint conditions have an impact on the slenderness ratio. As can be seen, the stability of a structure is affected by its end restraints.

Under Eurocode 6: Design of Masonry Structures, clause 5.5.1.4: Slenderness ratio of masonry walls defines the slenderness ratio and places a limit of 27 on the slenderness ratio where walls are subject to mainly vertical loading.

If the restraint conditions at the end of the wall are rigidly fixed, then the slenderness ratio is lowered and the wall is less likely to fail. Eurocode 6: Design of Masonry Structures, clause 5.5.1.2 proposes that walls can be restrained by floors, roofs, cross-walls or other rigid elements to which the wall is connected.

This can be simply demonstrated using a plastic ruler. If you stand this on a desk, pushing with the palm of your hand will cause the ruler to deflect. However, if you clasp the ruler at each end and push down on the ruler much more force is required to cause the ruler to bend.

Figure 8.3 shows a photograph of two pieces of dowel of the same cross-section with a central dowel which is a datum for the vertical line. A similar load is applied to both

Table 8.4: Load capabilities of a strut depending on end restraint

End	Load (N)
Pinned	743.05
Fixed	2970
Pin and fix	1516
Free	185.76

Figure 8.3: Photograph showing a fixed-end and pinned dowel.

dowels, however one is glued and therefore considered to be a fixed end and the second is pointed and pushes into the base plate as a pin. The deflection in the photograph demonstrates the difference between the dowel bars depending on the restraint of the end, and is representative of the loads that can be sustained.

The deflection due to a fixed end, as shown on the "left", was less than that of a pinned end, which can be seen on the "right" of the photograph. This experiment demonstrates that the restraint at the end of the structure can have an impact on its strength.

The effective height/length is taken as the point between the restraints multiplied by a coefficient depending on the end restraints. Reducing the effective height therefore reduces the slenderness ratio. For example, if a wall is restrained at mid-height this reduces its effective height and subsequently this can prevent buckling of the wall. Walls can be restrained by the use of lateral restraint straps, at first floor and eaves level, as specified under Part A of the Building Regulations, Structure.

Furthermore, clause 8.5.1.1 of Eurocode 6: Design of Masonry Structures provides details on the transfer of loads to the floor or roof where the wall is considered to be restrained by the floor or roof. The floor must be covered and the wall has to be connected using straps to transfer the load via a diaphragm action. Clause 8.5.1.2: Connection by straps provides further information on the use of the transfer of loads using straps.

Lateral restraint straps are galvanised steel, 1200 mm × 30 mm × 5 mm heavy-duty straps with a tensile strength of 8 kN/mm^2 in compliance with BS 5268 Part 3 and BS EN 845-1. In cases where the floor joists are parallel to the wall, the straps are placed at a maximum 2.0 m centres carried across and fixed to at least three joists by nails or screws along a noggin line. In cases where the wall is at right angles to the floor joist the strap is secured against the side of the floor joist. The strap is located and tightly secured against the masonry wall in the cavity. Note that the spacing of the straps is decreased to 1.25 m for buildings over four storeys.

As discussed above, a wall can be stiffened using a stiffening wall running at right angles to the wall. Under Eurocode 6: Design of Masonry Structures, a stiffening wall should have a length of at least 1/5th of the clear height and have a thickness of at least 0.3 times the effective thickness of the wall to be stiffened. Clause 5.5.1.2: Effective height of masonry walls gives clear directions on the criteria for stiffening walls.

Let us consider the example of the difference the slenderness of a wall can make if we take a single-skin wall 100 mm thick, 4.8 m long and 2.6 m high and apply a dead load of 30 kN/m and an imposed load of 122 kN/m. If the wall is constructed using a clay brick of compressive strength 10 N/mm^2 and a mortar designation (ii), we find that the wall fails due to the allowable compressive stress which equates to 0.577 N/mm^2, being less than the actual compressive stress which equates to 0.612 N/mm^2.

If, however, we stiffen the wall by the introduction of piers at 3.0 m centres, 450 mm wide, protruding 100 mm from the wall, the slenderness ratio of the wall is improved through its reduction and the wall is much stronger. The allowable compressive stress now becomes 0.870 $N/mm^2 > 0.612\ N/mm^2$.

A similar effect would occur if we thickened the wall along its entire length, thus reducing the slenderness ratio.

Walls also have to sustain lateral loads and masonry is not as strong laterally as it is in compression, having different physical properties vertically and horizontally. Lateral loads can be applied causing bending horizontally and vertically. In other words, the wall is non-isotropic. Hence, the strength and failure modes are different in the vertical and horizontal directions. In vertical bending the failure and cracking will develop along the length of the bedding joints. In horizontal bending the failure will develop along the vertical joints. Horizontal bending offers more resistance than vertical bending. However, as the vertical load increases from above, the strength of a wall along the bedding joints (that is the vertical bending resistance) will increase.

Under BS 5628 Part 1: 1992 and Eurocode 6: Design of Masonry Structures, the characteristic flexural strength of masonry is provided in Tables 3 and 6, respectively (which is the same table in both codes). The value of the moment of resistance of the wall is a function of the flexural strength of the masonry. As can be seen in the tables, the flexural strength for plains of failure perpendicular to the wall or contributing to the horizontal bending is much greater than for parallel or horizontal plains.

The moment of resistance is a function of the section modulus of the wall and the flexural strength. Hence, for a wall of similar mortar designation the characteristic flexural strength will remain unchanged but if the thickness is altered then the moment of resistance increases due to an increase in the section modulus. The section modulus is given by $Z = bd^2/6$ (b = breadth under consideration, d = depth or thickness), as seen in equation (2.26) of Chapter 2.

The shape of the wall can determine the lateral strength and to demonstrate this, if we consider a free-standing wall of different plan shapes, calculations can be undertaken to determine the ratio of the maximum value of the lateral load to the cross-sectional area of masonry.

Let us consider a free-standing garden wall 2.0 m high with special manufacturing control and normal construction control. The characteristic strength of brickwork is taken as 18 kN/m^3 with a mortar designation (ii). Let us consider the wall in four methods of construction and examine the lateral wind loading only with no other characteristic lateral imposed load (Qk). Thus, Qk in clause 36.5.2 of BS 5628 is considered to be equal to zero. The construction types considered are a solid wall 100 mm thick and a staggered

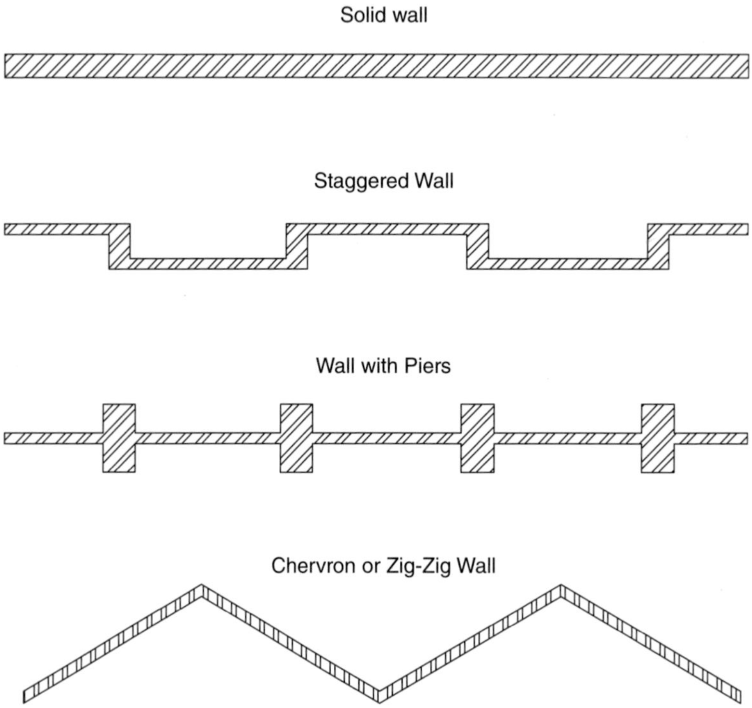

Figure 8.4: Four types of masonry wall to be analysed.

Table 8.5: Comparison of lateral moment of resistance for different wall types exposed to lateral wind loading

Wall type	Moment of resistance (kN·m/m)
Solid wall	0.21
Staggered wall	1.26
Wall with piers	1.76
Chevron wall	2.40

wall, chevron wall and single-brick wall with piers all 100 mm thick as shown in Figure 8.4.

Through the application of BS 5628: Part 1 we can analyse a 1.8-m length of each type of wall. The moment of resistance is calculated and comparisons obtained as in Table 8.5. The walls have been independently analysed by Geomex Architectural Design and Structural Engineering Ltd.

Interestingly, if a wall is constructed in a chevron pattern the wall is made much stronger. This is due to the average section modulus being increased with the angle, and thus the design resistance of the moment increases.

Lateral loads are applied to the wall primarily through wind, but other lateral loads such as the thrust from an arched structure may occur. It is assumed in the design of walls that such forces are carried to the floor acting as diaphragms or other walls acting as buttresses. These loads are ultimately transferred to the foundations.

Generally, masonry walls are weaker in their capacity to carry lateral loads than vertical loads. However, cross-walls, buttresses or piers are used to offer lateral stability and bracing to the overall structure. To this end the Building Regulations 2010 recommend minimum lengths, heights and thicknesses for walls to ensure lateral stability and the structural integrity of the overall structure. Under Part A of the Building Regulations 2010 the length of the buttress should be at least 1/6th of the overall height of the wall and be securely bonded to the wall.

Annex F of Eurocode 6: Design of Masonry Structures applies limiting height and length to the thickness ratio for walls depending on the restraint conditions at the edges. This information is provided in figures F1, F2 and F3 of the Annex in graphical form, with h/t on the y-axis and l/t on the x-axis.

Under Part A, Structure of the Building Regulations 2010 it is recommended that wall lengths do not exceed 12 m for buildings constructed using brick and blockwork up to three storeys.

Interestingly, the limiting factor for a panel dimension as used in clause 32.3 of BS 5628: 2005 is

$$\text{Limiting factor} = 50\,\text{tef} \tag{8.14}$$

where tef is the effective thickness of the wall. For a cavity wall this is taken as the thickness of the individual skin of wall or

$$\text{tef} = 2/3(\text{t}1 + \text{t}2) \tag{8.15}$$

Here t1 and t2 are the thicknesses of each individual skin of the wall.

If we use tef from equation (8.15) and substitute it into equation (8.14), the effective thickness becomes $200 \times 2/3 = 133$ mm, hence the limiting dimension is $50 \times \text{tef} = 50 \times 133 = 6666$ mm.

Leaning walls and stability

The stability of a wall can be determined by the following method and relies on the centre of gravity of the wall lying within the middle third of the wall. If this is the case then the wall is considered stable since no tension is experienced in the wall or foundation and the resultant force lies within the middle third of the structure. We cover the middle-third rule later in this chapter. It should be noted that the middle-third rule as demonstrated is undertaken on a free-standing wall and if the wall is subject to additional loads such as floors, then the stability position is much reduced.

Movement joints

All walls experience movement due to thermal and moisture changes, consequently movement joints are required to accommodate these movements. Different materials will expand and contract at different rates depending on the thermal coefficient of expansion and changes due to moisture. Thus, if two materials are placed together – such as a concrete or a steel lintel over an opening in a brick wall – opposing forces result in cracking.

Changes due to temperature changes

All the materials used in buildings are affected by changes in temperature, but to differing degrees depending on their coefficient of thermal expansion. The amount of movement can also be affected by restraints and the temperature range to which the material is exposed.

Let us take a calcium silicate brick wall, 6 m long with a thermal expansion coefficient of $\acute{\alpha} = 8 \times 10^{-6} - 14 \times 10^{-6}$ and a temperature range of 50°C.

The equation for the change in size is as follows:

$$\Delta L = \acute{\alpha} L t \tag{8.16}$$

Thus with the lower thermal expansion coefficient:

$$\Delta L = 8 \times 10^{-6} \times 6000(\text{mm}) \times 50 = 2.4\,\text{mm}$$

With the higher thermal expansion coefficient, ΔL = 4.2 mm.

Hence, the change in length of a 6-m-long brick wall can range between 2.4 and 4.2 mm, with a temperature range of 50°C.

Under BS 5628: Part 3, recommendations are established for the provision of movement joints in brick and blockwork. These are a maximum of 15.0 m for brickwork and 6.0 m for a concrete block wall. Aerated blocks should have movement joints in accordance with the manufacturer's specifications.

If we undertake the above calculations again for a 9.0-m-long wall in clay brick, with $\acute{\alpha} = 8 \times 10^{-6}$ as an upper value and a temperature range of 70°C, we find that the change in length is 5.04 mm.

Using the same temperature range as in the example above, for a 6.0-m-long concrete block wall with a thermal expansion coefficient of 12, the change in length of the wall becomes 5.04 mm.

Hence we can see that to restrict the change in length to approximately 5 mm, the length of the clay brick and concrete wall will vary due to the coefficient of thermal expansion.

If we apply the same formula to an aerated concrete block used on the internal skin of a wall with a coefficient of thermal expansion of 8×10^{-6}, we see that this is 5 mm for a wall length of 9.0 m. However, the range of temperatures on the internal skin of a wall may be greater and particularly behind radiators, where the temperature range may be 80°C. Thus, to sustain a limit of 5.0 mm, the length of the wall without a movement joint becomes 7.8 m.

Changes due to moisture changes

In addition to being affected by temperature changes, some materials are also affected by moisture changes and this can lead to cracking. Some materials experience reversible changes by expanding and shrinking when exposed to moisture and then drying out. Others show irreversible changes – this is where the material changes size and does not return to its original size. An example of this is cement: as it cures it shrinks and this shrinkage or movement can be experienced over a long period of time, diminishing as time proceeds.

Information is available from the Building Research Establishment which shows the moisture-induced size changes for materials, both reversible and irreversible.

Some materials can exhibit both reversible and irreversible movements; for example an assessment can be made on how much concrete will shrink as it cures, which is an irreversible movement. At the same time, if the material becomes wet, then a reversible change may also take place – returning to its original size as it dries – and these coefficients are different.

For example, if we consider a cement mortar or fine concrete, the reversible moisture movement is 0.02–0.06% of its size. However, the irreversible change due to moisture is -0.04–0.1%. The minus sign indicates contraction or shrinkage.

If we examine these percentage changes we can see that the irreversible moisture movement for brick shows that it expands, whereas concrete shrinks or contracts. Thus, over time, if these two materials are put together there will be cracking between them due to irreversible moisture movements.

Traditional design of walls

Traditionally, masonry walls were designed using empirical slenderness ratios handed down the generations by masons. Typical examples would be as follows:

- Slenderness ratio $h/t < 20$ for solid walls or fully grouted walls.
- Slenderness ratio $h/t < 18$ for other external walls not fully grouted or solid walls and non-load-bearing walls.

This was the case for brick walls, and a simplistic slenderness ratio of $h/t < 10$ appears to be correct for stone structures.

If we take a stone cottage of typical wall height 4.5 m, the wall thickness would be 450–500 mm. For a brick structure of two storeys, say 4.5 m high, the thickness would be 225 mm – thus a slenderness ratio of 20. The height of 4.5 m is measured from the ground floor to the wall plate. If the walls went higher than this, say to another storey, then the brick wall would be thickened. For example, another storey would give an overall height of 6.5 m. This would mean that the solid brick wall would be thickened to 325 mm. This includes 10 mm thickness for each mortar joint, thus maintaining a slenderness ratio no greater than 20.

The early Masons would not understand stresses, since this is a relatively modern concept, so they would not understand that the height of a wall is dependent on design stress as well as buckling capacity. Also, it was unlikely that these parameters or ratios would be committed to writing.

In 1875 the Public Health Act made it incumbent on local authorities to improve the structural stability of houses, with the particular focus on preventing fires. In 1878 the Building Act gave details on wall and foundation thicknesses, as can be seen in Figure 7.1 of Chapter 7.

Middle-third rule

We have now examined walls and foundations, and it is worth noting a key parameter concerning the design of these structural elements. Many wall constructions – particularly walls constructed in stone, brick and cob – will not accommodate tension forces very well but will accommodate compression forces. For this reason the forces acting on a wall and foundation should be primarily in compression, avoiding tension forces where possible to ensure the effective efficiency of the design.

The middle-third rule states that the resultant force should be directed towards the middle-third section of the cross-section of the wall or foundation, to ensure the forces are kept in compression. A simplistic demonstration using a free-standing wall can be seen in Figure 8.5. The centre of gravity of the wall is determined, and a vertical line drawn to the base. If the vertical line passes through the middle third of the wall at its base, the wall is stable.

Clearly, floors and roof structures will alter the centre of gravity and therefore shift the resulting force and centre of gravity of the wall.

If we consider the wall of a house, the loads are predominantly vertical but lateral forces from the roof can cause horizontal forces, with a concomitant shift in the resultant force as it progresses down the structure and the load path down the wall. Therefore, steep roofs with the majority of the force being vertical rather than horizontal will be better in this context than flatter roofs which have a larger horizontal force, potentially causing the resultant force outside the middle third of the wall.

In the case of a retaining wall there is a lateral horizontal force acting approximately a third of the height up the wall from the soil it retains, and a vertical force from the load of the wall and the weight of the soil on the heel. See Figure 8.6.

The resultant force R obtained from the horizontal load H and the vertical load V has to act within the middle third of the foundation otherwise tension forces develop and the

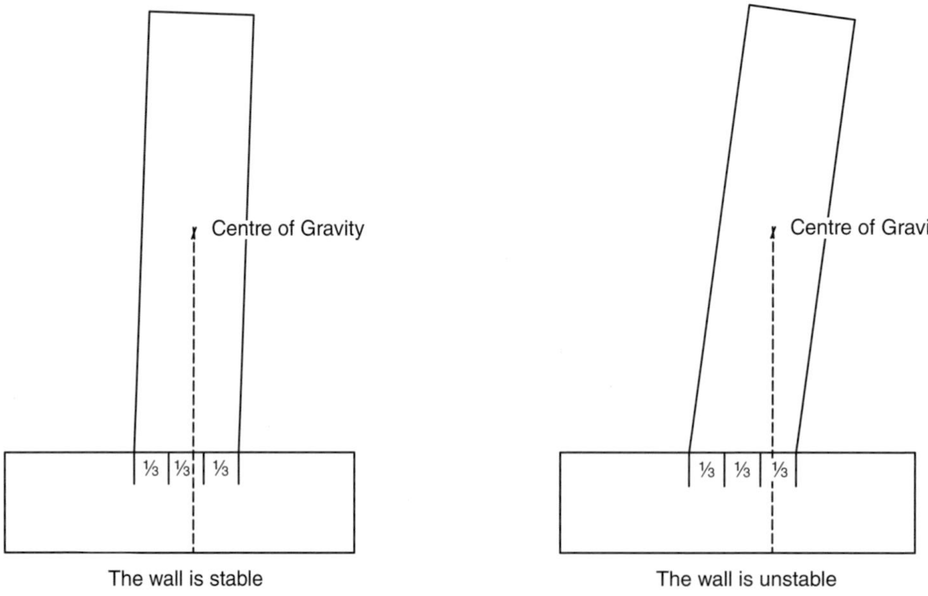

Figure 8.5: Showing how to determine the stability of a free-standing wall.

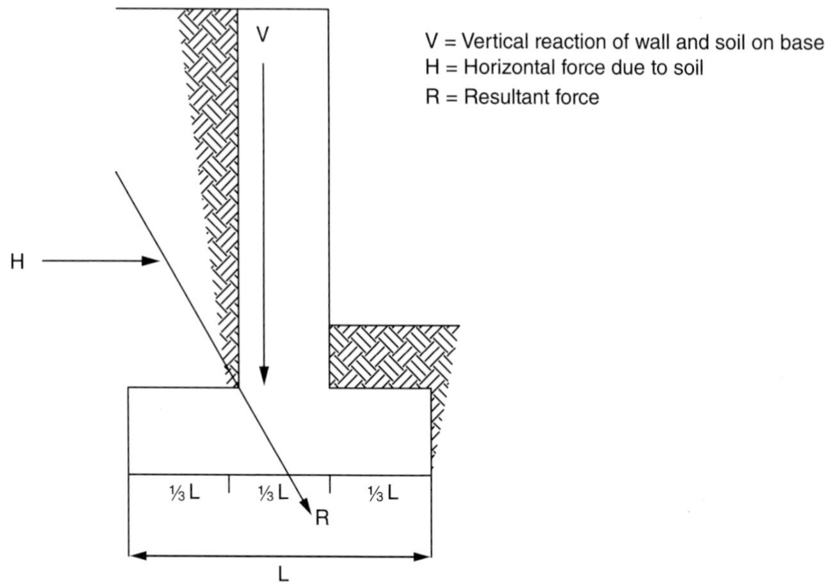

Figure 8.6: Middle third of the retaining structure.

base is likely to fail. To accommodate this, the base would have to be thickened or the vertical load increased, which has an impact on the bearing pressure under the foundation. Hence, these two parameters have to be considered.

A wall may be thickened to ensure the resultant force continues to fall within the middle third. Hence if we examine walls, particularly large structures, we can see piers or walls which are stepped. As we travel from the top to the bottom of the wall, the loads increase and in an attempt to ensure the resultant force continues to be directed through the middle third of the wall, the wall or piers are stepped out – thickening the wall to make the base wider.

If we consider a large structure such as a cathedral, these buttresses or wall thickenings can often be seen. Flying buttresses transfer the load from walls out and down the structure, ensuring this middle-third concept is maintained. This was particularly useful for cathedrals in the 1200s to 1300s, which were being constructed with higher and flatter ceilings and larger windows to allow the passage of light through the building. At this time the fashion was to introduce light, which was considered a spiritual concept, thus the more light the better. These tall, open structures now had to sustain large lateral loads, with the problem of pushing the resultant force outside the middle third of the wall. The flying buttress accommodated such loads and one of the first such uses of the flying buttress was at Durham Cathedral. If we examine the shape of the buttress, it directs the force downwards and at the same time introduces an interesting architectural feature to the building – sometimes passing over the nave.

Timber frame walls and raking

We discussed the construction of timber frame walls in Chapter 3, and these walls are the load-bearing elements in modern construction – usually lined with an outer skin of brick. However, some modern timber constructions – particularly in America – employ timber cladding rather than brick or masonry to protect the timber frame from the elements. As with traditional timber frame, all timber structures are prone to lateral movement called raking, and this movement is resisted by sheathing on one or both sides of the wall, which then acts as a diaphragm. The key to effective raking resistance is the fixings which provide the majority of the resistance to the lateral forces. The frame is also secured by bolts or nails along the sole plate, which prevents the overturning of the timber panel.

Thus, diaphragms provide the necessary resistance to raking and overturning and therefore have to be designed under the relevant code of practice. Eurocode 5: Design of Timber Structures provides two methods for calculating the raking resistance and the National Annex for the UK recommends method B. Timber frames must be sheathed using a timber panel such as plywood or oriented strand board (OSB). The new Eurocodes do not permit the use of plasterboard as a sheathing material to provide raking resistance.

Under Eurocode 5, the following criterion has to be met:

$$h/b < 4 \tag{8.17}$$

where

h = panel height
b = panel width

Hence, for a panel 2.4 m high and 1.2 m wide, the ratio = 2 < 4 therefore the criterion of equation (8.17) is met. This criterion ensures the in-plane strength of the wall is provided by the panel.

Clearly, the spacing and type of connections of the sheathing to the frame are important, and for this reason the fixings have to be screws or nails and the spacing of such has to meet specified criteria. They have to be equally spaced around the sheathing perimeter, and the spacing of the internal sheathing fastenings cannot exceed twice the distance of those on the external perimeter of the frame. The limiting factors are 150 mm spacing for nails and 200 mm for screws.

In this type of construction many panels are secured together to form a wall, and further criteria are set for the jointing and arrangement of the panels:

- The anchorage or loads to the sole plate or base of the panels must be adequate to prevent overturning and sliding.
- The vertical connection strength between the panels should be a minimum of 2.5 kN/m. This means that where panels are connected, the fixing connecting them must be able to sustain a vertical shear force of 2.5 kN/m.
- The tops of the panels must be continuous and linked by a member such as a continuous header plate or wall plate.

If we consider a wall made from a number of timber-sheathed panels, the raking resistance taken from EC5 is as follows:

$$\text{FvRd(raking resistance)} = \sum \text{FivRd} \tag{8.18}$$

(The raking strength sum of all the panels thus provides the raking strength of the wall.)

The sum of the raking strength FvRd calculated as above has to be greater than the horizontal design force on the wall:

$$\text{FvRd} > \text{FH} \tag{8.19}$$

FivRd, which is the raking strength of the panel, can be determined from EC5 and is equal to

$$\frac{\text{FfRdbi}}{\text{So}}\text{Kd Ki, q Ks Kn} = \text{FivRd} \tag{8.20}$$

where
FfRd = design capacity of each fastener in the lateral shear
bi = length of the wall
So = fastener spacing, provided by the equation below:

$$\text{So} = \frac{9700\text{d}}{\rho\text{k}} \tag{8.21}$$

where
pk = density of wall framing timber
d = diameter of the fastener

The following factors are determined from EC5:

Kd = dimension factor, determined from knowing the ratio of the length of the wall to its panel height

Kiq = uniformly distributed load factor; this takes all the loads on the wall and derives the equivalent uniformly distributed load to the wall (in kN/m)

Ks = fastener spacing; this is based on the spacing of the fasteners around the perimeter of the panel

Kn = sheathing material factor; Kn = 1 if sheathing is used on one side only; if, however, the sheathing is on both sides, EC5 provides a formula to calculate this value

As shown above, when a wall is subject to a lateral load this is resisted by the raking resistance of the diaphragm. The raking resistance is the sum of the raking resistance of each individual panel making up the wall, and this value has to exceed the horizontal force acting on the building or structure. The overturning formula is also provided in EC5 and ensures the wall panel does not lift when subjected to a horizontal force.

Chapter 9 Floors

The history of floors

Originally ground floors were constructed using compacted earth, but by the 1600s materials were being placed on the compacted earth such as brick, stone and tile. In some circumstances the floors were screeded using a plaster or lime, and these types of floor are of significant architectural importance.

Later, in Victorian times and around the 1900s, clay tiles or stone slabs were supported on ash or compacted earth. This gave a spongy feel, and this type of floor can sometimes be felt when walking across it or dropping both heels on the floor from a tip-toe position. Be careful whilst undertaking this procedure, as particularly springy floors may cause furniture to bounce and expensive ornaments to fall to the floor!

Suspended floors on the ground floor became prevalent in the 1700s, but the timber was laid directly onto the earth and this led to decay of the timber members. An example of a floor resting on the ground can be seen in Figure 9.1.

Suspended floors were used to prevent damp at the wall/floor interface. These types of floor became more fashionable in the Edwardian and Victorian era, and were supported off dwarf walls or sleeper walls. These walls comprised little or no foundations and were usually constructed of brick to the underside of the joists. Air bricks or vents with metal cast iron covers can give an external indication if the floor is suspended, but as a note of caution the ventilation was usually poor and damp can be a problem in older houses.

The introduction of damp-proof courses in the 1900s meant that the decay of timber ground-floor joists was reduced and dwarf or sleeper walls were constructed using staggered brickwork, which introduced gaps and allowed air to flow across the floor, thus reducing damp.

Although a typical Victorian terraced house would have a solid concrete floor in the hallway and kitchen, the remaining ground floors were suspended. However, by the 1950s solid concrete floors became more widely used on the ground floor and employed bituminous membranes as a barrier against damp. In the 1970s damp-proof plastic membranes were also introduced and this prevented damp rising through the floor. The added advantage was that the damp-proof membrane also reduced contact between the concrete and soil, thus inhibiting the likelihood of sulphate attack on the concrete.

Vinyl tiles were used in the 1950s to cover concrete floors, and the Health and Safety Executive advise that these products may contain asbestos and care should be taken if removing such floor coverings.

Structural Design of Buildings, First Edition. Paul Smith.

Figure 9.1: Photograph showing a timber floor laid directly onto earth in a Georgian property.

Traditionally, suspended first floors used boards laid over the joists secured by nails approximately 1 inch or 25–30 mm from the edge. The size of the joist was usually 50 mm (2 inches) wide and the depth calculated using the following formula:

$$\text{Span of the room in } \frac{\text{feet}}{2} + 2 \text{ inches}$$

For example, a room of 12-foot span would require a 12/2 + 2 = 8-inch joist. Typically the joists were spaced between 300 and 400 mm.

During the 1900s–1920s a typical floor joist in a terraced house would be 50 mm × 200 mm deep, positioned at 300–400 mm centres. The joists would be built into the walls of the property or in some cases corbelled using brickwork or wrought iron. After 1930 joist hangers were introduced to support the joists, and these were originally cantilevered from the wall.

In some late-Victorian houses it is possible to find a suspended first floor constructed using reinforced concrete slabs. These are strong structures and typically over-reinforced.

Modern solid floors

Solid floors are constructed on the ground floor using concrete poured over a compacted hardcore. The floors are usually 100 mm to 150 mm thick and thickened beneath internal load-bearing walls. The introduction of a damp-proof membrane is important for two reasons: first to ensure that damp does not migrate through the floor surface; second to prevent sulphates, magnesium or acid attacking the concrete which can cause structural failure of the floor structure. This will be discussed in more detail in a later chapter.

Insulation is also used to ensure the thermal qualities as recommended in the Building Regulations Part L 2010.

Insulation can be placed under or over the floor slab, but if the insulation is placed over the floor slab a 75 mm screed is recommended to prevent cracking. The concrete used in floors is usually a Grade ST2 or Gen 1 mix.

Large areas of concrete will require reinforcement to prevent cracking. BS 8110 recommends that the surface crack width should not exceed 0.3 mm, with the primary intention of preventing the ingress of moisture and thus corrosion to the reinforcement steel. Cracking is controlled by using minimum areas of reinforcement. Under BS 8110 this is taken as 0.13% bh for slabs, where b is the span and h is the depth of the slab. In addition to this the standard places maximum and minimum distances for the spacing of reinforcement.

In Eurocode 2: Design of Concrete Structures, the cross-sectional area of reinforcement should not be less than

$$A_{sl} = 0.26\,\frac{f_{ctm}}{f_{yk}}\,b_t d > 0.0013\,b_t d \tag{9.1}$$

where

A_{sl} = longitudinal tensile reinforcement
b_t = breadth of the member
d = depth of the member
f_{ctm} = tensile strength of concrete
f_{yk} = characteristic yield stress of reinforcement (500 N/mm^2)

Suspended floors and engineered floor joists

Modern suspended floors normally use timber joists at 400, 450 or 600 mm centres, but in recent times manufactured, engineered floor joists have been introduced to the market – such as ply web or joists with metal webs. These joists have a top and bottom chord or flange constructed of timber. This type of joist allows services to pass through the joist without the need for notches, thus avoiding weakening the joists. They also have the advantage of spanning greater distances than traditional floor joists. Typically, a 254 mm deep joist using Grade TR26 timber at 400 mm centres with a flange of 72 mm × 47 mm can span 5.65 m.

Holes and notches in floor joists

TRADA recommends that notches in floor joists should not be closer to a support than 0.7 times the span of the joist and not further away than 0.25 times the span. The maximum depth of the notch should be no greater than 0.125, or 1/8th of the joist depth.

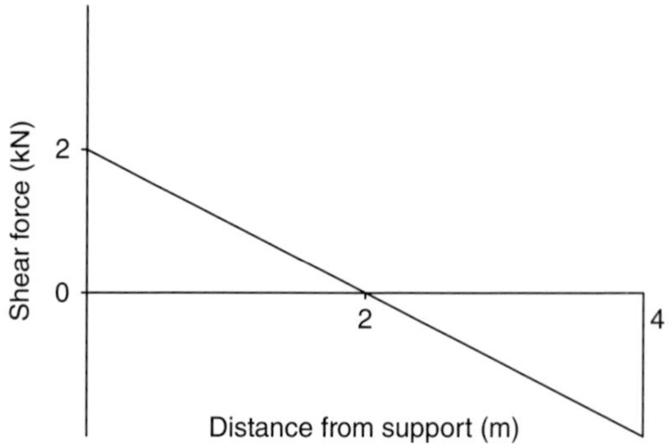

Figure 9.2: Shear force diagram for a typical joist.

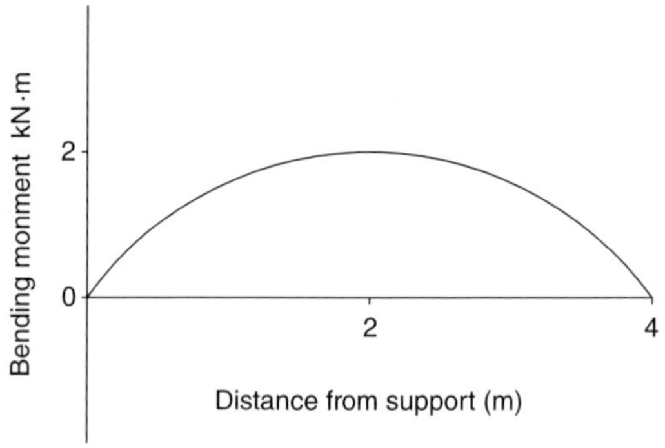

Figure 9.3: Bending moment diagram for a typical joist.

For holes in joists the maximum diameter should be 0.25 times the joist depth. In addition to this the holes should not be less than three diameters (centre to centre) apart, and located between 0.25 and 0.4 times the span from the support.

In clause 2.10.4 of BS 5268 the recommended maximum notch depth is half the depth of the joist, but this has to be shown using structural calculations to ensure the permissible shear stress is greater than the applied shear stress.

If we consider the bending moment and shear force diagrams for a uniformly distributed load along a timber floor joist and the stress distribution across the depth of the beam, we can make sense of the criteria for holes and notches as described above. See Figures 9.2 and 9.3.

As we can see from the shear force and bending moment diagrams in Figures 9.2 and 9.3, the shear force is greatest at the supports and the bending moment is zero. Therefore, we want to avoid reducing the size of a section near the supports where the resistance to the shear force is to be maintained (and thus the criteria for the length of the notch and the closeness of the notch to the bearing).

If we consider the criteria for holes as discussed above, these are placed in the centre of the joist where the neutral axis lies and the bending stresses are zero.

If we consider the permissible shear stress in a joist this is provided under BS 5268 as

$$\Gamma_{adm} = \Gamma_g k_3 k_8 \tag{9.2}$$

where

Γ_{adm} = permissible shear stress
Γ_g = grade shear stress of timber parallel to the grain
k_3 = duration of loading
k_5 = he/h, where he = remaining notch depth and h = depth of joist
k_8 = load-sharing coefficient

For a Grade C16 timber:

$\Gamma_g = 0.67$
$k_3 = 1$
k_5 = N/A
$k_8 = 1.1$ assuming the joists are spaced at no greater than 600 mm, then the value of 1.1 can be used

By substituting into equation (9.2) the permissible shear stress becomes $0.67 \times 1 \times 1.1 = 0.737$ N/mm^2.

If we consider a notched joist the formula becomes

$$\Gamma_{adm} = \Gamma_g k_3 k_5 k_8 \tag{9.3}$$

Let us consider a 200 mm joist and a 75 mm notch, then the value of $k_5 = 125/200 = 0.625$. Hence, substituting into equation (9.3) the permissible shear stress becomes $0.67 \times 1 \times 0.625 \times 1.1 = 0.46$ N/mm^2.

We can see that a notch of 75 mm reduces the shear strength by approximately 37%, which is the same as the notch compared with the original depth of the joist.

Earlier in Chapter 2, we discussed that where a beam is subjected to a load or bending moment along the length of the beam, then the fibres in the lower section of the beam are extended and those in the upper part of the beam are compressed. If a notch is cut out of the end of the beam, weakening the section due to a reduction in its permissible shear stress, the fibres can be strengthened by compressing them closer together. Experimentally it has been shown that by the addition of screws to the side of the notch the fibres are compressed, making the notch much stronger.

Below is a floor-joist calculation using a Grade C16 timber. The floor joists are 47 mm × 195 mm at 450 mm centres, spanning 3500 mm. The imposed load is taken as 1.5 kN/m^2, which is required under BS 6399 for residential and domestic dwellings. The dead load will vary depending on the floor and ceiling structure, but if we assume that the joists support the loads shown below – which are typical – we can use a load of not greater than 0.5 kN/m^2.

Dead loads:	
Plaster and ceiling	12 kg/m^2
Floorboards	17 kg/m^2
Fibreglass insulation	2 kg/m^2
Total	**31 kg/m^2 (0.31 kN/m^2)**

These calculations have been undertaken using structural engineering software and we gratefully acknowledge Tekla (UK) Ltd for their approval in the use of this software.

TIMBER JOIST DESIGN (BS 5268-2: 2002).

Joist details

Joist breadth;	b = **47** mm
Joist depth;	h = **195** mm
Joist spacing;	s = **450** mm
Timber strength class;	**C16**
Service class of timber;	**1**

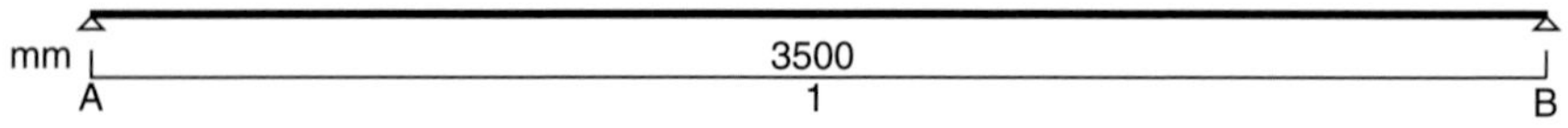

Span details

Number of spans;	N_{span} = **1**
Length of bearing;	L_b = **50** mm
Effective length of span;	L_{s1} = **3500** mm

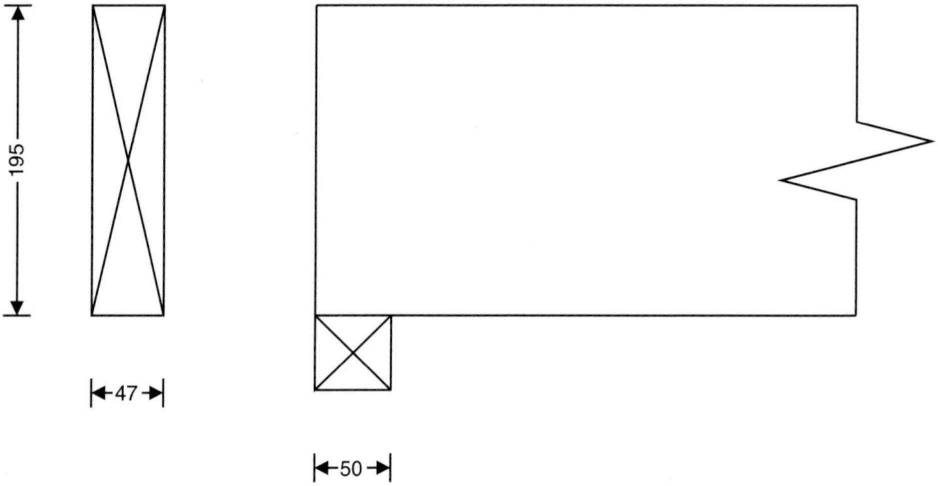

Section properties

Second moment of area;	$I = b \times h^3/12 = \mathbf{29041594}\ mm^4$
Section modulus;	$Z = b \times h^2/6 = \mathbf{297863}\ mm^3$

Loading details

Joist self-weight;	$F_{swt} = b \times h \times \rho_{char} \times g_{acc} = \mathbf{0.03}\ kN/m$
Dead load;	$F_{d\_udl} = \mathbf{0.50}\ kN/m^2$
Imposed UDL (long term);	$F_{i\_udl} = \mathbf{1.50}\ kN/m^2$
Imposed point load (medium term);	$F_{i\_pt} = \mathbf{1.40}\ kN$

Modification factors

Service class for bending parallel to grain;	K_{2m} = **1.00**
Service class for compression;	K_{2c} = **1.00**
Service class for shear parallel to grain;	K_{2s} = **1.00**
Service class for modulus of elasticity;	K_{2e} = **1.00**
Section depth factor;	K_7 = **1.05**
Load sharing factor;	K_8 = **1.10**

Consider long-term loads

Load duration factor; K_3 = **1.00**
Maximum bending moment; M = **1.421** kNm
Maximum shear force; V = **1.624** kN
Maximum support reaction; R = **1.624** kN
Maximum deflection; δ = **7.432** mm

Check bending stress

Bending stress; σ_m = **5.300** N/mm^2
Permissible bending stress; $\sigma_{m\_adm} = \sigma_m \times K_{2m} \times K_3 \times K_7 \times K_8$ = **6.113** N/mm^2
Applied bending stress; σ_{m_max} = M/Z = **4.770** N/mm^2

PASS – Applied bending stress within permissible limits

Check shear stress

Shear stress; τ = **0.670** N/mm^2
Permissible shear stress; $\tau_{adm} = \tau \times K_{2s} \times K_3 \times K_8$ = **0.737** N/mm^2
Applied shear stress; $\tau_{max} = 3 \times V/(2 \times b \times h)$ = **0.266** N/mm^2

PASS – Applied shear stress within permissible limits

Check bearing stress

Compression perpendicular to grain (no wane); σ_{cp1} = **2.200** N/mm^2
Permissible bearing stress; $\sigma_{c\_adm} = \sigma_{cp1} \times K_{2c} \times K_3 \times K_8$ = **2.420** N/mm^2
Applied bearing stress; $\sigma_{c\_max} = R/(b \times L_b)$ = **0.691** N/mm^2

PASS – Applied bearing stress within permissible limits

Check deflection

Permissible deflection; $\delta_{adm} = \min(L_{s1} \times 0.003, 14 \text{ mm})$ = **10.500** mm
Bending deflection (based on E_{mean}); $\delta_{bending}$ = **7.094** mm
Shear deflection; δ_{shear} = **0.338** mm
Total deflection; $\delta = \delta_{bending} + \delta_{shear}$ = **7.432** mm

PASS – Actual deflection within permissible limits

Consider medium-term loads

Load duration factor; K_3 = **1.25**
Maximum bending moment; M = **1.612** kNm
Maximum shear force; V = **1.843** kN
Maximum support reaction; R = **1.843** kN
Maximum deflection; δ = **7.210** mm

Check bending stress

Bending stress; σ_m = **5.300** N/mm^2
Permissible bending stress; $\sigma_{m\_adm} = \sigma_m \times K_{2m} \times K_3 \times K_7 \times K_8$ = **7.641** N/mm^2
Applied bending stress; σ_{m_max} = M/Z = **5.413** N/mm^2

PASS – Applied bending stress within permissible limits

Check shear stress

Shear stress; τ = **0.670** N/mm^2
Permissible shear stress; $\tau_{adm} = \tau \times K_{2s} \times K_3 \times K_8$ = **0.921** N/mm^2
Applied shear stress; $\tau_{max} = 3 \times V/(2 \times b \times h)$ = **0.302** N/mm^2

PASS – Applied shear stress within permissible limits

Check bearing stress

Compression perpendicular to grain (no wane); σ_{cp1} = **2.200** N/mm^2
Permissible bearing stress; $\sigma_{c\_adm} = \sigma_{cp1} \times K_{2c} \times K_3 \times K_8$ = **3.025** N/mm^2
Applied bearing stress; $\sigma_{c\_max} = R / (b \times L_b)$ = **0.784** N/mm^2

PASS – Applied bearing stress within permissible limits

Check deflection	
Permissible deflection;	δ_{adm} = min(L_{s1} × 0.003, 14 mm) = **10.500** mm
Bending deflection (based on E_{mean});	$\delta_{bending}$ = **6.826** mm
Shear deflection;	δ_{shear} = **0.384** mm
Total deflection;	$\delta = \delta_{bending} + \delta_{shear}$ = **7.210** mm
	PASS Actual deflection within permissible limits

Tables 9.1 and 9.2 use similar calculations to calculate the spans for domestic floor joists at various centres. The calculations include dead loads of 0.25 kN/m^2, 0.5 kN/m^2 and 1.25 kN/m^2 and an imposed load of 1.5 kN/m^2. The tables assume a 0.9-kN point load acting on the joist where it can cause maximum moment, shear or deflection. Table 9.1 is for C16 grade timber and Table 9.2 for C24 grade timber. These tables are for guidance only and a suitably qualified person will be required to assess the loads and ensure the correct size joists are used and no liability is admitted for any problems associated with using the tables.

Typically the loads are as follows:

- 0.25 kN/m^2 for suspended ground floors.
- 0.5 kN/m^2 for first floors.
- 1.25 kN/m^2 for party floors (which contain additional sound thermal insulation and additional fire elements).

We recommend that two additional joists are provided under bathrooms and lightweight partitions. We also recommend that a noggin line is placed at 1/3rd intervals along the span. Where partitions run at right angles to the joists, the spans should be reduced by 10%. The bearing of the joists is assumed to be 40–50 mm.

Limecrete

These floors consist of a mixture of lime and sand and can incorporate insulation to meet current-day Building Regulations. The advantage of this type of floor is that it can breathe and also offer resistance to the passage of moisture by the use of a breathable membrane. A typical diagram of this type of floor can be seen in Figure 9.4.

Care must be taken not to seal the floor by adding carpets or floor coverings which trap moisture in the floor, preventing it from being able to breathe.

The use of plaster and lime ash floors

These floors are traditional floors which originated in Medieval times and were used for many years into the 1900s. These floors comprised gypsum or plaster, which was burnt and mixed with aggregate, lime and ash. Usually these types of floor were employed at first-floor level as a floating floor over timber laths, reeds or timber.

Table 9.1: Floor joist spans for C16 grade timber (clear spans in metres)

Size of joist: breadth × depth	Dead loads not more than 0.25 kN/m^2: spacing of joist (mm)			Dead load not more than 0.5 kN/m^2: spacing of joist (mm)			Dead load not more than 1.25 kN/m^2: spacing of joist (mm)		
	400	450	600	400	450	600	400	450	600
38 × 97	1.76	1.66	1.43	1.64	1.55	1.35	1.43	1.35	0.71
38 × 120	2.36	2.23	1.94	2.18	2.07	1.80	1.86	1.77	1.55
38 × 145	2.85	2.74	2.48	2.68	2.58	2.32	2.33	2.22	1.96
38 × 170	3.33	3.20	2.90	3.14	3.02	2.70	2.74	2.63	2.32
38 × 195	3.81	3.67	3.30	3.59	3.45	3.10	3.14	3.01	2.65
38 × 220	4.29	4.13	3.70	4.05	3.89	3.45	3.53	3.39	2.95
47 × 97	1.95	1.84	1.60	1.81	1.72	1.50	1.57	1.49	1.31
47 × 120	2.54	2.44	2.15	2.39	2.27	2.00	2.04	1.94	1.71
47 × 145	3.06	2.94	2.67	2.88	2.77	2.51	2.52	2.42	2.15
47 × 170	3.58	3.44	3.12	3.37	3.24	2.94	2.95	2.83	2.56
47 × 195	4.09	3.94	3.57	3.86	3.71	3.36	3.38	3.24	2.93
47 × 220	4.60	4.43	4.02	4.34	4.18	3.79	3.80	3.65	3.25
50 × 97	2.00	1.89	1.65	1.87	1.77	1.54	1.61	1.53	1.34
50 × 120	2.59	2.49	2.22	2.44	2.34	2.05	2.09	1.99	1.75
50 × 145	3.12	3.00	2.72	2.94	2.83	2.56	2.57	2.47	2.21
50 × 170	3.65	3.51	3.19	3.44	3.31	3.00	3.01	2.89	2.61
50 × 195	4.17	4.02	3.65	3.94	3.79	3.44	3.45	3.31	3.00
50 × 220	4.70	4.52	4.11	4.43	4.26	3.87	3.88	3.73	3.38
75 × 120	2.96	2.85	2.59	2.79	2.69	2.44	2.45	2.35	2.09
75 × 145	3.56	3.43	3.12	3.37	3.24	2.94	2.95	2.84	2.57
75 × 170	4.16	4.01	3.65	3.93	3.79	3.44	3.45	3.32	3.01
75 × 195	4.75	4.58	4.17	4.49	4.33	3.94	3.95	3.80	3.45
75 × 220	5.34	5.15	4.70	5.05	4.87	4.43	4.45	4.28	3.88

Table independently compiled by structural calculations undertaken by Geomex Ltd, Structural Engineers and Architectural Design.

Table 9.2: Floor joist spans for C24 grade timber (clear spans in metres)

Size of joist: breadth × depth	Dead loads not more than 0.25 kN/m^2: spacing of joist (mm)			Dead load not more than 0.5 kN/m^2: spacing of joist (mm)			Dead load not more than 1.25 kN/m^2: spacing of joist (mm)		
	400	450	600	400	450	600	400	450	600
38 × 97	2.05	1.94	1.68	1.91	1.80	1.57	1.64	1.56	1.37
38 × 120	2.63	2.53	2.26	2.48	2.38	2.09	2.13	2.02	1.78
38 × 145	3.17	3.05	2.77	2.99	2.87	2.60	2.61	2.51	2.25
38 × 170	3.71	3.57	3.24	3.50	3.36	3.05	3.06	2.94	2.65
38 × 195	4.25	4.08	3.71	4.00	3.85	3.49	3.50	3.36	3.04
38 × 220	4.75	4.60	4.17	4.51	4.33	3.93	3.95	3.79	3.42
47 × 97	2.24	2.14	1.87	2.10	1.99	1.74	1.80	1.71	1.51
47 × 120	2.83	2.72	2.47	2.67	2.56	2.31	2.32	2.21	1.96
47 × 145	3.40	3.27	2.97	3.21	3.09	2.80	2.81	2.70	2.44
47 × 170	3.98	3.83	3.48	3.76	3.61	3.28	3.29	3.16	2.86
47 × 195	4.55	4.38	3.98	4.30	4.13	3.75	3.77	3.62	3.27
47 × 220	5.00	4.87	4.48	4.83	4.65	4.23	4.24	4.08	3.69
50 × 97	2.32	2.20	1.92	2.15	2.04	1.79	1.85	1.76	1.55
50 × 120	2.88	2.77	2.52	2.72	2.62	2.37	2.38	2.27	2.01
50 × 145	3.48	3.34	3.04	3.28	3.15	2.86	2.87	2.76	2.50
50 × 170	4.06	3.91	3.55	3.83	3.69	3.35	3.36	3.23	2.92
50 × 195	4.64	4.47	4.07	4.38	4.22	3.83	3.85	3.69	3.35
50 × 220	5.10	4.95	4.58	4.93	4.75	4.32	4.33	4.16	3.77
75 × 120	3.29	3.17	2.88	3.11	2.99	2.72	2.73	2.62	2.38
75 × 145	3.96	3.81	3.48	3.74	3.60	3.28	3.29	3.16	2.87
75 × 170	4.62	4.45	4.06	4.37	4.21	3.83	3.85	3.70	3.36
75 × 195	5.13	5.00	4.64	4.90	4.81	4.38	4.40	4.23	3.85
75 × 220	5.60	5.45	5.10	5.40	5.29	4.93	4.95	4.76	4.33

Table independently compiled by structural calculations undertaken by Geomex Ltd, Structural Engineers and Architectural Design.

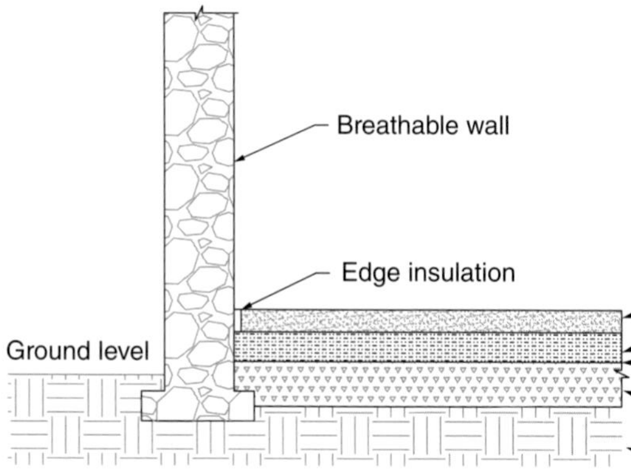

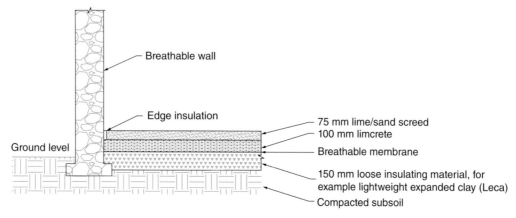

Figure 9.4: Diagram showing limecrete floor.

Beam and block suspended floors and hollow core floors

Beam and block floors comprise prestressed reinforced concrete beams with concrete blocks or polystyrene blocks laid between. A screed is then applied over the floor.

Hollow core flooring comprises prestressed reinforced concrete slabs spanning between walls or beams which support the edges of the slabs. The slabs are 750 mm or 1200 mm wide and laid interlocking or jointed.

Typically loadings based on occupancy and usage are forwarded to the manufacturer, who then provides drawings and calculations to design the floors and prove that the floors can sustain the proposed loads. The loadings are calculated based on Eurocode EC2 and are accredited by an independent body such as BSI or CE monitoring.

Damp

Dampness in floors can be as a result of traditional floors, which were laid directly on the earth, being prevented from breathing. A tile or slab would have been laid directly on the earth using lime, and if this type of floor is overlaid with a cement screed or with carpets and floor coverings the moisture becomes trapped, hence the floor begins to smell damp.

In other cases, traditional floors are replaced with solid floors in the belief that they can offer protection against damp and improve thermal qualities. Unfortunately this results in serious problems to the fabric and structural integrity of a building. Moisture originally escaped through the floor, but the sealing or impervious nature of the new construction results in the damp migrating to the walls, where it rises up. In some types of construction this can have disastrous effects on the construction of the wall. The problems associated with dampness in walls have been discussed earlier.

Moisture penetrating solid walls such as brickwork can affect timber joists fixed into the wall. This is particularly problematic where suspended floors are used on the ground floor and fixed into the inner brick skin of the wall. Damp can cause wet-rot decay and a

Table 9.3: Span tables for ceiling joists using C24 grade timber (clear spans in metres)

Size of joist: breadth × depth	Dead loads not more than 0.25 kN/m^2: spacing of joist (mm)			Dead load not more than 0.5 kN/m^2: spacing of joist (mm)		
	400	450	600	400	450	600
47 × 72	1.43	1.41	1.38	1.35	1.33	1.28
47 × 97	2.13	2.10	2.02	2.00	1.96	1.87
47 × 120	2.81	2.77	2.65	2.65	2.56	2.42
47 × 145	3.56	3.50	3.40	3.30	3.25	3.03
47 × 170	4.35	4.25	4.10	4.00	3.87	3.63
47 × 195	5.10	4.97	4.70	4.64	4.55	4.24
47 × 220	5.83	5.70	5.38	5.32	5.20	4.85

Table independently compiled by structural calculations undertaken by Geomex Ltd, Structural Engineers and Architectural Design.

loss of bearing of the joist, thus leading to collapse in extreme cases. Houses built before the 1900s were particularly prone to this, as discussed above.

Early floor joists were constructed into the wall and were prone to dampness, although other methods of securing joists to walls included corbelling and wrought iron hangers, which attempted to reduce the effect of damp. By the turn of the 19th century into the 20th century, steel hangers were introduced into cavity walls which meant the effect of damp on timber joists was minimised. In some cases a damp-proof membrane is wrapped around the joist end to prevent moisture causing decay, but care has to be taken not to cause condensation which will have a similar effect.

Salts

Salts are usually visible by white furring or staining, and naturally may attract damp. To remove salt patches they should be brushed off; under no circumstances should they be washed, as this re-dissolves the salts which may migrate back into the floor.

Sulphate attack

Sulphates are found in mortars and bricks and react with moisture to form a chemical reaction, which results in the mortar returning to its original constituents – for example sand and cement. This causes the mortar to expand, hence in a chimney this can cause the chimney to lean, in floors the concrete domes and in bedding joints the joints expand.

In chimneys this can be as a result of hot flue gases passing through the chimney condensing before they reach the top, thus water condenses on the colder side of the chimney and a sulphate attack can result. This is characterised by a leaning chimney.

In floors the floor rises in the middle and is evidenced by a domed surface, which is sometimes accompanied by star cracking or map cracking. Until the 1970s damp-proof membranes were not widely used in floor construction, and this meant that the subsoil or materials beneath the floor were in direct contact with the concrete.

Black ash from power stations was plentiful in the 1950s and was used beneath concrete floors as a sub-base to regulate the floor level. Red shale is a by-product of coal mining and this was also used for the same purpose. Both of these products were used as a form of hardcore and contained soluble sulphates – usually sodium, potassium or magnesium – which react with the tricalcium aluminates constituent of the cement, causing a sulphate reaction. Another source of sulphates is from plaster.

Following the introduction of damp-proof membranes, which prevented direct contact with the sub-base material, sulphate attacks became less frequent.

For sulphate reaction to take place, the following conditions have to be prevalent:

- Tricalcium aluminates present in the cement.
- A significant amount of moisture also present.

In floor slabs the moisture can result from leaking pipes or natural ground water. The resulting doming effect of the floor can cause the walls to be forced out below the damp-proof course.

In the case of brickwork, the sulphates are found in the brick and for the joints to suffer sulphate attack the sulphate content of the brick has to be greater than 0.5% and the mortar has to be permeable, allowing the moisture to pass between the mortar and brickwork.

In cases where renders are applied to brickwork this makes the situation worse, since the moisture is trapped between the brick and render.

Affected areas of mortar or masonry can be sent for analysis to confirm the findings.

Ceilings

Table 9.3 provides the span tables for ceiling joists which act as a tension member resisting the horizontal forces of the roof, thus offering triangulation to the roof frame but also supporting any storage load in the roof frame. The minimum bearing of the joists is assumed to be 35 mm. The tables are for C24 grade timber and the imposed load is assumed to be 0.25 kN/m^2. The dead load is variable, depending on the materials used in the ceiling construction, and we have assumed a dead load of either 0.25 kN/m^2 or 0.5 kN/m^2.

It is strongly recommended that professional advice is sought to ascertain spans for ceiling joists and the determination of timber strengths and loads likely to be encountered.

Chapter 10 Roofs

Trussed and cut roofs

Essentially there are two types of roof: trussed roofs and cut roofs. In the context of a truss roof, we are not talking about a traditional truss roof but the modern truss roof. Modern truss roofs were introduced in America in the late 1940s, but they were not introduced in the UK until after 1964. Truss roofs are constructed using fabricated structural members held together using plates nailed to the timber members. The weakest part of the truss is arguably the steel plate. Conversely, a cut roof is one where the members are all "cut" from individual timber members, screwed, nailed or bolted together.

Modern truss roofs

There are a number different types of truss used in construction, and these are manufactured by specialist companies who calculate the loadings, spacing of the trusses and member sizes. A diagram of the different types can be seen in Figure 10.1, but the most common is a fink or "W" truss.

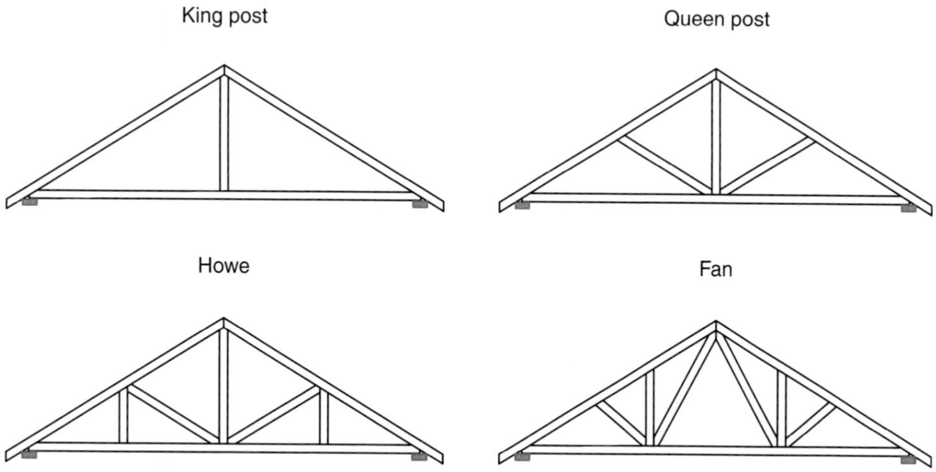

Figure 10.1: Different modern truss types.

Structural Design of Buildings, First Edition. Paul Smith.

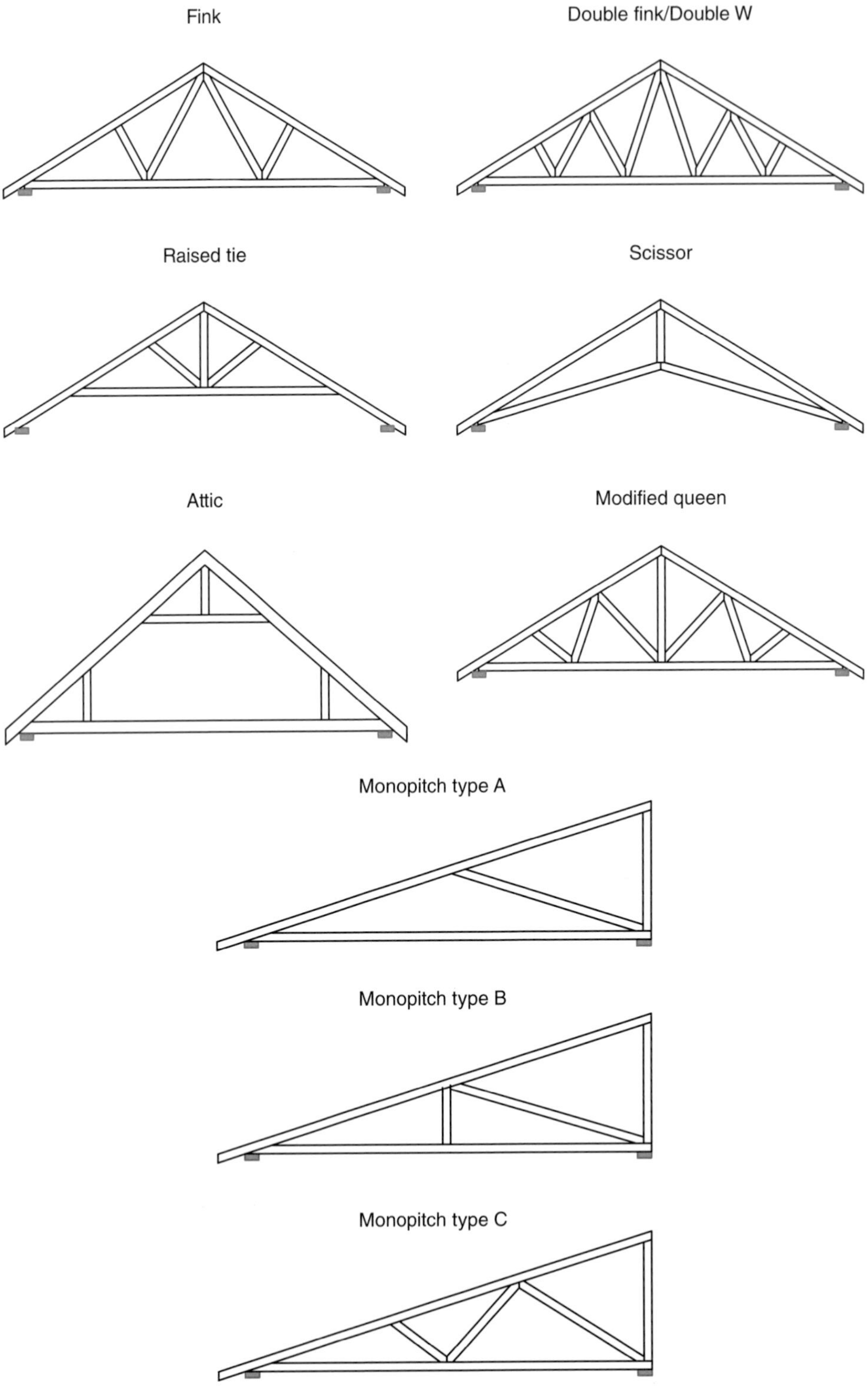

Figure 10.1: (Continued)

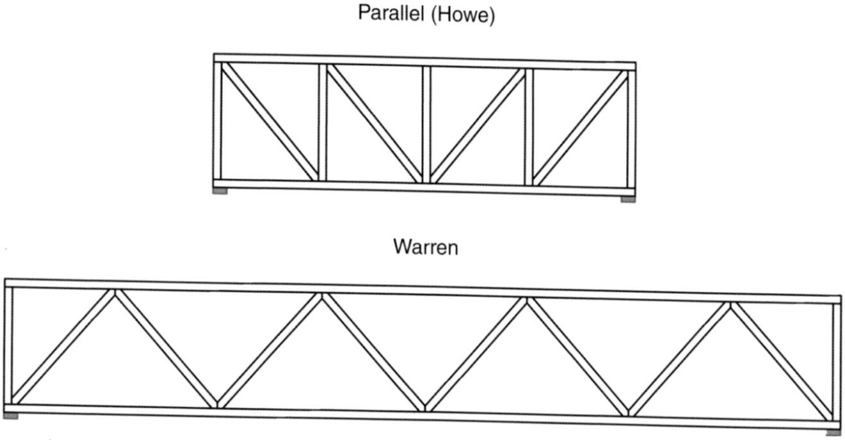

Figure 10.1: (Continued)

Cut roofs

There are different types of cut roofs, and diagrams of these can be seen below.

Coupled roof

A coupled roof has two members supported by the walls. See Figure 10.2.

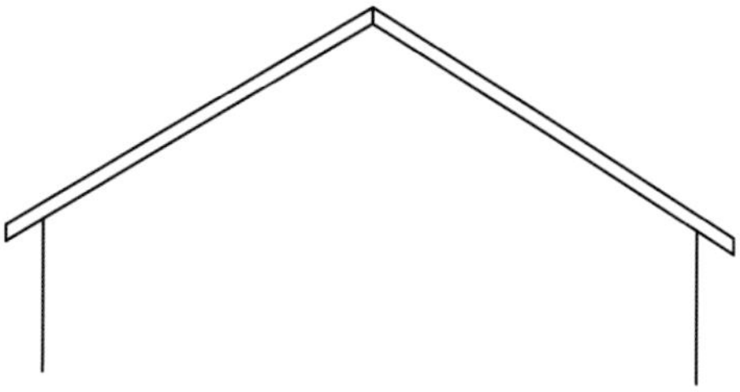

Figure 10.2: Coupled roof.

Close coupled roof

A close coupled roof has a ceiling joist ensuring triangulation of the roof frame. See Figure 10.3.

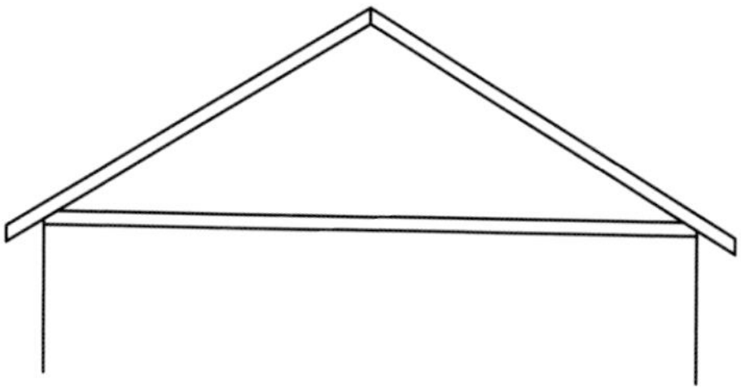

Figure 10.3: Close coupled roof.

Collared roof

A collared roof has a joist but this is secured higher than the eaves and to be effective the collar is placed between a half and a third of the vertical height of the distance between the eaves and the ridge. See Figure 10.4.

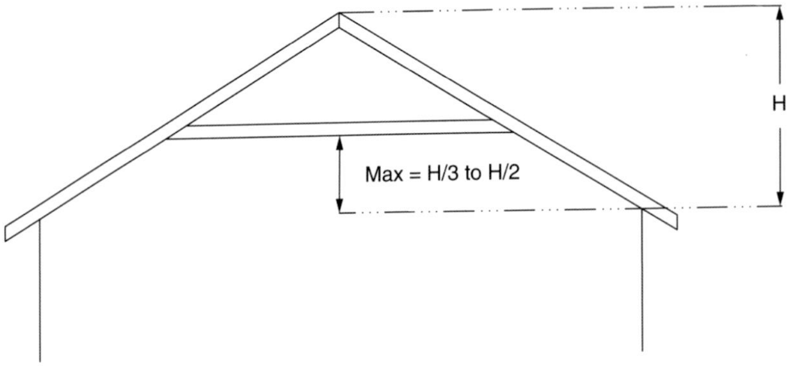

Figure 10.4: Collared roof.

Lean to roof

As the name suggests, this type of roof leans against a wall which continues above the lean to roof. See Figure 10.5.

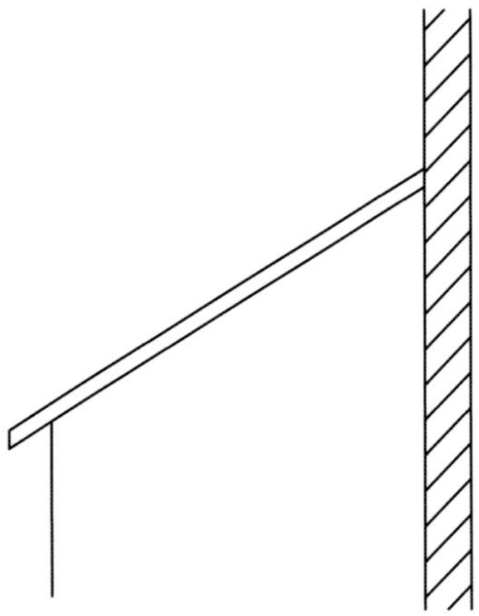

Figure 10.5: Lean to roof.

Monopitch roof

In this situation the roof slopes in one direction only and the wall at the highest slope does not continue. See Figure 10.6.

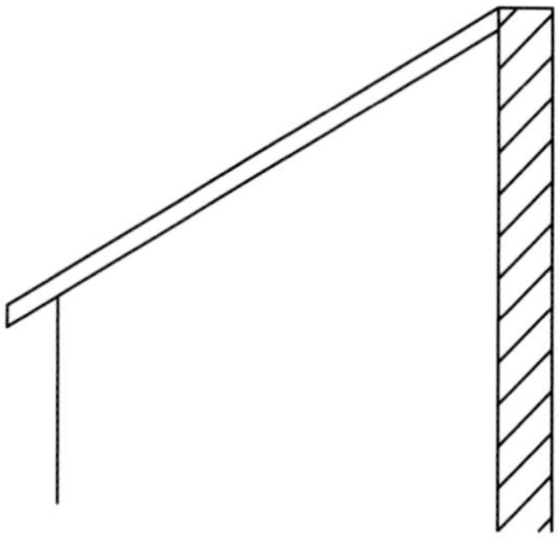

Figure 10.6: Monopitch roof.

Mansard roof

This type of roof is more common in the west of the country, in such places as Cornwall and Gloucestershire, and in countries such as Holland and America. This is a cut roof and for small spans can accommodate additional rooms in the roof. This type of roof structure is also known as a Dutch or gambrel roof. See Figure 10.7.

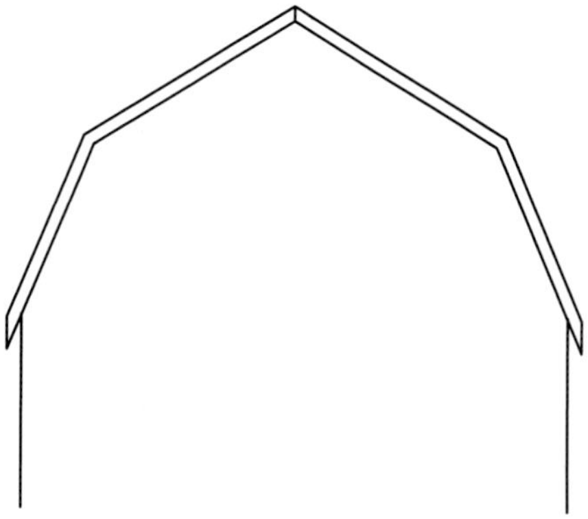

Figure 10.7: Gambrel roof.

Hipped roof

To prevent the height of a roof appearing excessive, a hip can be placed on the end or both ends of the roof. This may be necessary as a planning condition to maintain the style and character of roofs in the area, or so as not to dominate surrounding features and architecture.

Roof components

The primary purpose of a roof is to ensure that moisture does not penetrate into the building, but at the same time it has to be strong enough to sustain the imposed, dead and wind loads to which it is exposed.

Rafters

The rafters support battens to which the tiles or slates of the roof covering are attached. The rafter size will depend on the loads, and one of the first failures is deflection. Although the roof does not structurally fail, it may sustain large deflections and undulations along its length if the loading, particularly the dead load, is too heavy for the rafter section used. Rafters are subject to compressive and bending stresses and their size and spacing is important in accommodating such stresses. Care has to be taken when changing a roof covering to ensure the rafter size, span and spacing are still able to accommodate the new load.

The rafters connect to a wall plate which is secured to the wall, and the principle is to direct as much of the force vertically to the walls from the rafters as possible and minimise the horizontal force, since walls act better in compression and can sustain vertical loads far more resiliently than lateral or horizontal loads. Therefore, roofs are designed to ensure that loads are directed vertically. A steeper-pitched roof will be more efficient in directing forces vertically and minimising the horizontal thrust on walls. A bird mouth cut on the rafter where it adjoins the wall plate also tries to achieve the maximum likelihood of vertical rather than horizontal thrusts.

An example of a rafter design can be seen below. Let us consider 47 mm × 125 mm rafters at 400 mm centres with an imposed load of 0.75 kN/m^2 and a dead load as measured on the slope of 0.70 kN/m^2. The roof has a proposed slope of 30° and we will use a C16 grade structural timber.

These calculations have been undertaken using structural engineering software and we gratefully acknowledge Tekla (UK) Ltd for their approval in the use of this software.

TIMBER RAFTER DESIGN (BS 5268-2: 2002).

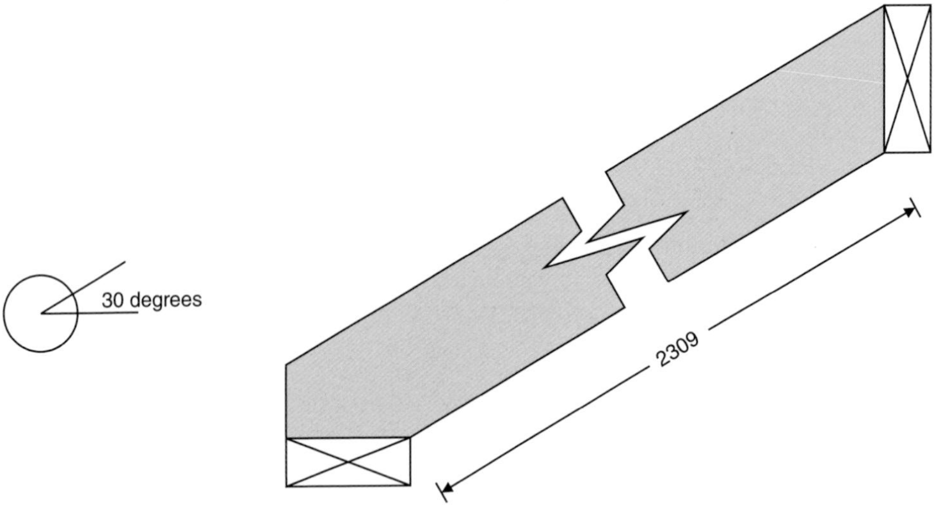

Rafter details

Breadth of timber sections;	$b = \mathbf{47}$ mm
Depth of timber sections;	$h = \mathbf{125}$ mm
Rafter spacing;	$s = \mathbf{400}$ mm
Rafter slope;	$\alpha = \mathbf{30.0}$ deg
Clear span of rafter on horizontal;	$L_{clh} = \mathbf{2000}$ mm
Clear span of rafter on slope;	$L_{cl} = L_{clh}/\cos(\alpha) = \mathbf{2309}$ mm
Rafter span;	**Single span**
Timber strength class;	**C16**

Section properties

Cross-sectional area of rafter;	$A = b \times h = \mathbf{5875}$ mm^2
Section modulus;	$Z = b \times h^2/6 = \mathbf{122396}$ mm^3
Second moment of area;	$I = b \times h^3/12 = \mathbf{7649740}$ mm^4
Radius of gyration;	$r = \sqrt{(I / A)} = \mathbf{36.1}$ mm

Loading details

Rafter self-weight;	$F_j = b \times h \times \rho_{char} \times g_{acc} = \mathbf{0.02}$ kN/m
Dead load on slope;	$F_d = \mathbf{0.70}$ kN/m^2
Imposed load on plan;	$F_u = \mathbf{0.75}$ kN/m^2
Imposed point load;	$F_p = \mathbf{0.90}$ kN

Modification factors

Section depth factor;	$K_7 = (300\ \text{mm}/h)^{0.11} = \mathbf{1.10}$
Load sharing factor;	$K_8 = \mathbf{1.10}$

Consider long-term load condition

Load duration factor;	$K_3 = \mathbf{1.00}$
Total UDL perpendicular to rafter;	$F = F_d \times \cos(\alpha) \times s + F_j \times \cos(\alpha) = \mathbf{0.258}$ kN/m
Notional bearing length;	$L_b = F \times L_{cl}/[2 \times (b \times \sigma_{cp1} \times K_8 - F)] = \mathbf{3}$ mm
Effective span;	$L_{eff} = L_{cl} + L_b = \mathbf{2312}$ mm

Check bending stress

Bending stress parallel to grain;	$\sigma_m = \mathbf{5.300}$ N/mm^2
Permissible bending stress;	$\sigma_{m\_adm} = \sigma_m \times K_3 \times K_7 \times K_8 = \mathbf{6.419}$ N/mm^2
Applied bending stress;	$\sigma_{m\_max} = F \times L_{eff}^2/(8 \times Z) = \mathbf{1.408}$ N/mm^2

PASS – Applied bending stress within permissible limits

Check compressive stress parallel to grain

Compression stress parallel to grain;	$\sigma_c = \mathbf{6.800}$ N/mm^2
Minimum modulus of elasticity;	$E_{min} = \mathbf{5800}$ N/mm^2
Compression member factor;	$K_{12} = \mathbf{0.63}$
Permissible compressive stress;	$\sigma_{c\_adm} = \sigma_c \times K_3 \times K_8 \times K_{12} = \mathbf{4.697}$ N/mm^2
Applied compressive stress;	$\sigma_{c\_max} = F \times L_{eff} \times (\cot(\alpha) + 3 \times \tan(\alpha))/(2 \times A) =$ **0.176** N/mm^2

PASS – Applied compressive stress within permissible limits

Check combined bending and compressive stress parallel to grain

Euler stress;	$\sigma_e = \pi^2 \times E_{min}/\lambda^2 = \mathbf{13.944}$ N/mm^2
Euler coefficient;	$K_{eu} = 1 - (1.5 \times \sigma_{c\_max} \times K_{12}/\sigma_e) = \mathbf{0.988}$
Combined axial compression and bending check;	$\sigma_{m\_max}/(\sigma_{m\_adm} \times K_{eu}) + \sigma_{c\_max}/\sigma_{c\_adm} =$ **0.259**; **< 1**

PASS – Combined compressive and bending stresses are within permissible limits

Check shear stress

Shear stress parallel to grain;	$\tau = \mathbf{0.670}$ N/mm^2
Permissible shear stress;	$\tau_{adm} = \tau \times K_3 \times K_8 = \mathbf{0.737}$ N/mm^2
Applied shear stress;	$\tau_{max} = 3 \times F \times L_{eff} / (4 \times A) = \mathbf{0.076}$ N/mm^2

PASS – Applied shear stress within permissible limits

Check deflection

Permissible deflection; $\delta_{adm} = 0.003 \times L_{eff} = \mathbf{6.936}$ mm

Bending deflection; $\delta_b = 5 \times F \times L_{eff}^4/(384 \times E_{mean} \times I) = \mathbf{1.426}$ mm

Shear deflection; $\delta_s = 12 \times F \times L_{eff}^2/(5 \times E_{mean} \times A) = \mathbf{0.064}$ mm

Total deflection; $\delta_{max} = \delta_b + \delta_s = \mathbf{1.490}$ mm

PASS – Total deflection within permissible limits

Consider medium-term load condition

Load duration factor; $K_3 = \mathbf{1.25}$

Total UDL perpendicular to rafter; $F = [F_u \times \cos(\alpha)^2 + F_d \times \cos(\alpha)] \times s + F_j \times \cos(\alpha) =$ **0.483** kN/m

Notional bearing length; $L_b = F \times L_{cl} / [2 \times (b \times \sigma_{cp1} \times K_8 - F)] = \mathbf{5}$ mm

Effective span; $L_{eff} = L_{cl} + L_b = \mathbf{2314}$ mm12

Check bending stress

Bending stress parallel to grain; $\sigma_m = \mathbf{5.300}$ N/mm^2

Permissible bending stress; $\sigma_{m\_adm} = \sigma_m \times K_3 \times K_7 \times K_8 = \mathbf{8.024}$ N/mm^2

Applied bending stress; $\sigma_{m\_max} = F \times L_{eff}^2/(8 \times Z) = 2.642$ N/mm^2

PASS – Applied bending stress within permissible limits

Check compressive stress parallel to grain

Compression stress parallel to grain; $\sigma_c = \mathbf{6.800}$ N/mm^2

Minimum modulus of elasticity; $E_{min} = \mathbf{5800}$ N/mm^2

Compression member factor; $K_{12} = \mathbf{0.59}$

Permissible compressive stress; $\sigma_{c\_adm} = \sigma_c \times K_3 \times K_8 \times K_{12} = \mathbf{5.510}$ N/mm^2

Applied compressive stress; $\sigma_{c\_max} = F \times L_{eff} \times (\cot(\alpha) + 3 \times \tan(\alpha)) / (2 \times A) =$ **0.330** N/mm^2

PASS – Applied compressive stress within permissible limits

Check combined bending and compressive stress parallel to grain

Euler stress; $\sigma_e = \pi^2 \times E_{min} / \lambda^2 = \mathbf{13.916}$ N/mm^2

Euler coefficient; $K_{eu} = 1 - (1.5 \times \sigma_{c\_max} \times K_{12} / \sigma_e) = \mathbf{0.979}$

Combined axial compression and bending check; $\sigma_{m\_max} / (\sigma_{m\_adm} \times K_{eu}) + \sigma_{c\_max} / \sigma_{c\_adm} = \mathbf{0.396}$; **< 1**

PASS – Combined compressive and bending stresses are within permissible limits

Check shear stress

Shear stress parallel to grain; $\tau = \mathbf{0.670}$ N/mm^2

Permissible shear stress; $\tau_{adm} = \tau \times K_3 \times K_8 = \mathbf{0.921}$ N/mm^2

Applied shear stress; $\tau_{max} = 3 \times F \times L_{eff} / (4 \times A) = \mathbf{0.143}$ N/mm^2

PASS – Applied shear stress within permissible limits

Check deflection

Permissible deflection; $\delta_{adm} = 0.003 \times L_{eff} = \mathbf{6.943}$ mm

Bending deflection; $\delta_b = 5 \times F \times L_{eff}^4/(384 \times E_{mean} \times I) = \mathbf{2.680}$ mm

Shear deflection; $\delta_s = 12 \times F \times L_{eff}^2/(5 \times E_{mean} \times A) = \mathbf{0.120}$ mm

Total deflection; $\delta_{max} = \delta_b + \delta_s = \mathbf{2.800}$ mm

PASS – Total deflection within permissible limits

Consider short-term load condition

Load duration factor; $K_3 = \mathbf{1.50}$

Total UDL perpendicular to rafter; $F = F_d \times \cos(\alpha) \times s + F_j \times \cos(\alpha) = \mathbf{0.258}$ kN/m

Notional bearing length; $L_b = [F \times L_{cl} + F_p \times \cos(\alpha)] / [2 \times (b \times \sigma_{cp1} \times K_8 - F)] = \mathbf{6}$ mm

Effective span; $L_{eff} = L_{cl} + L_b = \mathbf{2315}$ mm

Check bending stress

Bending stress parallel to grain; σ_m = **5.300** N/mm^2

Permissible bending stress; $\sigma_{m\_adm} = \sigma_m \times K_3 \times K_7 \times K_8$ = **9.629** N/mm^2

Applied bending stress; $\sigma_{m_{max}} = F \times L_{eff}^2/(8 \times Z) + F_p \times \cos(\alpha) \times L_{eff}/(4 \times Z)$ = **5.099** N/mm^2

PASS – Applied bending stress within permissible limits

Check compressive stress parallel to grain

Compression stress parallel to grain; σ_c = **6.800** N/mm^2

Minimum modulus of elasticity; E_{min} = **5800** N/mm^2

Compression member factor; K_{12} = **0.55**

Permissible compressive stress; $\sigma_{c\_adm} = \sigma_c \times K_3 \times K_8 \times K_{12}$ = **6.182** N/mm^2

Applied compressive stress; $\sigma_{c\_max} = F \times L_{eff} \times (\cot(\alpha) + 3 \times \tan(\alpha)) / (2 \times A) + F_p \times \sin(\alpha)/A$ = **0.253** N/mm^2

PASS – Applied compressive stress within permissible limits

Check combined bending and compressive stress parallel to grain

Euler stress; $\sigma_e = \pi^2 \times E_{min} / \lambda^2$ = **13.902** N/mm^2

Euler coefficient; $K_{eu} = 1 - (1.5 \times \sigma_{c\_max} \times K_{12} / \sigma_e)$ = **0.985**

Combined axial compression and bending check; $\sigma_{m\_max} / (\sigma_{m\_adm} \times K_{eu}) + \sigma_{c\_max} / \sigma_{c\_adm}$ = **0.578**; **< 1**

PASS – Combined compressive and bending stresses are within permissible limits

Check shear stress

Shear stress parallel to grain; τ = **0.670** N/mm^2

Permissible shear stress; $\tau_{adm} = \tau \times K_3 \times K_8$ = **1.106** N/mm^2

Applied shear stress; $\tau_{max} = 3 \times F \times L_{eff} / (4 \times A) + 3 \times F_p \times \cos(\alpha) / (2 \times A)$ = **0.275** N/mm^2

PASS – Applied shear stress within permissible limits

Check deflection

Permissible deflection; $\delta_{adm} = 0.003 \times L_{eff}$ = **6.946** mm

Bending deflection; $\delta_b = L_{eff}^3 \times (5 \times F \times L_{eff} / 384 + F_p \times \cos(\alpha)/ 48) / (E_{mean} \times I)$ = **4.429** mm

Shear deflection; $\delta_s = 12 \times L_{eff} \times (F \times L_{eff} + 2 \times F_p \times \cos(\alpha)) / (5 \times E_{mean} \times A)$ = **0.232** mm

Total deflection; $\delta_{max} = \delta_b + \delta_s$ = **4.660** mm

PASS – Total deflection within permissible limits

Let us take the same example and ascertain the effect of changing the pitch from 30° to 40°, with all other criteria remaining the same.

These calculations have been undertaken using structural engineering software and we gratefully acknowledge Tekla (UK) Ltd for their approval in the use of this software.

TIMBER RAFTER DESIGN (BS 5268-2: 2002).

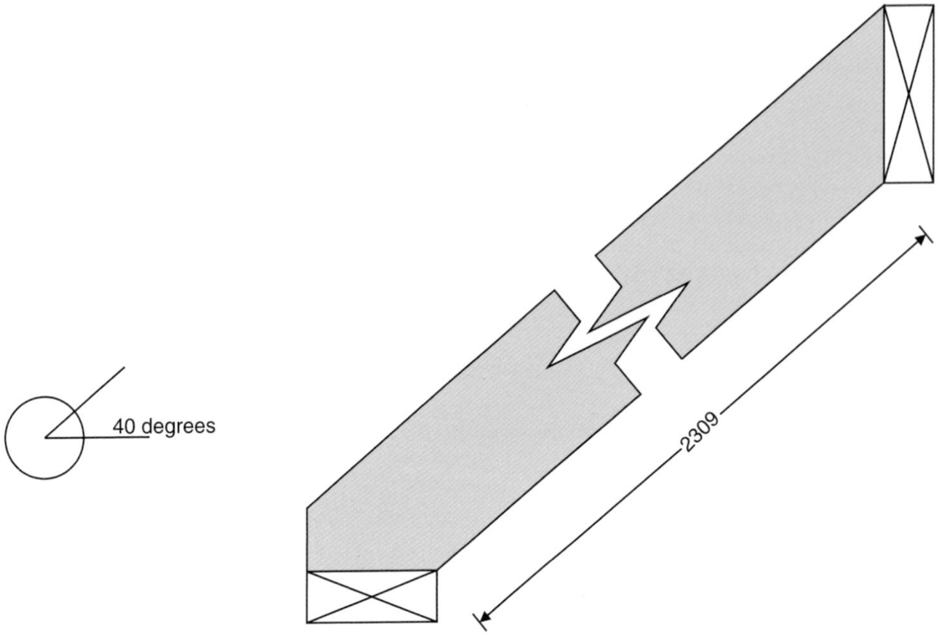

Rafter details

Breadth of timber sections;	$b = \mathbf{47}$ mm
Depth of timber sections;	$h = \mathbf{125}$ mm
Rafter spacing;	$s = \mathbf{400}$ mm
Rafter slope;	$\alpha = \mathbf{40.0}$ deg
Clear span of rafter on horizontal;	$L_{clh} = \mathbf{1769}$ mm
Clear span of rafter on slope;	$L_{cl} = L_{clh}/\cos(\alpha) = \mathbf{2309}$ mm
Rafter span;	**Single span**
Timber strength class;	**C16**

Section properties

Cross-sectional area of rafter;	$A = b \times h = \mathbf{5875}$ mm^2
Section modulus;	$Z = b \times h^2/6 = \mathbf{122396}$ mm^3
Second moment of area;	$I = b \times h^3/12 = \mathbf{7649740}$ mm^4
Radius of gyration;	$r = \sqrt{(I / A)} = \mathbf{36.1}$ mm

Loading details

Rafter self-weight;	$F_j = b \times h \times \rho_{char} \times g_{acc} = \mathbf{0.02}$ kN/m
Dead load on slope;	$F_d = \mathbf{0.70}$ kN/m^2
Imposed load on plan;	$F_u = \mathbf{0.75}$ kN/m^2
Imposed point load;	$F_p = \mathbf{0.90}$ kN

Modification factors

Section depth factor;	$K_7 = (300 \text{ mm}/h)^{0.11} = \mathbf{1.10}$
Load sharing factor;	$K_8 = \mathbf{1.10}$

Consider long-term load condition

Load duration factor;	$K_3 = \mathbf{1.00}$
Total UDL perpendicular to rafter;	$F = F_d \times \cos(\alpha) \times s + F_j \times \cos(\alpha) = \mathbf{0.228}$ kN/m
Notional bearing length;	$L_b = F \times L_{cl}/[2 \times (b \times \sigma_{cp1} \times K_8 - F)] = \mathbf{2}$ mm
Effective span;	$L_{eff} = L_{cl} + L_b = \mathbf{2312}$ mm

Check bending stress

Bending stress parallel to grain; σ_m = **5.300** N/mm^2

Permissible bending stress; $\sigma_{m\_adm} = \sigma_m \times K_3 \times K_7 \times K_8$ = **6.419** N/mm^2

Applied bending stress; $\sigma_{m\_max} = F \times L_{eff}^2/(8 \times Z)$ = **1.245** N/mm^2

PASS – Applied bending stress within permissible limits

Check compressive stress parallel to grain

Compression stress parallel to grain; σ_c = **6.800** N/mm^2

Minimum modulus of elasticity; E_{min} = **5800** N/mm^2

Compression member factor; K_{12} = **0.63**

Permissible compressive stress; $\sigma_{c\_adm} = \sigma_c \times K_3 \times K_8 \times K_{12}$ = **4.698** N/mm^2

Applied compressive stress; $\sigma_{c\_max} = F \times L_{eff} \times (\cot(\alpha) + 3 \times \tan(\alpha)) / (2 \times A)$ = **0.166** N/mm^2

PASS – Applied compressive stress within permissible limits

Check combined bending and compressive stress parallel to grain

Euler stress; $\sigma_e = \pi^2 \times E_{min} / \lambda^2$ = **13.949** N/mm^2

Euler coefficient; $K_{eu} = 1 - (1.5 \times \sigma_{c\_max} \times K_{12} / \sigma_e)$ = **0.989**

Combined axial compression and bending check; $\sigma_{m\_max} / (\sigma_{m\_adm} \times K_{eu}) + \sigma_{c\_max} / \sigma_{c\_adm}$ = **0.232**; < **1**

PASS – Combined compressive and bending stresses are within permissible limits

Check shear stress

Shear stress parallel to grain; τ = **0.670** N/mm^2

Permissible shear stress; $\tau_{adm} = \tau \times K_3 \times K_8$ = **0.737** N/mm^2

Applied shear stress; $\tau_{max} = 3 \times F \times L_{eff} / (4 \times A)$ = **0.067** N/mm^2

PASS – Applied shear stress within permissible limits

Check deflection

Permissible deflection; $\delta_{adm} = 0.003 \times L_{eff}$ = **6.935** mm

Bending deflection; $\delta_b = 5 \times F \times L_{eff}^4/(384 \times E_{mean} \times I)$ = **1.260** mm

Shear deflection; $\delta_s = 12 \times F \times L_{eff}^2/(5 \times E_{mean} \times A)$ = **0.057** mm

Total deflection; $\delta_{max} = \delta_b + \delta_s$ = **1.317** mm

PASS – Total deflection within permissible limits

Consider medium-term load condition

Load duration factor; K_3 = **1.25**

Total UDL perpendicular to rafter; $F = [F_u \times \cos(\alpha)^2 + F_d \times \cos(\alpha)] \times s + F_j \times \cos(\alpha)$ = **0.404** kN/m

Notional bearing length; $L_b = F \times L_{cl} / [2 \times (b \times \sigma_{cp1} \times K_8 - F)]$ = **4** mm

Effective span; $L_{eff} = L_{cl} + L_b$ = **2313** mm

Check bending stress

Bending stress parallel to grain; σ_m = **5.300** N/mm^2

Permissible bending stress; $\sigma_{m\_adm} = \sigma_m \times K_3 \times K_7 \times K_8$ = **8.024** N/mm^2

Applied bending stress; $\sigma_{m\_max} = F \times L_{eff}^2/(8 \times Z)$ = **2.209** N/mm^2

PASS – Applied bending stress within permissible limits

Check compressive stress parallel to grain

Compression stress parallel to grain; σ_c = **6.800** N/mm^2

Minimum modulus of elasticity; E_{min} = **5800** N/mm^2

Compression member factor; K_{12} = **0.59**

Permissible compressive stress; $\sigma_{c\_adm} = \sigma_c \times K_3 \times K_8 \times K_{12}$ = **5.512** N/mm^2

Applied compressive stress; $\sigma_{c\_max} = F \times L_{eff} \times (\cot(\alpha) + 3 \times \tan(\alpha)) / (2 \times A)$ = **0.295** N/mm^2

PASS – Applied compressive stress within permissible limits

Check combined bending and compressive stress parallel to grain

Euler stress; $\sigma_e = \pi^2 \times E_{min} / \lambda^2 = \mathbf{13.927}$ N/mm^2

Euler coefficient; $K_{eu} = 1 - (1.5 \times \sigma_{c\_max} \times K_{12} / \sigma_e) = \mathbf{0.981}$

Combined axial compression and bending check; $\sigma_{m\_max} / (\sigma_{m\_adm} \times K_{eu}) + \sigma_{c\_max} / \sigma_{c\_adm} =$ **0.334**; < 1

PASS – Combined compressive and bending stresses are within permissible limits

Check shear stress

Shear stress parallel to grain; $\tau = \mathbf{0.670}$ N/mm^2

Permissible shear stress; $\tau_{adm} = \tau \times K_3 \times K_8 = \mathbf{0.921}$ N/mm^2

Applied shear stress; $\tau_{max} = 3 \times F \times L_{eff} / (4 \times A) = \mathbf{0.119}$ N/mm^2

PASS – Applied shear stress within permissible limits

Check deflection

Permissible deflection; $\delta_{adm} = 0.003 \times L_{eff} = \mathbf{6.940}$ mm

Bending deflection; $\delta_b = 5 \times F \times L_{eff}^4 / (384 \times E_{mean} \times I) = \mathbf{2.239}$ mm

Shear deflection; $\delta_s = 12 \times F \times L_{eff}^2 / (5 \times E_{mean} \times A) = \mathbf{0.100}$ mm

Total deflection; $\delta_{max} = \delta_b + \delta_s = \mathbf{2.340}$ mm

PASS – Total deflection within permissible limits

Consider short-term load condition

Load duration factor; $K_3 = \mathbf{1.50}$

Total UDL perpendicular to rafter; $F = F_d \times \cos(\alpha) \times s + F_j \times \cos(\alpha) = \mathbf{0.228}$ kN/m

Notional bearing length; $L_b = [F \times L_{cl} + F_p \times \cos(\alpha)] / [2 \times (b \times \sigma_{cp1} \times K_8 - F)]$ = **5** mm

Effective span; $L_{eff} = L_{cl} + L_b = \mathbf{2315}$ mm

Check bending stress

Bending stress parallel to grain; $\sigma_m = \mathbf{5.300}$ N/mm^2

Permissible bending stress; $\sigma_{m\_adm} = \sigma_m \times K_3 \times K_7 \times K_8 = \mathbf{9.629}$ N/mm^2

Applied bending stress; $\sigma_{m_{max}} = F \times L_{eff}^2 / (8 \times Z) + F_p \times \cos(\alpha) \times L_{eff} / (4 \times Z) = \mathbf{4.508}$ N/mm^2

PASS – Applied bending stress within permissible limits

Check compressive stress parallel to grain

Compression stress parallel to grain; $\sigma_c = \mathbf{6.800}$ N/mm^2

Minimum modulus of elasticity; $E_{min} = \mathbf{5800}$ N/mm^2

Compression member factor; $K_{12} = \mathbf{0.55}$

Permissible compressive stress; $\sigma_{c\_adm} = \sigma_c \times K_3 \times K_8 \times K_{12} = \mathbf{6.184}$ N/mm^2

Applied compressive stress; $\sigma_{c\_max} = F \times L_{eff} \times (\cot(\alpha) + 3 \times \tan(\alpha)) / (2 \times A) + F_p \times \sin(\alpha) / A = \mathbf{0.265}$ N/mm^2

PASS – Applied compressive stress within permissible limits

Check combined bending and compressive stress parallel to grain

Euler stress; $\sigma_e = \pi^2 \times E_{min} / \lambda^2 = \mathbf{13.913}$ N/mm^2

Euler coefficient; $K_{eu} = 1 - (1.5 \times \sigma_{c\_max} \times K_{12} / \sigma_e) = \mathbf{0.984}$

Combined axial compression and bending check; $\sigma_{m\_max} / (\sigma_{m\_adm} \times K_{eu}) + \sigma_{c\_max} / \sigma_{c\_adm} =$ **0.519**; < 1

PASS – Combined compressive and bending stresses are within permissible limits

Check shear stress

Shear stress parallel to grain; $\tau = \mathbf{0.670}$ N/mm^2

Permissible shear stress; $\tau_{adm} = \tau \times K_3 \times K_8 = \mathbf{1.106}$ N/mm^2

Applied shear stress; $\tau_{max} = 3 \times F \times L_{eff} / (4 \times A) + 3 \times F_p \times \cos(\alpha) / (2 \times A) = \mathbf{0.243}$ N/mm^2

PASS – Applied shear stress within permissible limits

Check deflection

Permissible deflection; $\delta_{adm} = 0.003 \times L_{eff} = \mathbf{6.944}$ mm

Bending deflection; $\delta_b = L_{eff}^3 \times (5 \times F \times L_{eff}/384 + F_p \times \cos(\alpha)/48)/(E_{mean} \times I) = \mathbf{3.913}$ mm

Shear deflection; $\delta_s = 12 \times L_{eff} \times (F \times L_{eff} + 2 \times F_p \times \cos(\alpha))/(5 \times E_{mean} \times A) = \mathbf{0.205}$ mm

Total deflection; $\delta_{max} = \delta_b + \delta_s = \mathbf{4.118}$ mm

PASS – Total deflection within permissible limits

Table 10.1 shows a comparison of the results with short-term loading.

We can see from this simple analysis that the greater the pitch, the more efficient the timber becomes since the force is directed vertically where the timber acts more efficiently as a compression member.

Purlin

The purlin acts like a beam and is primarily to prevent deflection in the rafters by reducing the span. In older, more traditional roofs the rafters were bird-mouthed over the purlin to ensure the force was directed vertically rather than horizontally.

In traditional Victorian and Edwardian houses, where hipped roofs were fashionable, the purlin can be seen to extend along each roof and in some cases is unsupported. The theory is that the force or load from the hip and the side of the roof is equal and opposite to the force or load from the other side. Consequently, the purlin acts as a continuous beam pushing against itself, with the loads and forces equalising out.

Joists

Ceiling joists carry the ceiling, which usually comprises lath and plaster or in more modern constructions plasterboard and skim. The key to ensuring these brittle finishes do not crack is to ensure the limits on deflection as outlined under the relevant Eurocode Standards and previous British Standards are observed. The deflection can be reduced by inserting ceiling binders across the joist, which reduce the span. However, the second function of the joist is to triangulate the roof structure and ensure the roof does not spread by acting as a tie between the rafters.

Table 10.1: Comparison of rafter loading for different roof pitches

Structural	Pitch at 30°	Pitch at 40°
Shear stress (N/mm^2)	0.275	0.243
Combined axial bending and compression	0.578	0.519
Deflection (mm)	4.66	4.118

On a flat roof, joists are used to carry the roof dead and imposed loads. The dead loads will be dependent on the type of materials used and some examples include felt, fibre glass and metal sheets. The imposed loads on a flat roof will be greater than those on a pitched roof, and consideration should also be given for additional imposed loads due to access and maintenance.

Joists and collars to triangulate the forces in the roof structure

Traditionally it was recognised that to prevent roof spread at the eaves and ensure the roof did not induce horizontal forces on the wall, triangulation of the roof frame was necessary to ensure equal and opposite forces were exhibited across the frame. This was achieved by the use of a collar or a ceiling joist acting as a tie. However, a collar placed too high will not respond in the desired way by acting in tension, but act in compression – thus allowing the base of the rafter to spread. Hence there are limitations on the height at which a collar should be placed. For collars to be effective, it is normally the case that they are located at a height of one-third to one-half of the height measured from the eaves to the ridge.

Failing this, and if vaulted or open roofs are desired, the roof frame has to employ some other means of resistance to ensure the roof cannot spread and these are described below.

As shown in Figure 10.2, a couple roof does not benefit from a collar or ceiling joist providing the necessary triangulation and tie. In this situation the roof has to be sufficiently designed that the load is mainly directed vertically down the walls, hence the roof has to have a steeper pitch. The resulting horizontal load must be able to be sustained by the walls, and this will depend on the construction, thickness and stiffness of the walls. Interestingly, lateral restraint straps are added to modern houses not only to afford restraint to the wall, but also to spread the load of the lateral force at wall plate level over a deeper section of wall. In cathedrals, flying buttresses are employed for this very purpose – to direct horizontal loads from roofs and ceilings vertically to the ground.

Wind bracing

Trussed and cut roofs will need to be braced against wind loading, which is absorbed by the roof frame. For truss roofs the truss manufacturer will normally specify the bracing on the truss diagram that accompanies the trusses when they are delivered. BS 5268: Part 3 and Eurocode 5 set out the requirements for bracing of roof structures for truss and cut roofs.

Normally, a series of diagonal and longitudinal timber members (nominally 22 mm × 97 mm) will be secured to each of the rafters and ceiling joists along the length of the building. The design standards referred to above set out the lapping requirements, fixing requirements and details for the bracing depending on the wind speed, pitch and length of the roof.

Roof spread

Failure to reduce the lateral or horizontal loads can result in roof spread, whereby the wall is unable to sustain the horizontal thrust and the wall plate is forced from its original position – or worse still the wall exhibits bowing and leaning. Cracking can also occur on the gable wall, and the photograph in Figure 10.8 is an example of roof spread where the gable wall has suffered tapered cracking from being pushed out by the purlins.

In domestic buildings the following methods are usually employed to prevent roof spread if ceiling joists or collars are not used.

Ridge beam

In this scenario a beam is placed at ridge level and is designed to sustain a minimum of half the loading on the roof. The rafters and roof frame hang from the beam and in order for the roof to spread, the roof would have to drop vertically. Thus, if this vertical movement is prevented, then roof spread cannot ensue. A photograph of a typical ridge beam can be seen in Figure 10.9. The critical thing is to make sure there is a good connection of the rafters to the ridge beam, to ensure they cannot slip from the beam. Consequently, a timber plate is normally bolted to the top flange of the ridge beam and the rafters are bird-mouthed over the timber plate. The rafters are skew nailed or screwed, and a small collar

Figure 10.8: Photograph showing a property with roof spread.

Figure 10.9: Photograph of a steel ridge beam.

placed under the beam across the rafters using a 47 mm × 100 mm timber. Ridge beams can be of steel or timber construction.

Ring beam

A ring beam is positioned at eaves level and the rafters connect firmly and securely onto the beam. As the name suggests, the beam is a ring – usually of concrete or steel although sometimes of timber – and for the roof to spread the ring beam must be pushed out. The beam and cross-members on the gable wall at the end prevent this from occurring. It can be seen that the connections at the ends of the beams on the wall plate provide the resistance to tension forces. Whether in concrete or steel, the strength of the connections is paramount.

In traditional solid brick buildings timber ties were connected across the gable walls connecting the wall plates in an attempt to offer a cross-tie to the side walls and provide some lateral restraint to the roof structure (with a similar principle to the ring beam). It is not uncommon for property developers, sometimes under instruction from warranty organisations, to request these to be removed. This is usually due to the idea that timber in a wall will be subject to wet rot and may cause a liability under the warranty. Without the introduction of another form of triangulation or lateral restraint mechanism this can compromise the structural integrity of the roof, walls or both.

Once roof spread has occurred, the impact on the remaining structure has to be considered. For example, care has to be taken to ensure the walls remain stable and that the load from the roof does not induce an excessive eccentric loading on the wall and foundation, thus making the wall unstable.

Hip beams and dragon ties

Hip beams carry the rafters on the hip of a roof and the rafters are connected to the hip beam, which carries the load to the walls. However, these beams impart a concentrated load on the corner of a building and this can result in the beam sliding over the wall plate or pushing out the corner of the building. Clearly some lateral restraint or restraint mechanism is required to overcome this lateral horizontal thrust.

One way to do this is with a dragon tie, which is a mechanism that carries the load from the hip beam in a triangular frame usually constructed of timber. The sides of the triangle rest on the wall and are at right angles to each other. Along the centre of the frame is a structural member attached to the apex of the triangle on the corner and attached at the other end to a tie.

The hip beam is attached to the central member, which pulls on the tie member, thus causing tension in the tie. The tie acting in tension transfers the load to the side members over the wall, which are then in compression. This ensures the triangulation of the forces on the corner of the wall. The forces are directed along the timbers over each wall, and where they meet on the corner the forces are horizontally equal and opposite, thus cancelling each other out with no resultant force causing a thrust on the wall.

Overloading of roof members

In some circumstances it may be found that roof members such as rafters and ceiling joist are under-sized, particularly if new roof coverings have been employed. For example, the removal of thatch at approximately 41.5 kg/m^2 and replacement with a heavier clay tile at approximately 78 kg/m^2 can make a difference of approximately 36.5 kg/m^2. This equates to an 88% increase in the load, and this type of exchange of materials was not uncommon in traditional timber frame housing; the effects in some parts of the UK can still be seen today. Figure 10.10 is a photograph of a house with a traditional thatched roof, but with part of the roof covered in a heavier clay tile. The clay tile was presumably changed later than the original thatch. If other finishes such as plasterboard and plaster are added, one can easily understand how rafters can begin to fail under the additional load if they are not strengthened. Not only does the load to the rafter increase, but some

Figure 10.10: Photograph showing a house with a part thatched and part tiled roof.

consideration of the increase in horizontal thrust at wall plate level is also required to ensure that roof spread does not result.

One way to combat this overloading is to identify the size of the rafter required and thicken the existing rafter by screwing and gluing an additional thickness of timber to the bottom.

Wind loading can cause problems to roofs and in traditional timber frame buildings which are constructed using trusses, supporting purlins and rafters, raking can occur which causes the roof to move laterally across the length of the building.

In truss roofs and modern cut roofs wind bracing is employed to comply with current-day design standards, but traditionally this was not always the case. Consideration needs to be given to wind bracing during the conversion of old barns and buildings. This can be resolved by the addition of wind bracing or ply in-fill panels between the rafters at the end of the building. However, you are advised to seek professional advice from a structural engineer if this is deemed necessary, since an assessment of the roof structure may be necessary to ensure the structural integrity has not been compromised and the wind loading will vary depending on the location and topology.

Alterations to roof structures

A roof frame is an intricate structure and alterations undertaken must be on guidance from suitably qualified persons. Trusses have a series of members that act in compression and tension, forming an integral structure. The removal or alteration of one member can distribute loads to other members that are not able to sustain these additional loads.

If alterations to a cut roof are being undertaken, then the order of the alterations is probably as important as the strengthening of the remaining members.

The alteration of a truss may result in a member changing from being in compression to being in tension, which can impact not only on the structure itself but also on the surrounding structure and fabric of the building.

Traditional timber frame building trusses

There are a number of types of truss found in traditional timber frame buildings, and examples and sketches of these can be seen in Figure 10.11.

There are many examples where the members have been removed to facilitate openings or access arrangements for newly constructed floors at roof level. In some circumstances these modifications have been undertaken without full knowledge and understanding of the resulting impact on the roof structure. Once the roof structure has deformed, it is very difficult and often expensive to recover the situation – even though the movement may be limited, the distortion will remain. In addition to this it is worth noting at this conjecture that the removal of such members in traditional buildings, undertaken in an attempt to accommodate modern-day living, seeks to destroy our significant architectural heritage and in listed buildings this is illegal without Listed Building Consent approvals from the Local Authority.

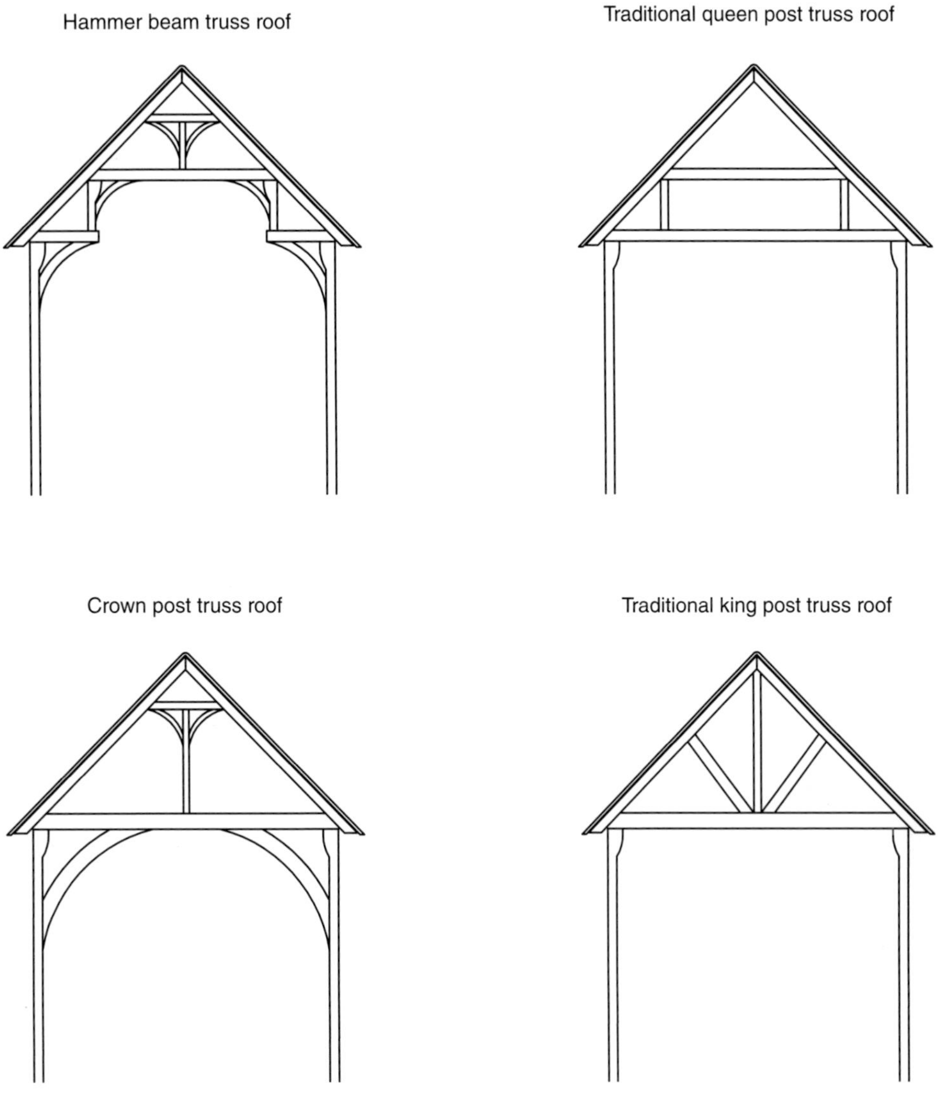

Figure 10.11: Traditional timber truss types.

Modern rafter design

Tables 10.2–10.4 provide the more common sizes of rafters for a given pitch of roof with dead and imposed loadings for Grade C24 timber. The imposed snow loading will vary across the country, and the geographical position in the country and exposure will determine the imposed loads to be used. In these tables an imposed load of 0.75 kN/m^2 has been used. The dead load will depend on the roof covering as well as any plasterboard or timber boarding used to line the underside of the roof. An understanding of the loads must be assumed before using the tables. These tables are for guidance only, and professional advice should be sought for final designs to be implemented. The tables provide clear spans for rafters.

Table 10.2: Rafter span for pitch of roof between 15° and 22.5°
Grade C24 timber
Imposed load 0.75 kN/m^2
Point load 0.9 kN

	Dead load 0.5 kN/m^2			**Dead load 0.75 kN/m^2**			**Dead load 1.0 kN/m^2**		
Size of rafter: breadth × depth (mm^2)	**Spacing of rafter 400 mm**	**Spacing of rafter 450 mm**	**Spacing of rafter 600 mm**	**Spacing of rafter 400 mm**	**Spacing of rafter 450 mm**	**Spacing of rafter 600 mm**	**Spacing of rafter 400 mm**	**Spacing of rafter 450 mm**	**Spacing of rafter 600 mm**
47 × 100	2.12	2.17	2.08	2.08	2.04	1.93	1.98	1.93	1.82
47 × 125	2.93	2.88	2.74	2.74	2.68	2.56	2.60	2.52	2.37
47 × 150	3.67	3.59	3.40	3.40	3.32	3.12	3.23	3.12	2.91
47 × 195	5.00	4.88	4.54	4.61	4.49	4.21	4.38	4.20	3.90

Table independently compiled by structural calculations undertaken by Geomex Ltd, Structural Engineers and Architectural Design.

Table 10.3: Rafter span for pitch of roof between 22.5° and 30°
Grade C24 timber
Imposed load 0.75 kN/m^2
Point load 0.9 kN

	Dead load 0.5 kN/m^2			**Dead load 0.75 kN/m^2**			**Dead load 1.0 kN/m^2**		
Size of rafter: breadth × depth (mm^2)	**Spacing of rafter 400 mm**	**Spacing of rafter 450 mm**	**Spacing of rafter 600 mm**	**Spacing of rafter 400 mm**	**Spacing of rafter 450 mm**	**Spacing of rafter 600 mm**	**Spacing of rafter 400 mm**	**Spacing of rafter 450 mm**	**Spacing of rafter 600 mm**
47 × 100	2.25	2.23	2.12	2.12	2.08	1.97	2.01	1.97	1.85
47 × 125	2.99	2.93	2.79	2.79	2.73	2.57	2.64	2.57	2.81
47 × 150	3.75	3.68	3.47	3.47	3.38	3.18	3.26	3.18	2.97
47 × 195	5.09	4.98	4.64	4.69	4.57	4.27	4.39	4.27	3.97

Table independently compiled by structural calculations undertaken by Geomex Ltd, Structural Engineers and Architectural Design.

Table 10.4: Rafter span for pitch of roof between 30° and 45°
Grade C24 timber
Imposed load 0.75 kN/m^2
Point load 0.9 kN

	Dead load 0.5 kN/m^2			**Dead load 0.75 kN/m^2**			**Dead load 1.0 kN/m^2**		
Size of rafter breadth × depth (mm^2)	**Spacing of rafter 400 mm**	**Spacing of rafter 450 mm**	**Spacing of rafter 600 mm**	**Spacing of rafter 400 mm**	**Spacing of rafter 450 mm**	**Spacing of rafter 600 mm**	**Spacing of rafter 400 mm**	**Spacing of rafter 450 mm**	**Spacing of rafter 600 mm**
47 × 100	2.32	2.28	2.18	2.18	2.14	2.02	2.07	2.02	1.91
47 × 125	3.08	3.02	2.87	2.87	2.80	2.64	2.71	2.64	2.47
47 × 150	3.84	3.77	3.56	3.56	3.47	3.26	3.35	3.26	3.04
47 × 195	5.22	5.10	4.78	4.81	4.68	4.38	4.50	4.38	4.07

Table independently compiled by structural calculations undertaken by Geomex Ltd, Structural Engineers and Architectural Design.

Flat roof construction

Table 10.5 provides joist spans for a flat roof construction for the loads identified using Grade C16 timber.

The table has an assumed imposed load of 1.02 kN/m^2 and a concentrated load of 0.9 kN with a nominal bearing of 40 mm.

Table 10.5: Flat joist spans using Grade C16 timber (clear spans in metres)

Size of joist: breadth × depth	Dead loads not more than 0.5 kN/m^2: spacing of joist (mm)			Dead load not more than 0.75 kN/m^2: spacing of joist (mm)			Dead load not more than 1.0 kN/m^2: spacing of joist (mm)		
	400	**450**	**600**	**400**	**450**	**600**	**400**	**450**	**600**
38 × 97	1.64	1.61	1.55	1.55	1.52	1.45	1.48	1.45	1.37
38 × 120	2.17	2.13	2.04	2.04	2.00	1.89	1.94	1.89	1.78
38 × 145	2.77	2.72	2.58	2.58	2.53	2.38	2.44	2.38	2.23
38 × 170	3.38	3.31	3.09	3.13	3.06	2.87	2.95	2.87	2.68
38 × 195	4.00	3.90	3.54	3.69	3.59	3.29	3.46	3.37	3.00
38 × 220	4.56	4.39	3.80	4.24	4.10	3.65	3.97	3.86	3.40
47 × 97	1.81	1.78	1.70	1.71	1.68	1.60	1.63	1.60	1.51
47 × 120	2.40	2.35	2.25	2.24	2.19	2.08	2.12	2.08	1.95
47 × 145	3.00	2.98	2.82	2.83	2.76	2.60	2.67	2.60	2.44
47 × 170	3.70	3.61	3.32	3.42	3.33	3.10	3.21	3.13	2.92
47 × 195	4.35	4.18	3.80	4.00	3.90	3.54	3.76	3.66	3.34
47 × 220	4.90	4.70	4.28	4.57	4.39	3.99	4.31	4.15	3.76
75 × 120	2.91	2.86	2.72	2.72	2.66	2.51	2.58	2.52	2.36
75 × 145	3.66	3.60	3.32	3.41	3.33	3.10	3.21	3.13	2.92
75 × 170	4.41	4.25	3.88	4.09	3.98	3.62	3.84	3.74	3.42
75 × 195	5.00	4.85	4.43	4.71	4.54	4.14	4.46	4.30	3.91
75 × 220	5.50	5.35	4.98	5.30	5.11	4.66	5.02	4.83	4.40

Table independently compiled by structural calculations undertaken by Geomex Ltd, Structural Engineers and Architectural Design.

The table has an assumed imposed load of 1.02 kN/m^2 and a concentrated load of 0.9 kN with a nominal bearing of 40 mm.

Chapter 11 Arches and Columns

The history of arches

The arch construction was used in Mesopotamia in at least 200 BC, arguably 4000 BC, and was constructed from sundried bricks. Later the Egyptians developed the arch further to use stone and then later still the ancient Romans used arches, applying them to a wide range of structures. One of the oldest surviving arch bridges is the Mycenaean Arkdiko Bridge in Greece, which was constructed in 1300 BC.

Arches are constructed using wedge-shaped voussoirs with a keystone at the top and the sides restrained using an abutment or alternative restraint. Figure 11.1 is a photograph showing the keystones and voussoirs of the English Bridge at Shrewsbury.

Arch bridges are constructed over false work known as cantering, which is removed on completion of the arch. The arch acts in compression and the more weight that is added the stronger the arch becomes, since this reduces the tension. Masonry acts particularly well under compression, not tension, thus the design ensures that the arch is constantly under compression.

Inversion theory

If we consider a chain slung between two points, an inverted arch is formed which is effectively a tension structure for a uniformly distributed load. If the same profile is used for an arch, with the same uniformly distributed load, then the arch is in pure compression. Consequently, it would be possible to cut the arch into segments or short lengths without the structural integrity of the arch being compromised.

The photograph in Figure 11.2 shows an inverted chain suspended between two points.

The chain is in tension and if we now construct an arch with the same inverted shape using timber segments, with an additional weight added to represent the difference between the weight of the segments and the weight of the chain, we can see an inverted structure carrying the same weight. The photograph in Figure 11.3 demonstrates this.

The structure is in pure compression and this is known as the inversion technique. The technique has been used to design many structures. This means that if the timber elements are replaced by stone segments, then the arch will still stand since the forces are being directed to the adjacent element through compression and the structure has no tension. In 1675 Robert Hooke found that hanging a chord in tension could provide an inverted structure in compression.

Structural Design of Buildings, First Edition. Paul Smith.

Figure 11.1: Photograph showing the keystones and voussoirs running through the length of the English Bridge at Shrewsbury in the formation of the arch.

Figure 11.2: Photograph demonstrating inversion theory.

Figure 11.3: Photograph showing an inverted arch in compression.

Line of thrust

The thrust must be resisted by the weight of the abutment, otherwise movement will occur in the arch. The photograph in Figure 11.4 shows where an arch over a doorway in a barn has failed. The restraint being offered by the adjacent brickwork is inadequate to support the lateral thrust from the arch.

The middle-third rule plays an important part in the transfer of compression forces to the adjacent arch segment of voussoirs. Each voussoir has a horizontal force H and a vertical weight W. The resultant force R is passed from one voussoir to its adjacent neighbour, forming a line of thrust. Therefore, the thrust is a function of the weight of the voussoirs and is dependent on the flatness of the arch and the vertical weight.

The thrust line is the line through which the internal forces flow, and the line of the thrust has to pass through the middle third of the neighbouring voussoir and eventually through the middle third of the base foundation in order for the arch to remain stable. Rankine identified that the thrust would have to pass through the middle third of a cross-section of an arch to avoid tensile stresses and ensure the structural stability of the arch.

On the abutments of an arch there is a compressive thrust T acting along the line of the arch and a vertical weight, which is the vertical force of the abutment. The resultant force is R and this has to fall within the middle third of the abutment, otherwise the abutment will be unstable. See Figure 11.5, which shows the forces in an abutment. The middle-third rule is explained in more detail in Chapter 8. However, it can be seen that if the resultant force R falls outside the middle third of the base then the stability can be restored by increasing the weight. This effectively reduces the angle and moves the resultant force R to a more vertical position to fall within the middle third of the abutment.

Although the structure will still be stable if the resultant force passes through the base or the adjacent voussoir, clearly if the line of thrust lies within the middle third of the

Figure 11.4: Photograph showing failure of abutment due to arch thrust.

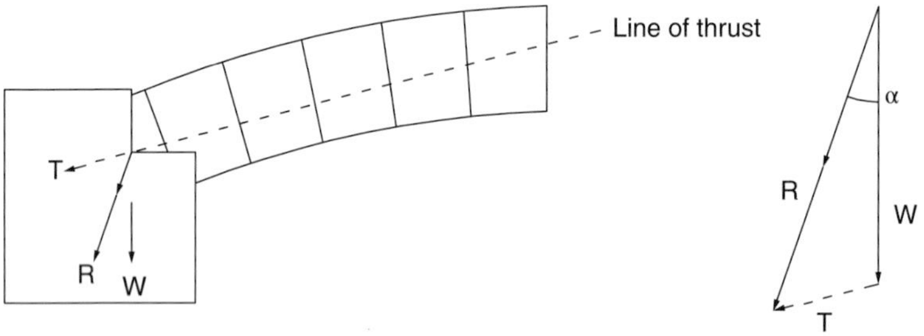

Figure 11.5: The forces in an arch abutment.

arch then a safety factor of three is seen to exist. This was considered good practice, rather than allowing the line of thrust to reach the extremities, since ground movements or settlement may cause a shift in the resultant force R, making the difference between stability and instability. Consequently, a change in the angle and direction of the resultant force can be accommodated if the structure has a safety factor included.

Formation of hinges

The problem in calculating the forces in an arch is that it is a statically indeterminate structure and the values of thrust and vertical forces cannot be determined through static equations. For this reason hinges are introduced to the arch to provide a means of making the structure statically determinate. If we consider a central point of thrust which is an assumed hinge, calculations can be undertaken in an otherwise statically indeterminate structure. A two-hinged arch, which has hinges at the abutments and a fixed hinge arch, cannot be solved using static equations. However, a three-hinged arch, which has hinges at the abutments and at the crown, can be solved. A hinge is a point where there is no moment; it can only transfer shear forces and axial forces. Thus, the summation of the moment is zero. A hinged connection allows the members to have different rotations but the same displacements.

Visible line of thrust

Stone is very good in compression but not tension, and therefore if the abutment moves the span of the arch will increase or decrease. If the span is increased, then the arch comes into tension and develops cracks or hinges. Small movements can occur when the temporary supports (false work) are removed from an arch, possibly due to some movement at the abutments caused by horizontal thrust, shrinkage of the mortar or some movement of the adjoining masonry. This movement will create the formation of hinges or cracks. This is not uncommon in arch bridges, and these structures are so strong and durable that they can withstand the formation of up to three hinges.

The position of the thrust line is essential for the stability of the arch and if this falls below the inner third of the arch, the arch will burst outwards and conversely if the line of thrust falls above the middle third, the arch will collapse inwards.

The line of thrust can be seen in Figure 11.2 as demonstrated in the inversion principle with the hanging chain. Thus in a simple masonry arch, if cracks are evident between the voussoirs and if a line is drawn joining the top of the cracks (that is, the position of the hinges), then this line shows the visible line of thrust in the arch since the thrust has to pass through the hinges. Figure 11.6 is a diagram of this concept. The arch will collapse

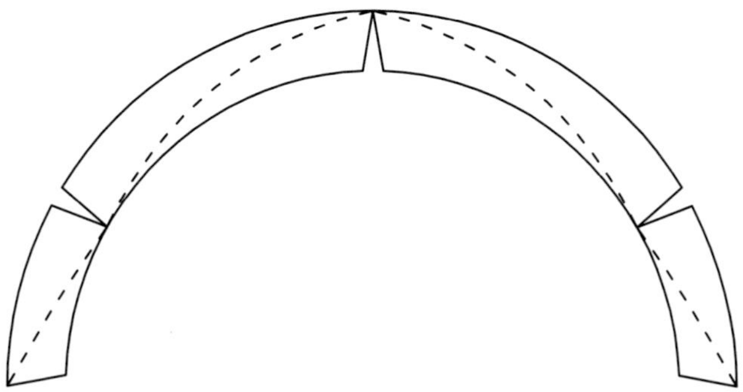

Figure 11.6: The visible line of thrust in an arch passing through the hinges.

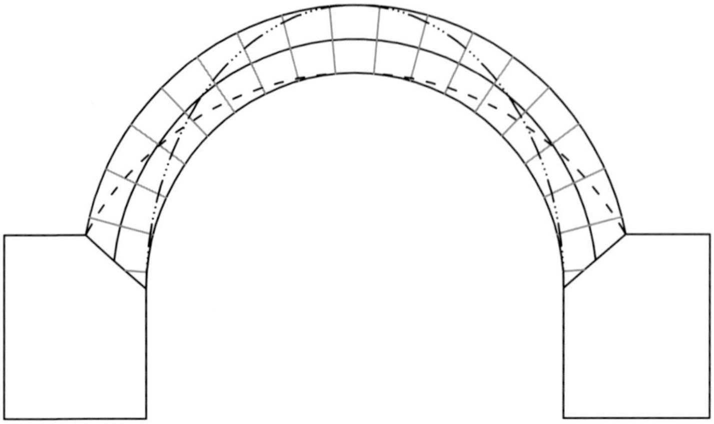

Figure 11.7: Diagram showing the maximum and minimum positions for the line of abutment thrust.

when sufficient hinges have formed and the line of thrust falls outside the boundaries of the arch. We see here that the arch is on the point of collapse. We can further demonstrate this by the inversion theory and, if the inverted chain supported at the abutments lies within the boundary of the arch, then the arch is stable and the chain will touch the arch at the hinges.

It can be seen that essentially the three-hinged arch is stable and it is only when additional hinges are formed – making the arch into a mechanism – that the arch fails.

In Figure 11.7 we show different positions for the thrust line and these lines will represent the limits between which all possible positions of the line of thrust must lie. The top line touches the extrados and is a line of resistance; the bottom line touches the intrados and is a line of compression. If we consider abutments supporting an arch and the abutments are moved closer together or further apart, the arch will sustain the formation of hinges at different parts of the arch depending on the directional movement of the abutment. For example, when the abutments move outwards there is a downward movement of the arch and tension results, causing cracking or the formation of a hinge. If the hinges are joined by a line, we can see in the diagram the steepest and flattest lines of thrust that result, which follow the same line as the chain in the inversion theory. The lines shown represent the minimum and maximum values of the horizontal component of the abutment thrust. The steepest line occurs when the abutments are moved closer together and the flattest line when the abutments are moved further apart. This was demonstrated by W.H. Barlow in 1846 to the Institution of Civil Engineers.

Pursuing this theory further, flat arches cannot have an arrangement of hinges that would allow for the line of thrust to fall outside the geometry limits or boundaries of the arch. Hence these types of arch can only fail when crushing of the overall structure occurs, or if one of the voussoirs slides out of position. Based on this theory the flat arch is considered to be very strong.

In the 14th and 15th centuries rib vaulting was introduced, which meant that the thrust and weight on the abutments could be reduced. This was achieved by introducing a series of beams or ribs constructed in stone, which supported the structure above and reduced the thickness of the material above and thus the weight having to be resisted by the abutment. The ribs supported the load and the in-fill between the ribs was constructed using an inferior quality stone, thus also reducing the time for construction and the cost of the structure.

Height and thickness of an arch

The height of an arch is called the sagitta and can be calculated from the radius and span using equation (11.1);

$$R = \frac{S}{2} + \frac{1D^2}{8S} \tag{11.1}$$

where S is the height of the arch (or sagitta)
D is the length of the arch
R is the radius of the arch

Clearly, for the middle-third rule to apply, there must be a minimum thickness of arch and this was shown to be approximately 1/20th of the span in early structures. The line of thrust falling within the line of the arch is also important in determining the thickness of the arch. If the line is within the middle third, then the arch is stable. The thickness/radius ratio varies between different structures. For example, for a chicken egg the ratio is 1:100, which is seen as an efficient structure compared with the Pantheon in Rome, for which the ratio is 1:18.

Gothic arch

The Gothic arch was introduced in the mid-12th century to the 13th century and was a pointed arch which was much stronger than the earlier circular Roman arches that had not changed since the Roman era. The Gothic arch originated in France but soon became accepted and used in British architecture, particularly in cathedrals, and one of the first cathedrals to use such architecture was Canterbury Cathedral. It allowed much wider openings and consequently allowed more light into the cathedrals, which had a significant meaning for religion.

This new style of arch was much stronger than the Roman circular arch, and was more flexible in its usage in terms of the shapes and widths it could span. The Gothic arch was more akin to the load path of a building, with loads moving down and out in a shallow curve rather than the wider Roman arch.

Domes

In arch construction each stone is laid resting on the one below, reaching the top of the arch where a keystone is introduced to form the arch by compressing the voussoirs together. The structure can be considered two-dimensional, and usually requires false work to support the structure until the keystone is laid. Domes are essentially three-dimensional arches and have the advantage that the adjacent stone to the side also adds lateral restraint to the dome ring as it increases in height. Thus, the rings of the dome support the structure as it rises.

Unlike an arch, if a load is applied to a dome it does not deform. Consider the example of the inverted arch demonstrated in Figures 11.2 and 11.3. If the load increases, the equilibrium of the tension member will change and consequently the shape of the string will change. Then the arch, the inverted compression shape of the structure, would also change. In a dome structure this cannot happen, since the shape cannot deform only form cracks if the supports move.

In conclusion, it can be seen that the stability and strength of an arch or dome is determined by the crushing strength of the members used in its construction; if the structural elements buckle or distort under the applied load, the structure will fail. The second mechanism by which these structures can fail is if the constraints supporting the structure move or distort. For example, if an abutment moves due to overwhelming horizontal forces then hinges develop in the structure, which are seen as cracks. If the line of thrust joining the top of the cracks falls outside the extremities of the arch or dome, then the structure will fail and collapse.

Columns

Structurally, a column acts in a similar manner to a wall. Ideally the column should be loaded axially and without the application of an eccentric load, which will induce a moment thus weakening the column. We have discussed before the use of stone columns in cathedrals, which are used to support the roof structures and ceilings, and the slenderness ratio can be seen to be less than 10 in most cases. This ensures that the columns would fail in crushing rather than in bending, for which the stone would have less resistance. The crushing strength of stone is very high, and this value of the slenderness ratio ensures the strength of these structures. Typically columns can sustain loads in excess of 100 tonnes/m^2 and loads in excess of 1000 tonnes are possible.

The height of a column is generally determined as a multiple of its width or diameter, and the ratios for the different types of column can be seen as follows:

Doric and Tuscan – seven to eight times the diameter
Ionic – nine times the diameter
Corinthian – nine and a half times the diameter

Although the individual structural elements such as columns can carry large loads, it is also important to recognise that the building overall has to be stable and the best results are achieved when the weight at the bottom of the building is higher than at the top. This ensures the centre of gravity is much lower and the building is stable. There have been examples in history where church buildings have collapsed through this lack of understanding. Although the stone columns had the capacity to carry heavy, tall roof structures, the centre of gravity of the building was high and consequently the overall stability of the building was compromised, leading to failure.

Columns can be constructed in timber, steel, concrete and masonry. The Romans and Greeks placed significant architectural significance on the column. The Greeks engaged three orders, or styles of architecture, and these were identified by the type of column used. The orders of columns are described below, and the top or "capital" of the column held significance in terms of identifying the order.

Doric

These are plain, short, stocky columns traditionally having 20 flutes or grooves in the column itself. The column is plain in appearance, with fluted shafts and rounded capitals (heads). These are meant to represent strength and masculinity, and thus in churches they portray male saints. Early examples were so stocky that they did not require a base. Figure 11.8 is a photograph of a Doric column.

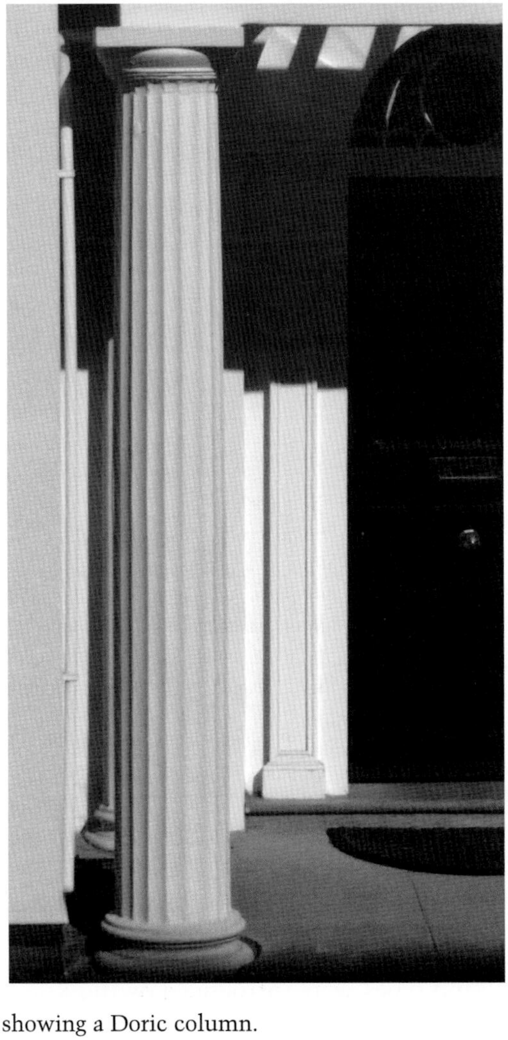

Figure 11.8: Photograph showing a Doric column.

Figure 11.9: Photograph showing an Ionic column.

Ionic

These are taller and more slender columns, with spiral scrolls on the capital. They represent femininity, wisdom and scholarship. In churches these types of column portray female saints. Figure 11.9 is a photograph of an Ionic column.

Corinthian

These columns are very decorative and fanciful, and have capitals with carved Acanthus Mollis leaves. This type of column was introduced later than the Ionic and Doric columns. The Romans added curves to the leaves and made the head of the column more decorative. These columns represent beauty and in churches they represent the Virgin Mary (Taylor, 2003). Figure 11.10 is a photograph of a Corinthian column.

Two further column types were introduced by the Romans and can be seen in house constructions particularly in the Georgian period.

Figure 11.10: Photograph showing a Corinthian column.

Tuscan

These columns are an architectural form based on ancient Italy; they have plain, smooth, round capitals and are very similar to Doric columns. However, the main differences are that unlike Doric columns, Tuscan columns are more slender and do not have grooves or flutes.

Composite

The composite column is a composite of the Ionic and Corinthian column orders. It dates back to Roman architecture of about 1 BC and is very intricate.

Chapter 12 Geology

The importance of understanding geology

The purpose of this chapter is to examine some of the more common geological terms and problems associated with soils, geological failures and geological features, which can have an adverse impact on buildings. The text is not an exhaustive appraisal of the geological problems likely to be encountered, but provides the reader with an appreciation of some possible problems.

Geological maps are available which provide information on the geology of an area, but a word of caution: the geology of an environment can change over a small area, thus there is no substitute for a geotechnical survey prior to commencing building work.

Most small-scale buildings are constructed without much regard for or research into the underlying soil conditions. Hence, on many occasions these features go undetected until problems arise and in other circumstances problems relating to the geology of the area are only detected once the construction has commenced, which can add unexpected costs to a project.

Sinkholes

Sinkholes occur notably in limestone areas that are cavernous, and where there is underground drainage. The damage caused by sinkholes can be severe and result in buildings being structurally damaged beyond repair. Figure 12.1 is a photograph showing the damage caused by a sinkhole to a property in Ripon, Yorkshire in 2014.

Essentially, there are four types of sinkhole.

Solution sinkhole

These are formed slowly, and can range from 1 to 1000 m in diameter and 1 to 100 m deep. The sinkholes are usually conical, cylindrical and irregular. These occur in limestone areas, where the limestone is exposed or covered in a thin layer of soil. Water dissolves the limestone at the surface as it permeates through the limestone or the thin layer of permeable material above, causing a sinkhole to open. The erosion of the limestone can be caused by acidic water, which dissolves the soluble carbonate in the limestone.

Collapse sinkhole

These occur when a natural cave below the surface collapses, thus opening the cave to the surface. These are rare, since most natural caves are stable.

Structural Design of Buildings, First Edition. Paul Smith.

Figure 12.1: Photograph showing damage caused by a sinkhole in Ripon, Yorkshire. By kind permission of The Press, York.

Buried or filled sinkhole

These are common in limestone and occur where the rock head depression is buried and the hole has been filled with sediment, resulting in no surface features. Hence, these can cause major problems if a building is constructed over the sinkhole due to differential settlement on the two types of material. If the rock surface is uneven and the rock head has pinnacles, it may contain closely spaced hollows. This is common in tropical limestone areas. Piles can be used to overcome this problem, but they may need to vary significantly in depth to accommodate the pinnacles.

Subsidence sinkholes

These are formed in unconsolidated sediment or clay above cavernous limestone by the downward wearing and washing away of the material above. In an incohesive material the migration of the particles is slow and dissipates in a similar manner to sand passing through an egg timer. The fissures beneath are filled with sediment and water washes away the sediment, causing holes to appear through the development of voids migrating to the surface.

Slow settlement occurs in sandy silts, but in clay or cohesive soils the drop-out can be instantaneous. The cohesion of the soil above retains its form, hiding the void beneath until it collapses.

The rate of removal of sediment can be enhanced by a drop in the water table. If the water table drops below the rock head, it will take with it sediment which is washed down the fissures more quickly. In Florida, for example, houses and roads have failed

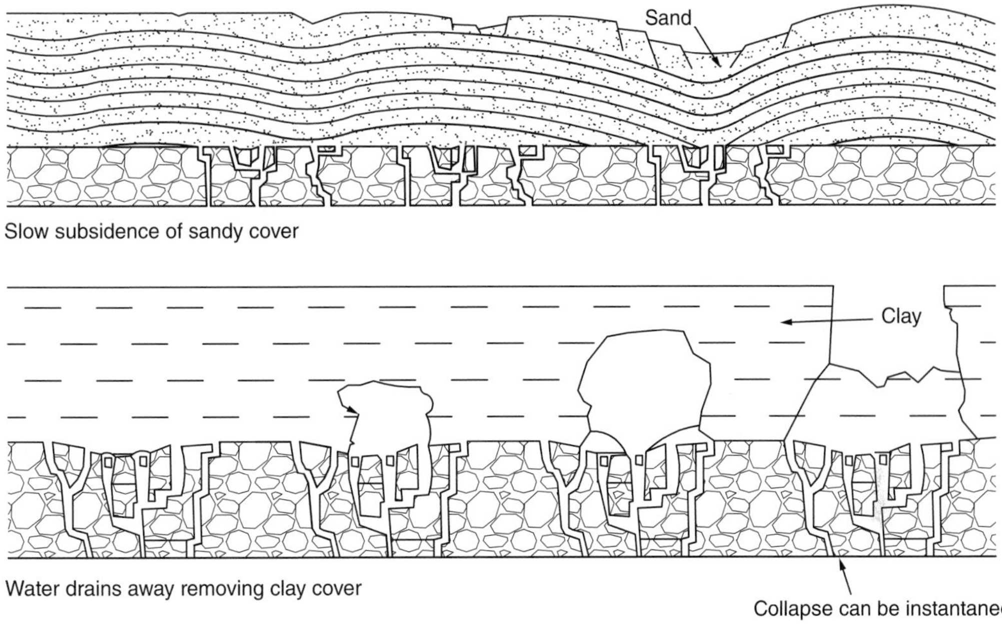

Figure 12.2: Progressive development of a sinkhole in sandy cover and clay cover.

in response to a drop in the water table where the soil is 2 to 20 m thick covering limestone.

Figure 12.2 shows the progressive development of subsidence sinkholes in sand and clay cover.

The illustrations in Figures 12.3–12.6 provide examples of the four different types of sinkhole described above.

Peat subsidence

Peat is a very weak organic material which impacts under foot and is extremely compressible. Building on peat is difficult due to major settlement or the material being washed away through poor drainage beneath the structure. Peat will flow quite easily and its combination with water can cause migration of the material.

The recommendation is to remove peat from beneath buildings, unless it is a very thick layer. Peat layers can be up to 10 m deep and on hills they are typically 2 to 5 m deep.

Usually building over peat will involve piling, but other solutions are to use peat blasting in which holes are bored into the peat and an explosive charge blasts gravel and sand into the peat. Peat can sustain very low loads and other options, including floating buildings over the material (such as on a raft-type foundation).

Tufa

This is a porous calcite deposited in a river or stream in limestone areas. The material is weak and is a cohesive rock, and care has to be taken for it not to be confused with the rock head. For example, the Les Cheurfas Dam in Algeria was constructed on a

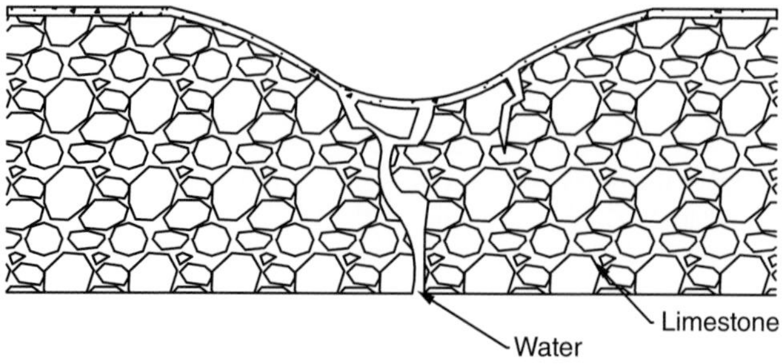

Figure 12.3: Illustrating a solution sinkhole.

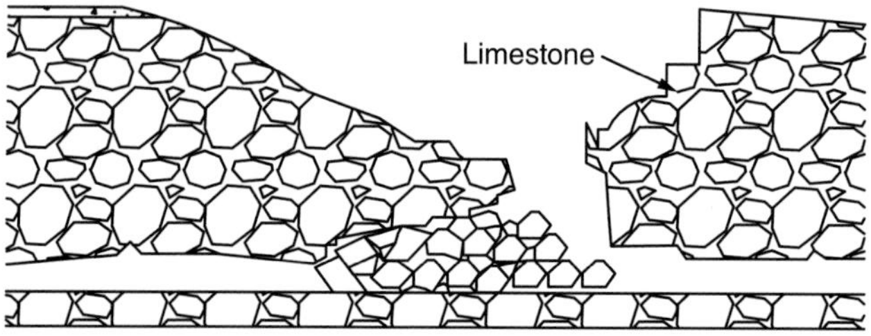

Figure 12.4: Illustrating a collapse sinkhole.

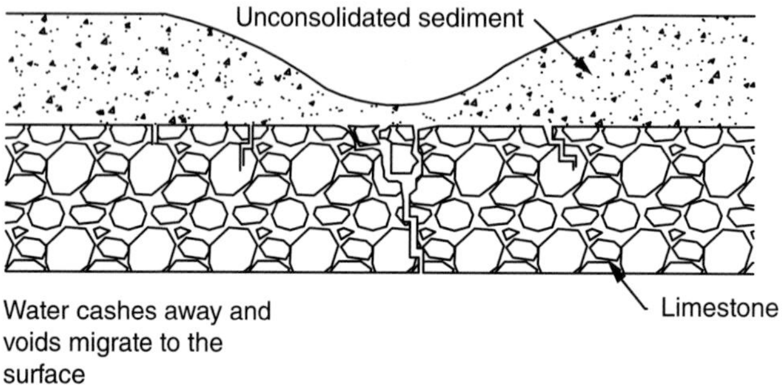

Figure 12.5: Illustrating a buried or filled sinkhole.

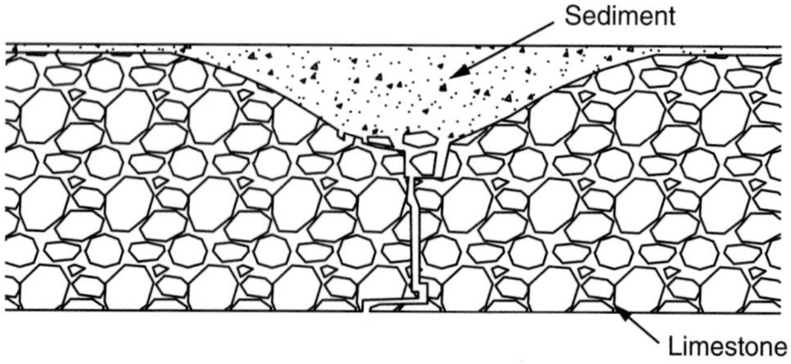

Figure 12.6: Illustrating a subsidence sinkhole.

quaternary conglomerate, made up of pebbles cemented with tufa, which was thought to be rock head.

Lake deposits (lacustrine)

These are lakes that are formed during de-glaciation in over-deepened valleys. When rivers run into the valleys the velocity slows, depositing silt that fills in the lake. Many of these are on boulder clay. The lake deposits appear as large, flat areas, similar in appearance to a flood plain. The lake sediment is similar to alluvium, except for the lack of coarser material, and is made up chiefly of clays and silts. Thus, these areas offer weak soils on which to build.

Alluvium

Alluvium is river-laid sediment comprising horizontal bedded or layered sediment, clay and larger parcels of sand and gravel. Alluvium is cemented together, but is loose and unconsolidated. Alluvium can vary considerably in consistency and strength across a site or within a small area, and it is strongly recommended that appropriate samples and tests be undertaken.

Salt subsidence

Salts are soluble in water and natural subsidence is common in salt areas such as Cheshire due to solution at the rock head where the salt is in contact with ground water.

Most subsidence is over brine streams, which are zones of concentrated ground water flowing into fissures, creating linear subsidence up to 5 km long, 300 m wide and 10 m deep.

Landslips

These are defined as the movement of earth material, rock or soil due to gravity. A photograph of a landslip can be seen in Figure 12.7.

Statham defines mass movement processes as those for which "large quantities of sediment move together in close grain to grain contact" as opposed to those in which single grains are dispersed through a fluid-transporting medium such as a river or stream.

The cause of slope instability is usually a sequence of events resulting in downward slope movement. Slope failures seldom have single causes. A slope will fail when the forces creating movement exceed those resisting it: shear stress > shear strength.

The following forces can lead to an increase in shear stress.

Removal of lateral underlying support or toe removal

This can occur as a result of water, such as the sea and rivers, or glacial ice movements removing lateral supports. The natural corrosion by sea or rivers, or man-made activity, resulting in the removal of the toe weight will reduce the resistance to a landslip since the rotational slip zones are afforded less load to prevent rotational movement.

Figure 12.7: Photograph showing a landslip.

Increased disturbing forces

Natural accumulations of snow, water or man-made pressures as a result of stockpiling materials on the top of a slope can cause instability. Tip heaps are a good example. The head loading places an additional load at the top of the embankment, and a slip can occur since the rotational forces are outweighed by the resistive forces preventing the slip.

Transitory earth stresses or vibration

Vibrations from traffic, pile drivers and earthquakes are all parameters that can destabilise embankments due to an increase in stress, causing them to slip.

Increased internal pressure

An increase in pore water pressure, for example in cracks (especially tension cracks), can cause instability.

The forces that can lead to a decrease in shearing resistance are seen below, namely materials, weathering changes and moisture.

Materials

If the geological beds experience a decrease in the shear stress due to an increase in water content, then slopes will exhibit instability. Examples of such materials are clays and shales.

Weathering changes

Weathering reduces the effective cohesion between materials. The slope angle α is the only control, and if this exceeds the critical value then the slope becomes unstable. This critical value is the angle of shearing ϕ, which is a measure of the inter-particle frictional strength. Failures are usually shallow and roughly parallel to the ground surface.

The absorption of water may cause fabric changes in clays, causing fissures and slaking.

Moisture

Increases in ground-water levels as a result of increased rain fall or even human interference can result in instability.

The addition of water will cause landslips due to the increase in pore water pressure, which reduces the effective vertical stresses in the soil and thus the vertical friction. An increase in rainfall can increase the likelihood of landslips, thus the period between December and April in the UK is when landslips are most likely to occur.

The pore water pressure is the pressure on the ground water within the soil. Water infiltrates between the particles of the soil and is under pressure due to external forces. The voids between the particles are referred to as the pores. The pore water pressure u is measured relative to atmospheric pressure and at a depth d below the saturated soil the pore water pressure is provided by equation (12.1):

$$u = \gamma_w d \tag{12.1}$$

where γ_w is the density of water.

Cohesionless materials

The only control in a cohesionless material such as sand is the slope angle, and if this exceeds a critical value then the slope becomes unstable. This critical value is the angle of shearing resistance, which is a measure of the inter-particle frictional strength. Stability in these types of material is independent of slope height or soil depth, and failures are usually shallow and roughly parallel to the ground surface.

Cohesive materials

Cohesive materials such as clays can be considered in two ways: firstly for vertical slopes such as cliffs and secondly for non-vertical slopes.

For a vertical slope, failure occurs due to the development of a shear plane, including through the soil forming a slip surface.

For non-vertical slopes the direction of the slip plane bends towards the slope surface and a rotational slip occurs. It is worth noting that rotational slips are often retrogressive.

Slip circles

Slip circles on slope failures only occur in homogeneous materials, but if the material comprises a number of soils then the failure plane is modified and may be elongated.

If clays are over-consolidated as a result of a higher overburden pressure or a thickness of ice, the degree of over-consolidation influences the stability of the slope. This results in an increased particle interlock and a decrease in porosity, thus the slope may be more stable. In the case of high ice content, when the ice melts the slope will become less stable.

Undrained saturated slope debris can move very easily, causing slip circles in the soil. These are known as landslips. The slumping of hillsides is known as solifluction, which is the downward slope movement of saturated debris or soil. These can be recognised by hummocks on a hill side.

Stabilising of slopes

The stability of slopes can be improved in the following ways:

- Improved drainage to maintain the pore water pressure.
- Rock bolts, usually 4–12 m long, which increase the friction between rocks and decrease the water flow.
- Improving the toe load, which also prevents rotational slips in cohesive soils.

Mining

Coal has been extracted in the UK for over a thousand years, beginning with shallow pits and adits. As technologies improved and demand became greater, particularly during the Industrial Revolution, mining became more commonplace. In the 12th and 13th centuries the early stages of coal mining began with the use of shafts. These were typically not more than 10 m below ground level, and coal was extracted via a shaft hand dug to access the coal seam. The spoil was usually placed at the side of the shaft and coal extracted by working along the seam. Such mine workings can still be seen in landscapes such as the Forest of Dean, where spoil heaps are still visible.

As the coal was extracted from the bottom of the shaft, the sides of the shaft could collapse and this additional debris or spoil was removed, resulting in a bell pit. Once this was exploited, another pit was commenced nearby and consequently a series of pits would result.

This type of mining is detected by a series of depressions and spoil heaps on the landscape. Sometimes the top of the shaft was covered with timber, which rots in time, and shafts have opened up – causing problems for development. Figure 12.8 is a photograph of a mining landscape taken in the Forest of Dean, Gloucestershire. Many of the depressions and spoil heaps are overgrown with vegetation, but the outline of the landscape can still be seen even a hundred years after mining was last undertaken.

As well as coal, other forms of mining have been undertaken in the UK and the Coal Authority should keep records of all mining activities. Historical records and old maps can also be consulted, since the mining records held by the Coal Authority may not cover the complete history of mining in an area. Internet searches can also reveal historical land use and areas prone to subsidence, which may be an indication of previous mining activity.

Pillar and stall mining

The second stage of coal mining was pillar and stall mining, which was used in the UK on a small scale but more extensively in America. This method was used to extract level beds of coal. The coal was extracted from the ground and pillars left in place, providing support to the ground above. Up to 75% extraction was undertaken using this method.

Figure 12.8: Photograph showing spoil heaps as a result of mining in the Forest of Dean, Gloucestershire.

Headings were driven away from a central shaft in a grid pattern. The amount of coal removed can be expressed by an extraction ratio r%:

$$r = \frac{2ab + b^2}{(a + b)^2} \times 100 \tag{12.2}$$

where
a = width of pillars
b = width of stall

Until the 19th century extraction ratios varied between 30% and 70%, where % = percentage voids.

These mines can fail in two ways: pillar failure and roof span failure. These are explained in more detail below.

Pillar failure

The load on the pillars is critical in assessing the stability of the mine, and thus how much overburden is carried on each pillar.

The stress on each pillar can be calculated using the cross-section of the pillar and comparing it with the compressive stress of coal. The stability of the pillars depends on their load, the ratio of the seam thickness to the pillar width (effectively the slenderness ratio), the depth of the overburden and the strength of the coal.

Pillars can also fail due to a weak floor, where the pillars sink into the ground supporting them. In addition, the pillars can become weathered and deteriorate. In some circumstances dry stone pillars were installed, and these can become unstable over time.

It should be noted that if one pillar fails then this disperses the load to adjacent pillars and can cause them also to become overloaded, thus leading to a progressive collapse within the mine. This happened in 1960 in the Coal Brook Mine in South Africa. The mine employed pillars which collapsed over an area of 3 km^2 within five minutes, killing 437 people.

Failure of these types of mine can affect large areas depending on the area of coal removed beneath the surface, but the vertical movement is relatively small.

Roof span failure

This occurs when an upward-migrating void extends from the roof level of the mine to the surface. The roof progressively caves in until it reaches the surface, causing a crown hole. As a rule of thumb the void migration will not produce a crown hole at the surface if the height from the seam void to the ground surface H is greater than 10t, where t is the seam thickness mined.

This type of failure usually occurs in weaker rocks and is rare for mines greater than 30 m in depth.

The roof may fail due to increased loads from buildings over the mine or weathering of the rock above.

Open-cast mining

This is a continuous operation; the surface is removed to expose the coal and then backfilled with unconsolidated waste. These types of mine require time to settle before building can be undertaken.

Open-cut mining

This is basically quarrying and the holes are left open, since nothing is used to backfill the extracted coal. Sometimes localised wall failures can occur.

Slope mining or caving

This creates large underground voids. Inclined slopes are excavated into the ground, extracting coal. The slopes can be inclined at an angle or vertical. Inclined slopes clearly create a wider hazard in terms of potential collapse. The voids are usually propped. The props can be made of timber, and as these deteriorate the void becomes susceptible to collapse.

Long-wall mining

This is considered to be the modern way to extract coal, but was developed in the 17th century and is a total extraction method. The roof support is temporary and the coal is extracted in panels. A panel is first isolated by tunnelling around it. The cutting machine moves forward towards the face of the coal along the isolated panel, taking with it the temporary jacks and causing the coal behind to collapse into the void, where the fallen debris or "goaf" is removed. On the surface subsidence occurs and a bowl effect results. The factors affecting the amount of ground movement are:

- The angle of draw, which is the angle taken from the goaf or void in the seam measured to the edge of the bowl on the surface, usually limited to 25–35° in the UK.
- The width of the panel.
- The depth of the panel.
- The seam thickness.
- The amount of support given to the goaf.

If the angles of draw intercept on or below the surface, then maximum subsidence occurs.

Subsidence is measured as a percentage of the seam thickness, and can typically be 75% but may vary between 40% and 90%.

Buildings on the surface undergo a series of movements. Firstly the building exhibits tension, then it tilts and experiences compression. The time taken for this subsidence to occur depends on the time taken for a face to be worked or the rate of advance. Subsidence occurs almost immediately (within days), equilibrium is usually reached within one year and a residual settlement of less than 5% is achieved within two years.

If a fault line occurs where the coal is mined, rather than the structure undergoing a gradual tilt, the subsidence follows an abrupt concentrated movement at the surface.

During this process most structures cannot cope with the tension, and most damage occurs when buildings are subjected to strains of 0.05–0.1%.

Precautionary measures can be undertaken to limit the impact of the movements caused by long-wall mining, with the main aim of achieving flexibility of the structure. Such precautions include constructing buildings on raft foundations laid on pea gravel, which reduces the friction between foundation material and foundation. This isolates the structure from the ground, reducing the friction.

Planned discontinuities can be built into the building with the introduction of expansion or flexible joints in the walls. Some buildings can be braced for controlled distortion – for example the braces of pin-jointed steel frame buildings can be set with springs so that they can distort. Jacking devices can also be employed, and on bridges flexible or hinged bearings and rockers can be used.

In compression zones trenching techniques can be used, where a hole is excavated around a structure below foundation level and backfilled with a compressible material. Thus, as the movement occurs, the compressible material absorbs the movements.

Structures subjected to tension can employ tie bolts between floors across the external walls.

In extreme cases, particularly where listed buildings or stately homes are concerned, the coal measure is sterilised and not worked in order to protect the building above. One example where this approach was undertaken was at Selby Abbey.

Where mining has already taken place, pile-jacked raft foundations may be employed with the piles extending below the bottom of the seam. In other cases foamed concrete can be used to fill the voids left by mining. In these circumstances extensive ground investigation programmes are required, and specialist geotechnical and engineering design solutions are needed.

Loess

Loess is wind-blown silt, which is fine and dry. Thus, when it is wetted for the first time, it compacts causing hydro-compaction. This is a problem in semi-desert areas when the ground is irrigated for the first time. In America this is called collapsing soil.

Quick sand

Quicksand is grainy soil, usually sand, that has been liquefied through saturation of water. The water in the sand is trapped and thus a liquid results, which has no strength and cannot support any load. The saturated sand appears to be quite solid until it is disturbed by a change in pressure or a shock. The sand forms a loose suspension and has a spongy, fluid texture.

Liquefaction is a specific condition relating to quick sand, where the pore water pressure increases and the liquefied sand/soil loses strength. It can result in subsidence and structures become unstable. This condition can be caused as the result of an earthquake.

Seismic activity

Movements due to seismic activity have to be accommodated in the design of buildings. Structures need to be able to absorb the energy from earthquakes, and this is achieved in larger buildings by the use of dampers. Design standards in different countries outline the requirements.

The world comprises six major plates and approximately ten minor ones. In transform plate boundary activity there is a sideways movement of the plate, for example the San Andreas Fault, and the fault movement is not smooth. When it moves there is a rapid release of energy in the form of an earthquake. Divergent plate boundaries, for example the Mid-Atlantic Ridge, are where basalt magma is made by fractionation forming a new plate, and this causes surface lifting and faulting with small earthquakes and the formation of small volcanoes. Finally, convergent plate boundaries are where plates move together, destroying the boundary. As the plates move over each other mountains are formed, examples being the Rocky Mountains and the Alps.

Drainage and the water table

Clay has a high porosity and a very open structure prone to deformation, and this can cause settlement under loading but may also cause subsidence. The amount of subsidence can relate to a number of factors, including differential loading, a loss of water pressure, porosity, the thickness of the clay bed (which may be differential over a sloping rock head) and the silt or sand content, which reduces the amount of subsidence.

The dominant control on clay subsidence is a water head decline or a loss of pore water pressure, and this is as a result of over-pumping. If there is an over-pumping of water from the ground, the clays can compact as the water table declines. In Santa Clara Valley, California some areas experienced a total of 4 m subsidence when water was pumped from the sand aquifers and inter-bedded clays compacted.

Chapter 13 Site Investigation

Site investigation

Site investigation is very important, and the amount of money invested in this part of a small development is usually minimal compared with the consequences of finding poor subsoil on which to build a structure. We have listed below some of the field survey and site investigation works that can be undertaken prior to commencing any building project.

Boreholes

These are essential for soil study and the process usually involves a shell and auger for soils. This method cannot drive through rock and normally ranges from 5 to 20 m depth, but stops at rocks or boulders. A similar borehole process can be used for investigation of rocks using a rock drill, which is a diamond- or tungsten-tipped cylinder.

Trenches

These need to be excavated to a depth of 3–5 m to expose a good section through the soil and down to the bed rock if possible. These are useful to trace faults and fractures within bedrock and essential for the exploration of the head material. This means of survey is cheaper than using boreholes, but the exploration is limited by its depth. Trial holes can be excavated, but these are limited to one area and the geology or soil may change over the area – thus a trench excavation is more useful.

Geophysics

This measures the physical properties of the ground from the surface. The main use is to identify the position of the rock head or to search for local anomalies such as mine shafts.

Gravity surveys

These measure small variations in the gravity, such as a void or a very porous or weak rock. These areas of anomaly need to be drilled when identified. The results show a series of contours but are impossible to interpret over voids less than 10 m in diameter, for example over mine shafts.

Structural Design of Buildings, First Edition. Paul Smith.

Magnetic surveys

These types of survey use a proton magnetometer to measure the earth's magnetic field. This is a one-person operation and is a walking-pace survey over the site. The survey searches for anomalies in readings and then drilling is undertaken to identify the reason for the change. Very useful for shaft searches.

Electromagnetic surveys

These are useful on non-contouring terrain and the machine is essentially a conductivity meter or metal detector. The apparatus measures the ground conductivity within an induced field to a depth of approximately 6 m. The survey will recognise the change in geology from sand to clay, for example, by measuring the resistivity and is good for shallow site investigation of soils. Boreholes can then be used to identify the anomalies.

Electrical surveys

These are resistivity surveys and are a traditional method of measuring ground electrical properties. They have been largely replaced by electromagnetic surveys, but these surveys have a greater depth potential.

Ground-penetrating radar

This requires specialist training to interpret the results, but essentially uses electromagnetic radiation pulses which are transmitted into the ground. The apparatus detects the reflected signal from the sub-surface material. The apparatus measures changes in the material, voids and cracks and is similar to seismic reflection surveys, but uses electromagnetic radiation rather than acoustic measurements. The effectiveness of the apparatus depends on the electrical conductivity of the ground material.

Seismic reflection surveys

This method of survey transmits shock waves through the ground and records the reflected vibrations. The method is useful for oil exploration. Traditionally, this method of exploration employs a truck which vibrates the ground or uses explosives and then measures the returned shock waves on a geophone. This survey method is difficult to use in site investigation as it is not easy to interpret the results for shallow surveys.

Seismic refraction

Refracted waves from shallow-depth investigations are measured using a hammer to produce a signal which is measured by the apparatus. This is useful in shallow site investigations to depths up to 15 m. The velocity of a shock wave through the material is measured, and therefore if that material overlies a material of similar transmittance then the results may be similar. For example, soil will provide a value of between 200 and 500 m/s, shale will produce a signal at 1000 m/s and a typical value for granite is 5000 m/s. The apparatus measures the velocity of the shock wave through the rock and this has a direct correlation with the strength of the rock. Thus, it is possible to detect stronger rocks under weaker ones and consequently ascertain the position of a rock head.

Made-up ground or fill

Great care has to be taken when building foundations on made-up ground or fill, since there can be extreme variations in the material. Boreholes may provide false readings and the results across the ground may vary considerably depending on the material directly below the area of inspection. The variation of material can lead to difficulty in the design of foundations, since the bearing capacity of the material will also vary. In addition to this the made-up ground may contain chemicals, gases and toxic materials and the ground may be prone to ignition or burning below the surface. The ground can be improved under strictly controlled conditions by the use of imported fill, laid and compacted in layers. Materials such as pulverised fuel ash can be laid, and BS 8004: 1986 provides guidance on the use of suitable fill material.

Walkover

A walkover is an invaluable method of understanding surface topology and landscape features, thus providing some indication of the history and underlying geology and soil conditions. This exercise could identify landslips, which may present themselves as ripples in the land surface or along slopes. Figure 13.1 is a photograph showing such an anomaly, where local dip variations could indicate back dips in a previous land slide.

A change in slope could indicate a change in ground conditions; boulder clay will present itself as bumpy ground, flat soft areas may indicate the presence of peat.

Drainage and water levels are also interesting to note and may provide some explanation of ground water levels, other geological features or anomalies. The water levels in ditches may indicate high ground water level. Ponding of water on the surface may indicate the presence of sinkholes.

A check of local rock exposures will provide an indication of the rock head and geology of the area. Quarries, natural exposures and old cuttings will provide information on the type and bedding planes of rock.

Historical land use can provide information on potential hazards likely to be encountered during construction. Gruffy hummocks in the ground may indicate the presence of mining in the area, where material has been excavated and left on the surface. Many shafts may have been sunk in a small area, and the ground may appear lumpy. Local knowledge is of great importance to the developer.

Examination of the state of existing buildings and structures in the area may reveal evidence of mining subsidence, cuttings will reveal slope stability issues and railway

Figure 13.1: Photograph showing dips, indicating a previous landslip.

embankments will indicate the stability of slopes and soil in the area, since these are traditionally built on the limit of their slope stability.

Archaeological remains may also be evident on the site of the proposed development, and research in local libraries and discussions with the Local Authority may reveal if there is any historic interest in the area and any impact this may have upon the proposed works. If any archaeological remains or artefacts are discovered once the works commence, then the developer or builder should seek the advice of the Local Authority before continuing.

Site access will be key to any development, particularly if the development is a new proposition. Access is very important, and not only ground access but also overhead – for example cables, low bridges and overhangs from buildings.

Another consideration is the site location and the ecological impact of the site. Badgers are a protected species, and evidence of badger sets may present problems for any potential development. Newts, bats and other protected species all need to be considered, and the cost of moving their habitat. These species can only be handled by licensed individuals and it is unlawful to disturb the habitat or interfere with such species.

Certain plants are also protected, and others can present further problems.

Japanese knotweed

This plant is a particular problem if found on a site or in a garden. Japanese knotweed was originally introduced by the Victorians as a fast-growing plant in their attempt to revolutionise garden design. Japanese knotweed can grow as much as 1.5 m in four weeks, growing to a maximum height of about 3–4 m during each season, and is the most aggressive weed in the UK. The plant has the appearance of bamboo, with hollow stems and nodules on the side. The leaves are broad ovals, approximately 10–14 cm long and 5–12 cm wide. The plant flowers in early autumn and late summer, producing small cream or white erect flowers. The weed can cause heave below materials such as concrete and tarmac, and can grow through such materials once they become damaged.

The Wildlife and Country Act 1981 confirms that it is an offence to plant or encourage the growth of this plant in the wild. Waste involving this material has to be managed and disposed of very carefully, and the Environment Agency must ensure that this is undertaken in accordance with an approved code of practice. The disposal of this plant is covered by the Environmental Protection Act 1990. To transport such a weed the contractor has to be registered with the Environment Agency. This plant is very difficult to control and remove, and treatment to eradicate the plant can take up to three years.

Buddleia

This plant thrives on lime and is prevalent in Victorian masonry structures, where lime mortar was used. The seeds attach and develop in the lime and the plant causes damage as it grows, by forcing mortars apart and destabilising structures.

As a wind-blown propagating plant these are particularly prevalent on railways, where there is an abundance of lime material on the track ballast and along the buildings constructed using lime mortar. The seeds are spurred along the track as trains pass. On bombsites after the war they thrived on the mortar rubble, thus earning the name of "the bombsite plant".

This plant attracts butterflies and is commonly known as the "butterfly bush". It has over 100 species, mostly shrubs which grow to a height of 5 m. The flowers can be blue, pink, white or yellow. Introduced into Britain in the 20th century (although the B. globosa variety was introduced as early as 1774), these soon became popular garden plants. The plant B. davidii is an invasive species, and in certain states in America is considered a noxious weed.

Desk study

A large amount of information is available by undertaking a desk study, and with the advent of the Internet a great deal of information can be assembled in a very short time and in an economic way.

Geological maps provide information on the geology of the area, and these are available at scales of 1:50 000 and 1:10 000. Map centres can provide local geological maps, which reveal important information on geological features – including areas of landslip.

Historical maps are also available online which provide important information on previous land use, possibly locating mines, and further useful information which may indicate the presence of structures, land contamination etc.

Aerial photographs can identify past land use and identify drift and landslip areas.

There are also websites available for identifying areas of subsidence and historical land use. Flooding details can be provided by the Environment Agency and mining information obtained from the British Coal website. The presence of radon in certain areas of the country has become more concerning over recent years, and the National Radiological Protection Board provides a "radon atlas" of England and Wales. The atlas identifies areas of concern and where provision needs to be made for the presence of radon gas in the construction of buildings and extensions.

Radon

Radon is a naturally occurring radioactive gas, which cannot be detected by smell. Specialist equipment is required to identify the presence of such an element.

Radon results from the decay of uranium, which is found in all rocks and soils. The amounts vary across the country, and the concentration of the gas is reduced when exposed to air. However, in some areas of the country concentrations can be significant enough to require preventative action. The action level recommended by the government is 200 Bqm^3.

Radon gas can cause health problems, including lung cancer, and is credited with being the second largest cause of lung cancer after smoking.

Protection can be afforded by using appropriate ventilation. Over the years properties have become increasingly draught-proofed to improve thermal efficiency, but unfortunately this has resulted in air flow being reduced and any gas escaping through the ground into the building becomes trapped – thus posing a health risk.

Chapter 14 Stability of Buildings

Disproportionate collapse

This chapter concentrates on disproportionate or progressive collapse and the overall stability of buildings. In America, this type of collapse has also acquired the name "pancake collapse" and describes a condition whereby the floors fall in on top of one another. Under the Building Regulations 2010 Approved Document A, it states that "The building shall be constructed so that in the event of an accident the building will not suffer collapse to an extent disproportionate to the cause."

In America, the National Institute of Standards and Technology provided a comprehensive set of building code changes following the collapse of the New York World Trade Center on 11th September 2001. These recommendations were approved by the International Code Council and provided substantial improvements to fire-proofing and fire resistance of substrates to protect buildings. This concentrates on the fire resistance of buildings and protecting the structural frame.

The UK regulations with regard to disproportionate collapse of buildings were tightly controlled some 30 years earlier, following the Ronan Point disaster. The Building Regulations were altered in 1970 to ensure that such a failure did not repeat itself. Ronan Point was a 22-storey building in east London which suffered a gas explosion on 16th May 1968, resulting in the collapse of the corner of the building and the death of four people.

The building was a Tower Hamlet, constructed using large-panel systems in reinforced concrete. These were fixed by bolting them together in construction. The building relied on these panels as load-bearing members and did not have a structural frame to support the panels. Thus, when the gas explosion on the 18th floor knocked out the corner of the building the wall above collapsed, since there was nothing now supporting it. This resulted in a progressive collapse up the building to the top floor.

The second phase of the collapse then unravelled, as the remaining floors below the initial blast became overloaded by the debris of the collapsed floors above, resulting in a progressive collapse through the floors below down to ground level.

Under the current Building Regulations a building has to be "sufficiently robust to sustain a limited extent of damage or failure depending upon the class of building". The class of building depends on type and occupancy – for example houses not exceeding four storeys and agricultural buildings are classed as 1.

In the case of a class 1 building, if it is designed and conforms to the criteria set out in Building Regulations 2010 Part A1 and A2 with respect to the structural aspects of the building, then it is considered acceptable and no additional measures are necessary. Buildings are classified as follows.

Structural Design of Buildings, First Edition. Paul Smith.

Class 1

- Houses not exceeding four storeys.
- Agricultural buildings.
- Buildings into which people rarely go, provided no part of the building is closer to another building, or area where people do go, than a distance of 1.5 times the building height.

Class 2A

- Five-storey houses.
- Hotels not exceeding four storeys.
- Flats, apartments and other residential buildings not exceeding four storeys.
- Offices not exceeding four storeys.
- Industrial buildings not exceeding three storeys.
- Retail premises not exceeding three storeys and with less than 2000 m^2 floor area on each storey.
- Single-storey educational buildings.
- All buildings not exceeding two storeys to which members of the public are entitled access and which contain floor areas not exceeding 2000 m^2 on each storey.

In the above cases, effective horizontal ties and/or anchors of suspended floors to walls are required. These are established in the relevant structural codes of practice for masonry, concrete and steel. The Building Regulations also establish the maximum distance between lateral supports as 2.25H, where H is the height of the building.

This also has a direct relationship to Ronan Point, since subsequent inspection of the flats as they were dismantled revealed that a strong wind or a fire could have had a similar effect. Indeed, the floor slabs were not securely fixed to the walls and the bolts were found to be loose. Although blast angle ties were inserted after the disaster between the floors and walls, the ties were largely ineffective.

Class 2B

- Hotels, flats, apartments and other residential buildings greater than four storeys but not exceeding 15 storeys.
- Educational buildings greater than one storey but not exceeding 15 storeys.
- Retail premises greater than three storeys but not exceeding 15 storeys.
- Hospitals not exceeding three storeys.
- Offices greater than four storeys but not exceeding 15 storeys.
- All buildings to which members of the public are admitted that contain floor areas exceeding 2000 m^2 but less than 5000 m^2 on each storey.
- Car parking not exceeding six storeys.

In these cases effective horizontal and vertical ties should be included in accordance with the design standards for concrete and steel structures. In addition to this an analysis of the building has to be undertaken, where each supporting column and each beam supporting one or more columns or a load-bearing wall on each floor is removed. The building must remain stable and the area of collapse be confined to 15% of the floor area of that storey or 70 m^2, whichever is smaller. The collapse also has to be confined to the immediate adjacent storeys. Where the above is not met and the structural member under consideration results in a larger collapse than that prescribed, then the structural member is considered to be a "key member" and consequently has to be able to sustain an accidental design load of 34 kN/m^2 applied in the horizontal and vertical directions – one direction at a time. This is in addition to a third of the normal characteristic loading (i.e., wind and imposed loading).

Class 3

- All buildings defined above as class 2A and 2B that exceed the limits on area and/or number of storeys.
- Grandstands accommodating more than 5000 spectators.
- Buildings containing hazardous substances and/or processes.

For a class 3 building: "A systematic risk assessment of the building should be undertaken taking into account all the normal hazards that may reasonably be foreseen, together with any abnormal hazards." Looking at the types of building included in this class it is not surprising that an individual risk assessment is required.

Chapter 15 Dimensions of Buildings

Building Regulations Part A

We have discussed the proportions of walls and other structural members within a building, and in this chapter we will examine the criteria set out in Part A of the Building Regulations 2010. The requirements under Part A of the Building Regulations 2010 are as follows.

Loading

A1. (1) *The building shall be constructed so that the combined dead, imposed and wind loads are sustained and transmitted by it to the ground:*

(a) *safely; and*

(b) *without causing such deflection or deformation of any part of the building, or such movement of the ground, as will impair the stability of any part of another building.*

(2) *In assessing whether a building complies with sub paragraph (1) regard shall be had to the imposed and wind loads to which it is likely to be subjected in the ordinary course of its use for the purpose for which it is intended.*

Ground movement

A2. The building shall be constructed so that ground movement caused by:

(a) *swelling, shrinkage or freezing of the subsoil; or*

(b) *landslip or subsidence (other than subsidence arising from shrinkage, in so far as the risk can be reasonably foreseen), will not impair the stability of any part of the building.*

Bearing this in mind, new builds and extensions have to conform to the above criteria. It should be noted that a change of use of the occupancy of the building may also increase the loading on a building, particularly the imposed loads as defined under BS 6339. For example, a building constructed as a residential property will have a smaller imposed load than an office building. Hence, any change of use may increase the loads on the building and cause some structural members – such as floor joists and beams – to become overloaded. This is also true for commercial premises, where a change of occupancy may result in a different commercial activity. A foundry, for example, will be susceptible to much greater loads than a garage. Therefore, the tolerances and loadings to a floor slab may be such that the floor has to be strengthened to accommodate the proposed change of use.

In designing buildings the Building Regulations provide guidance and limitations on the proportion and size of the building to ensure the stability, structural elements and fabric of the building are adequately connected with load paths directing loads through

Structural Design of Buildings, First Edition. Paul Smith.

the building to the foundation. The regulations also provide guidance on walls, floors and roofs to ensure the necessary lateral restraint is provided.

Originally the sizes of timber members were provided in the Building Regulations, but these are now available as a separate publication from the Timber Research and Development Association – known as the "TRADA Eurocode 5 span tables".

Slenderness ratio

We have discussed the slenderness ratio in Chapter 8, and how adding piers, thickening walls and the shape of walls can increase its strength. For this reason the Building Regulations place limits on the length, thickness and height of walls. These limits relate to walls not in excess of 15 m high.

The thicknesses of solid walls are provided and, for example, if we take a wall constructed using coursed brickwork this thickness is a minimum of 1/16th of the storey height. Thus, a storey 3.0 m high should have a minimum thickness of 187.5 mm. If the masonry is uncoursed, for example stone, the thickness of the wall is increased by 1.33 times the thickness above. Thus, for a similar wall in uncoursed stone the thickness becomes $1.33 \times 187.5 = 249.4$ mm. Hence, Section A1/2 of the Building Regulations provides a slenderness ratio for solid walls as seen under Section 2C5 to 2C10.

For cavity walls the Building Regulations provide limits on the width of cavities (maximum 509 mm) and the spacing of wall ties to ensure the two skins act together to achieve the desired slenderness. If these criteria are met, the thickness is defined as the combined thickness of the two leaves + 10 mm. Hence the wall is 210 mm thick, and this thickness should not be less than that for solid walls (i.e., 1/16th of the storey-height). Thus, a 210-mm-thick wall can stand a maximum storey height of 3360 mm.

Internal load-bearing walls also have limits on their thickness, and these are defined in Table 15.1. This provides the thicknesses for external, compartment and separating walls, and these values are reduced for internal walls using the following expression:

$$\frac{\text{Specified thickness from table}}{3} - 5\text{ mm} = \text{Internal load-bearing wall thickness}$$

Table 15.1: Minimum thickness of certain external walls, compartments walls and separating walls. Reproduced from Part A1/2 of the Building Regulations (table 3)

Height of wall	Length of wall	Minimum thickness of wall
Not exceeding 3.5 m	Not exceeding 12 m	190 mm
Exceeding 3.5 m but not exceeding 9.0 m	Not exceeding 9.0 m	190 mm
Exceeding 3.5 m but not exceeding 9.0 m	Exceeding 9.0 m but not exceeding 12.0 m	290 mm from base for the height of one storey and 190 mm for the rest of its height
Exceeding 9.0 m but not exceeding 12.0 m	Not exceeding 9.0 m	290 mm from base for the height of one storey and 190 mm for the rest of the height
Exceeding 9.0 m but not exceeding 12.0 m	Exceeding 9.0 m but not exceeding 12.0 m	290 mm from base for the height of one storey and 190 mm for the rest of the height

Table 15.1 has already established some limits on the dimensions for walls, and further sizes and proportions are provided in the Building Regulations in relation to height-to-width ratios to ensure stability. These criteria limit the height of residential buildings to 15 m, and diagrams one and two in Part A1/2 of the Building Regulations provide such details.

Let us consider a wall 3.50 m high, 11.5 m long and with a thickness of 190 mm. The wall will be loaded with a dead load of 23 kN/m and an imposed load of 12.5 kN/m. If we say that the first example represents a wall in Bristol with a typical wind load of 0.554 kN/m, the following results are obtained.

These calculations have been undertaken using structural engineering software and we gratefully acknowledge Tekla (UK) Ltd for their approval in the use of this software.

MASONRY WALL PANEL DESIGN (EN 1996-1-1: 2005).

In accordance with EN 1996-1-1: 2005 incorporating corrigenda February 2006 and July 2009 and the UK national annex

Masonry panel details
Unreinforced masonry wall without openings
Panel length; L = **11500** mm
Panel height; h = **3500** mm

Panel support conditions
All edges supported, right and left continuous

Effective height of masonry walls – Section 5.5.1.2
Reduction factor; ρ_2 = **1.000**
$\rho_4 = \rho_2 / (1 + [\rho_2 \times h / L]^2)$ = **0.915**
Effective height of wall – eq 5.2; $h_{ef} = \rho_4 \times h$ = **3203** mm

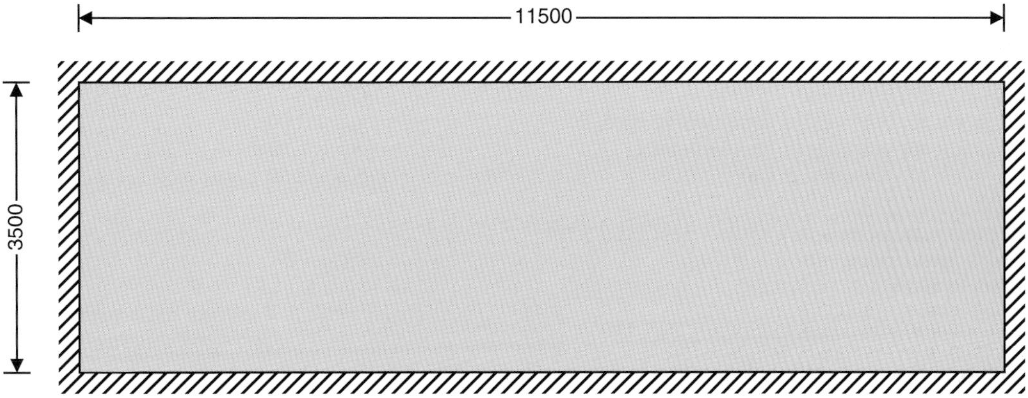

Single-leaf wall construction details
Wall thickness; t = **190** mm

Effective thickness of masonry walls – Section 5.5.1.3
Effective thickness; $t_{ef} = t$ = **190** mm

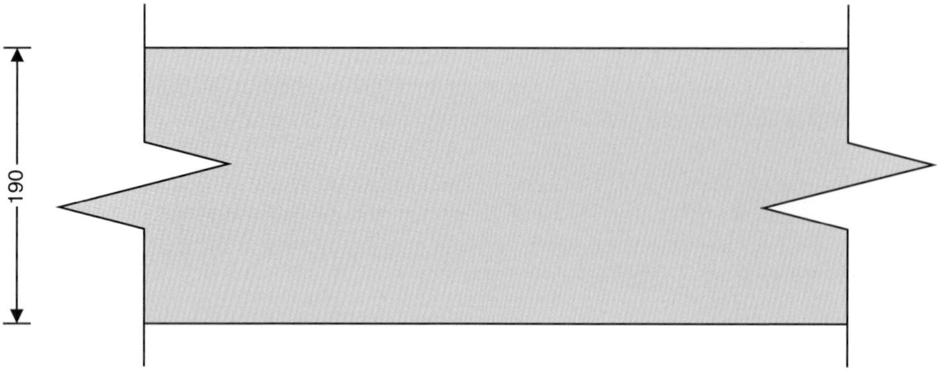

Masonry details

Masonry type;	**Clay with water absorption between 7% and 12% – Group 1**
Mean compressive strength of masonry unit;	f_b = **20** N/mm^2
Density of masonry;	γ = **18** kN/m^3
Mortar type;	**M4 – General-purpose mortar**
Compressive strength of masonry mortar;	f_m = **4** N/mm^2
Compressive strength factor – Table NA.4;	K = **0.50**
Characteristic compressive strength of masonry – eq 3.2	$f_k = K \times f_b^{0.7} \times f_m^{0.3} = \mathbf{6.17}$ N/mm^2
Characteristic flexural strength of masonry having a plane of failure parallel to the bed joints – Table NA.6	f_{xk1} = **0.4** N/mm^2
Characteristic flexural strength of masonry having a plane of failure perpendicular to the bed joints – Table NA.6	f_{xk2} = **1.1** N/mm^2

Lateral loading details

Characteristic wind load on panel;	W_k = **0.554** kN/m^2

Vertical loading details

Dead load on top of wall;	G_k = **23** kN/m
Imposed load on top of wall;	Q_k = **12.5** kN/m

Partial factors for material strength

Category of manufacturing control;	**Category II**
Class of execution control;	**Class 2**
Partial factor for masonry in compressive flexure;	γ_{Mc} = **3.00**
Partial factor for masonry in tensile flexure;	γ_{Mt} = **2.70**
Partial factor for masonry in shear;	γ_{Mv} = **2.50**

Slenderness ratio of masonry walls – Section 5.5.1.4

Allowable slenderness ratio;	SR_{all} = **27**
Slenderness ratio;	$SR = h_{ef}/t_{ef}$ = **16.9**

PASS – Slenderness ratio is less than maximum allowable

Unreinforced masonry walls subjected to lateral loading – Section 6.3

Limiting height and length-to-thickness ratio for walls under serviceability limit state – Annex F

Length-to-thickness ratio;	L/t = **60.526**
Limiting height-to-thickness ratio – Figure F.1;	44.868
Height-to-thickness ratio;	h/t = **18.421**

PASS – Limiting height-to-thickness ratio is not exceeded

Partial safety factors for design loads

Partial safety factor for variable wind load;	γ_{fW} = **1.50**
Partial safety factor for permanent load;	γ_{fG} = **1.00**

Design moments of resistance in panels

Self-weight at middle of wall;	$S_{wt} = 0.5 \times h \times t \times \gamma$ = **5.985** kN/m
Design compressive strength of masonry;	$f_d = f_k/\gamma_{Mc}$ = **2.057** N/mm^2
Design vertical compressive stress;	$\sigma_d = \min(\gamma_{fG} \times (G_k + S_{wt}) / t, 0.2 \times f_d)$ = **0.153** N/mm^2
Design flexural strength of masonry parallel to bed joints	$f_{xd1} = f_{xk1} / \gamma_{Mc}$ = **0.133** N/mm^2
Apparent design flexural strength of masonry parallel to bed joints	$f_{xd1,app} = f_{xd1} + \sigma_d$ = **0.286** N/mm^2
Design flexural strength of masonry perpendicular to bed joints	$f_{xd2} = f_{xk2} / \gamma_{Mc}$ = **0.367** N/mm^2
Elastic section modulus of wall;	$Z = t^2 / 6$ = **6016667** mm^3/m
Moment of resistance parallel to bed joints – eq 6.15	$M_{Rd1} = f_{xd1,app} \times Z$ = **1.72** kNm/m
Moment of resistance perpendicular to bed joints – eq 6.15	$M_{Rd2} = f_{xd2} \times Z$ = **2.206** kNm/m

Design moment in panels

Orthogonal strength ratio;	$\mu = f_{xd1,app} / f_{xd2}$ = **0.78**

Using yield line analysis to calculate bending moment coefficient

Bending moment coefficient;	α = **0.009**
Design moment in wall;	$M_{Ed} = \gamma_{fW} \times \alpha \times W_k \times L^2$ = **0.977** kNm/m

PASS – Resistance moment exceeds design moment

Unreinforced masonry walls subjected to mainly vertical loading – Section 6.1

Partial safety factors for design loads

Partial safety factor for permanent load;	γ_{fG} = **1.35**
Partial safety factor for variable imposed load;	γ_{fQ} = **1.50**

Check vertical loads

Reduction factor for slenderness and eccentricity – Section 6.1.2.2

Design bending moment at top or bottom of wall;	$M_{id} = \gamma_{fG} \times G_k \times e_G + \gamma_{fQ} \times Q_k \times e_Q$ = **0.0** kNm/m
Design vertical load at top or bottom of wall;	$N_{id} = \gamma_{fG} \times G_k + \gamma_{fQ} \times Q_k$ = **49.8** kN/m
Initial eccentricity – cl.5.5.1.1;	$e_{init} = h_{ef} / 450$ = **7.1** mm
Eccentricity due to horizontal load;	$e_h = M_{Ed} / N_{id}$ = **19.6** mm
Eccentricity at top or bottom of wall – eq 6.5;	$e_i = \max(M_{id} / N_{id} + e_h + e_{init}, 0.05 \times t)$ = **26.7** mm
Reduction factor at top or bottom of wall – eq 6.4;	$\Phi_i = \max(1 - 2 \times e_i / t, 0)$ = **0.719**
Design bending moment at middle of wall;	$M_{md} = \gamma_{fG} \times G_k \times e_G + \gamma_{fQ} \times Q_k \times e_Q$ = **0.0** kNm/m
Design vertical load at middle of wall;	$N_{md} = \gamma_{fG} \times G_k + \gamma_{fQ} \times Q_k + t \times \gamma \times h / 2$ = **55.8** kN/m
Eccentricity due to horizontal load;	$e_{hm} = M_{Ed} / N_{md}$ = **17.5** mm
Eccentricity at middle of wall due to loads – eq 6.7;	$e_m = M_{md} / N_{md} + e_{hm} + e_{init}$ = **24.6** mm
Eccentricity at middle of wall due to creep;	e_k = **0** mm
Eccentricity at middle of wall – eq 6.6;	$e_{mk} = \max(e_m + e_k, 0.05 \times t)$ = **24.6** mm
From eq G.2;	$A_1 = 1 - 2 \times e_{mk} / t$ = **0.741**
Short-term secant modulus of elasticity factor;	K_E = **1000**
Modulus of elasticity – cl.3.7.2;	$E = K_E \times f_k$ = **6170** N/mm^2
Slenderness – eq G.4;	$\lambda = (h_{ef}/t_{ef}) \times \sqrt{[f_k/E]}$ = **0.533**

From eq G.3; $u = (\lambda - 0.063) / (0.73 - 1.17 \times e_{mk} / t) = \mathbf{0.813}$
Reduction factor at middle of wall – eq G.1; $\Phi_m = \max(A_1 \times e_e^{-(u \times u)/2}, 0) = \mathbf{0.532}$
Reduction factor for slenderness and eccentricity; $\Phi = \min(\Phi_i, \Phi_m) = \mathbf{0.532}$

Verification of unreinforced masonry walls subjected to mainly vertical loading – Section 6.1.2
Design value of the vertical load; $N_{Ed} = \max(N_{id}, N_{md}) = \mathbf{55.785}$ kN/m
Design compressive strength of masonry; $f_d = f_k / \gamma_{Mc} = \mathbf{2.057}$ N/mm^2
Vertical resistance of wall – eq 6.2; $N_{Rd} = \Phi \times t \times f_d = \mathbf{208.023}$ kN/m
PASS – Design vertical resistance exceeds applied design vertical load

We can see from this simple analysis that the wall passes the structural analysis.

Let us now consider the same imposed and dead loading and the same wall parameters in Newcastle, and assess a typical wind load for this area of say 0.869 kN/m^2.

These calculations have been undertaken using structural engineering software and we gratefully acknowledge Tekla (UK) Ltd for their approval in the use of this software.

MASONRY WALL PANEL DESIGN (EN 1996-1-1: 2005).

In accordance with EN 1996-1-1: 2005 incorporating corrigenda February 2006 and July 2009 and the UK national annex

Masonry panel details
Unreinforced masonry wall without openings
Panel length; L = **11500** mm
Panel height; h = **3500** mm

Panel support conditions
All edges supported, right and left continuous

Effective height of masonry walls – Section 5.5.1.2
Reduction factor; $\rho_2 = \mathbf{1.000}$
$\rho_4 = \rho_2 / (1 + [\rho_2 \times h / L]^2) = \mathbf{0.915}$
Effective height of wall – eq 5.2; $h_{ef} = \rho_4 \times h = \mathbf{3203}$ mm

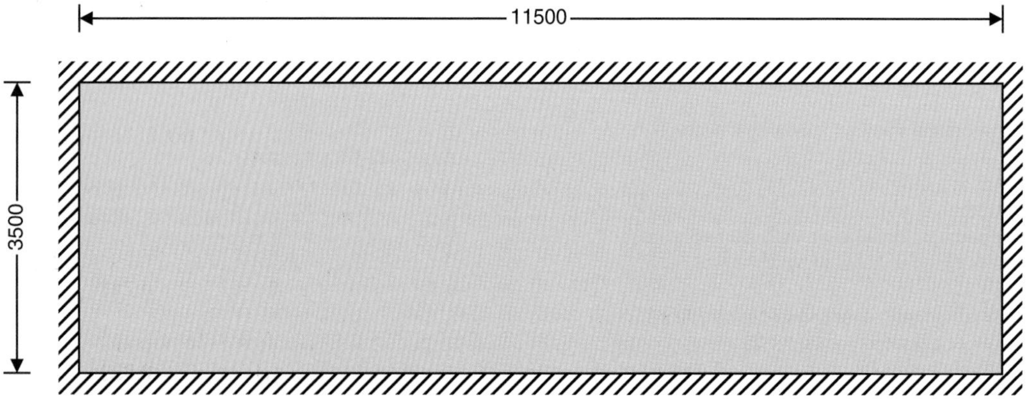

Single-leaf wall construction details
Wall thickness; $t = \mathbf{190}$ mm

Effective thickness of masonry walls – Section 5.5.1.3
Effective thickness; $t_{ef} = t = \mathbf{190}$ mm

Masonry details

Masonry type;	**Clay with water absorption between 7% and 12% – Group 1**
Mean compressive strength of masonry unit;	$f_b = \mathbf{20}$ N/mm^2
Density of masonry;	$\gamma = \mathbf{18}$ kN/m^3
Mortar type;	**M4 – General-purpose mortar**
Compressive strength of masonry mortar;	$f_m = \mathbf{4}$ N/mm^2
Compressive strength factor – Table NA.4;	$K = \mathbf{0.50}$
Characteristic compressive strength of masonry – eq 3.2	$f_k = K \times f_b^{0.7} \times f_m^{0.3} = \mathbf{6.17}$ N/mm^2
Characteristic flexural strength of masonry having a plane of failure parallel to the bed joints – Table NA.6	$f_{xk1} = \mathbf{0.4}$ N/mm^2
Characteristic flexural strength of masonry having a plane of failure perpendicular to the bed joints – Table NA.6	$f_{xk2} = \mathbf{1.1}$ N/mm^2

Lateral loading details
Characteristic wind load on panel; $W_k = \mathbf{0.869}$ kN/m^2

Vertical loading details
Dead load on top of wall; $G_k = \mathbf{23}$ kN/m;
Imposed load on top of wall; $Q_k = \mathbf{12.5}$ kN/m;

Partial factors for material strength

Category of manufacturing control;	**Category II**
Class of execution control;	**Class 2**
Partial factor for masonry in compressive flexure;	$\gamma_{Mc} = \mathbf{3.00}$
Partial factor for masonry in tensile flexure;	$\gamma_{Mt} = \mathbf{2.70}$
Partial factor for masonry in shear;	$\gamma_{Mv} = \mathbf{2.50}$

Slenderness ratio of masonry walls – Section 5.5.1.4
Allowable slenderness ratio; $SR_{all} = \mathbf{27}$
Slenderness ratio; $SR = h_{ef} / t_{ef} = \mathbf{16.9}$

PASS – Slenderness ratio is less than maximum allowable

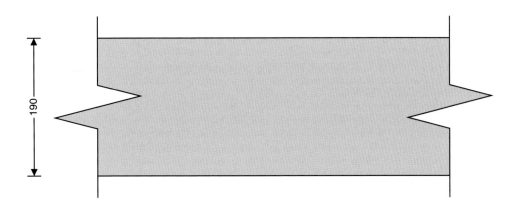

Unreinforced masonry walls subjected to lateral loading – Section 6.3

Limiting height and length-to-thickness ratio for walls under serviceability limit state – Annex F

Length-to-thickness ratio; $L / t = \mathbf{60.526}$

Limiting height-to-thickness ratio – Figure F.1; 44.868

Height-to-thickness ratio; $h / t = \mathbf{18.421}$

PASS – Limiting height-to-thickness ratio is not exceeded

Partial safety factors for design loads

Partial safety factor for variable wind load; $\gamma_{fW} = \mathbf{1.50}$

Partial safety factor for permanent load; $\gamma_{fG} = \mathbf{1.00}$

Design moments of resistance in panels

Self-weight at middle of wall; $S_{wt} = 0.5 \times h \times t \times \gamma = \mathbf{5.985}$ kN/m

Design compressive strength of masonry; $f_d = f_k / \gamma_{Mc} = \mathbf{2.057}$ N/mm^2

Design vertical compressive stress; $\sigma_d = \min(\gamma_{fG} \times (G_k + S_{wt}) / t, 0.2 \times f_d) =$ **0.153** N/mm^2

Design flexural strength of masonry parallel to bed joints $f_{xd1} = f_{xk1} / \gamma_{Mc} = \mathbf{0.133}$ N/mm^2

Apparent design flexural strength of masonry parallel to bed joints $f_{xd1,app} = f_{xd1} + \sigma_d = \mathbf{0.286}$ N/mm^2

Design flexural strength of masonry perpendicular to bed joints $f_{xd2} = f_{xk2} / \gamma_{Mc} = \mathbf{0.367}$ N/mm^2

Elastic section modulus of wall; $Z = t^2 / 6 = \mathbf{6016667}$ mm^3/m

Moment of resistance parallel to bed joints – eq 6.15 $M_{Rd1} = f_{xd1,app} \times Z = \mathbf{1.72}$ kNm/m

Moment of resistance perpendicular to bed joints – eq 6.15 $M_{Rd2} = f_{xd2} \times Z = \mathbf{2.206}$ kNm/m

Design moment in panels

Orthogonal strength ratio; $\mu = f_{xd1,app} / f_{xd2} = \mathbf{0.78}$

Using yield line analysis to calculate bending moment coefficient

Bending moment coefficient; $\alpha = \mathbf{0.009}$

Design moment in wall; $M_{Ed} = \gamma_{fW} \times \alpha \times W_k \times L^2 = \mathbf{1.532}$ kNm/m

PASS – Resistance moment exceeds design moment

Unreinforced masonry walls subjected to mainly vertical loading – Section 6.1

Partial safety factors for design loads

Partial safety factor for permanent load; $\gamma_{fG} = \mathbf{1.35}$

Partial safety factor for variable imposed load; $\gamma_{fQ} = \mathbf{1.50}$

Check vertical loads

Reduction factor for slenderness and eccentricity – Section 6.1.2.2

Design bending moment at top or bottom of wall; $M_{id} = \gamma_{fG} \times G_k \times e_G + \gamma_{fQ} \times Q_k \times e_Q =$ **0.0** kNm/m

Design vertical load at top or bottom of wall; $N_{id} = \gamma_{fG} \times G_k + \gamma_{fQ} \times Q_k = \mathbf{49.8}$ kN/m

Initial eccentricity – cl.5.5.1.1; $e_{init} = h_{ef} / 450 = \mathbf{7.1}$ mm

Eccentricity due to horizontal load; $e_h = M_{Ed} / N_{id} = \mathbf{30.8}$ mm

Eccentricity at top or bottom of wall – eq 6.5; $e_i = \max(M_{id} / N_{id} + e_h + e_{init}, 0.05 \times t) =$ **37.9** mm

Reduction factor at top or bottom of wall – eq 6.4; $\Phi_i = \max(1 - 2 \times e_i / t, 0) = \mathbf{0.601}$

Design bending moment at middle of wall; $M_{md} = \gamma_{fG} \times G_k \times e_G + \gamma_{fQ} \times Q_k \times e_Q = \mathbf{0.0}$ kNm/m

Design vertical load at middle of wall; $N_{md} = \gamma_{fG} \times G_k + \gamma_{fQ} \times Q_k + t \times \gamma \times h / 2 =$ **55.8** kN/m

Eccentricity due to horizontal load;	$e_{hm} = M_{Ed} / N_{md} = \mathbf{27.5}$ mm
Eccentricity at middle of wall due to loads – eq 6.7;	$e_m = M_{md} / N_{md} + e_{hm} + e_{init} = \mathbf{34.6}$ mm
Eccentricity at middle of wall due to creep;	$e_k = \mathbf{0}$ mm
Eccentricity at middle of wall – eq 6.6;	$e_{mk} = \max(e_m + e_k, 0.05 \times t) = \mathbf{34.6}$ mm
From eq G.2;	$A_1 = 1 - 2 \times e_{mk} / t = \mathbf{0.636}$
Short-term secant modulus of elasticity factor;	$K_E = \mathbf{1000}$
Modulus of elasticity – cl.3.7.2;	$E = K_E \times f_k = \mathbf{6170}$ N/mm^2
Slenderness – eq G.4;	$\lambda = (h_{ef} / t_{ef}) \times \sqrt{[f_k / E]} = \mathbf{0.533}$
From eq G.3;	$u = (\lambda - 0.063) / (0.73 - 1.17 \times e_{mk} / t) = \mathbf{0.909}$
Reduction factor at middle of wall – eq G.1;	$\Phi_m = \max(A_1 \times e_e^{-(u \times u)/2}, 0) = \mathbf{0.421}$
Reduction factor for slenderness and eccentricity;	$\Phi = \min(\Phi_i, \Phi_m) = \mathbf{0.421}$

Verification of unreinforced masonry walls subjected to mainly vertical loading – Section 6.1.2

Design value of the vertical load;	$N_{Ed} = \max(N_{id}, N_{md}) = \mathbf{55.785}$ kN/m
Design compressive strength of masonry;	$f_d = f_k / \gamma_{Mc} = \mathbf{2.057}$ N/mm^2
Vertical resistance of wall – eq 6.2;	$N_{Rd} = \Phi \times t \times f_d = \mathbf{164.365}$ kN/m

PASS – Design vertical resistance exceeds applied design vertical load

If we examine the two walls, the design bending moment to the wall increases from 0.977 kN·m/m in Bristol to 1.532 kN·m/m in Newcastle, where the wind load is greater.

If we extend the dimension of the wall beyond the limits prescribed in the Building Regulations for the Newcastle example, to 12.5 m long, we can see that the wall fails since the resistance moment is exceeded by the design moment. See below.

These calculations have been undertaken using structural engineering software and we gratefully acknowledge Tekla (UK) Ltd for their approval in the use of this software.

MASONRY WALL PANEL DESIGN (EN 1996-1-1: 2005)

In accordance with EN 1996-1-1: 2005 incorporating corrigenda February 2006 and July 2009 and the UK national annex

Masonry panel details

Unreinforced masonry wall without openings

Panel length;	$L = \mathbf{12500}$ mm
Panel height;	$h = \mathbf{3500}$ mm

Panel support conditions

All edges supported, right and left continuous

Effective height of masonry walls – Section 5.5.1.2

Reduction factor;	$\rho_2 = \mathbf{1.000}$
	$\rho_4 = \rho_2 / (1 + [\rho_2 \times h / L]^2) = \mathbf{0.927}$
Effective height of wall – eq 5.2;	$h_{ef} = \rho_4 \times h = \mathbf{3246}$ mm

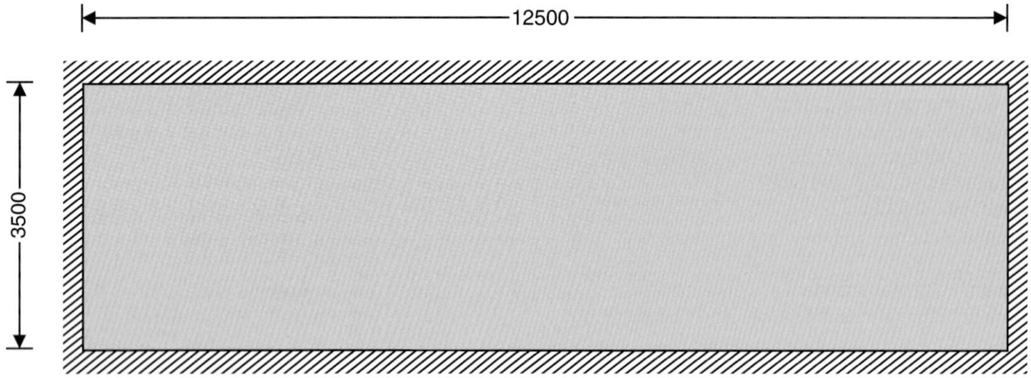

Single-leaf wall construction details

Wall thickness; $t = \mathbf{190}$ mm

Effective thickness of masonry walls – Section 5.5.1.3

Effective thickness; $t_{ef} = t = \mathbf{190}$ mm

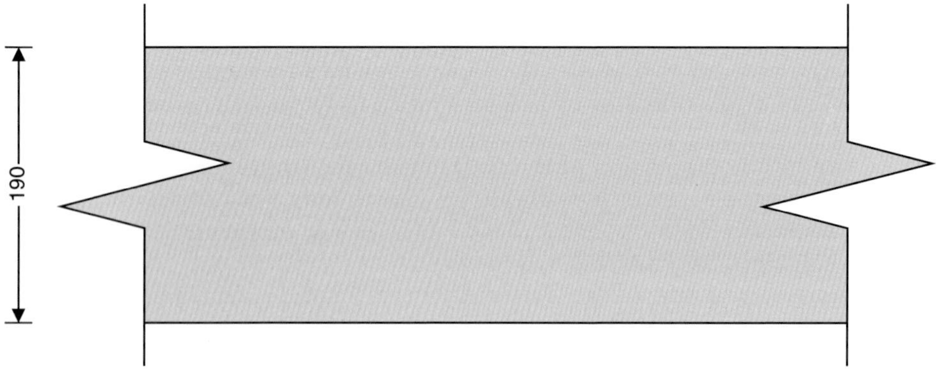

Masonry details

Masonry type;	**Clay with water absorption between 7% and 12% – Group 1**
Mean compressive strength of masonry unit;	$f_b = \mathbf{20}$ N/mm^2
Density of masonry;	$\gamma = \mathbf{18}$ kN/m^3
Mortar type;	**M4 - General purpose mortar**
Compressive strength of masonry mortar;	$f_m = \mathbf{4}$ N/mm^2
Compressive strength factor – Table NA.4;	$K = \mathbf{0.50}$
Characteristic compressive strength of masonry – eq 3.2	$f_k = K \times f_b^{0.7} \times f_m^{0.3} = \mathbf{6.17}$ N/mm^2
Characteristic flexural strength of masonry having a plane of failure parallel to the bed joints – Table NA.6	$f_{xk1} = \mathbf{0.4}$ N/mm^2
Characteristic flexural strength of masonry having a plane of failure perpendicular to the bed joints – Table NA.6	$f_{xk2} = \mathbf{1.1}$ N/mm^2

Lateral loading details

Characteristic wind load on panel; $W_k = \mathbf{0.869}$ kN/m^2

Vertical loading details

Dead load on top of wall; $G_k = \mathbf{23}$ kN/m;

Imposed load on top of wall; $Q_k = \mathbf{12.5}$ kN/m;

Partial factors for material strength

Category of manufacturing control;	**Category II**
Class of execution control;	**Class 2**
Partial factor for masonry in compressive flexure;	γ_{Mc} = **3.00**
Partial factor for masonry in tensile flexure;	γ_{Mt} = **2.70**
Partial factor for masonry in shear;	γ_{Mv} = **2.50**

Slenderness ratio of masonry walls – Section 5.5.1.4

Allowable slenderness ratio;	SR_{all} = **27**
Slenderness ratio;	$SR = h_{ef} / t_{ef}$ = **17.1**

PASS – Slenderness ratio is less than maximum allowable

Unreinforced masonry walls subjected to lateral loading – Section 6.3

Limiting height and length-to-thickness ratio for walls under serviceability limit state – Annex F

Length-to-thickness ratio;	L / t = **65.789**
Limiting height-to-thickness ratio – Figure F.1;	43.553
Height-to-thickness ratio;	h / t = **18.421**

PASS – Limiting height-to-thickness ratio is not exceeded

Partial safety factors for design loads

Partial safety factor for variable wind load;	γ_{fW} = **1.50**
Partial safety factor for permanent load;	γ_{fG} = **1.00**

Design moments of resistance in panels

Self-weight at middle of wall;	$S_{wt} = 0.5 \times h \times t \times \gamma$ = **5.985** kN/m
Design compressive strength of masonry;	$f_d = f_k / \gamma_{Mc}$ = **2.057** N/mm^2
Design vertical compressive stress;	$\sigma_d = \min(\gamma_{fG} \times (G_k + S_{wt}) / t, 0.2 \times f_d)$ = **0.153** N/mm^2
Design flexural strength of masonry parallel to bed joints	$f_{xd1} = f_{xk1} / \gamma_{Mc}$ = **0.133** N/mm^2
Apparent design flexural strength of masonry parallel to bed joints	$f_{xd1,app} = f_{xd1} + \sigma_d$ = **0.286** N/mm^2
Design flexural strength of masonry perpendicular to bed joints	$f_{xd2} = f_{xk2} / \gamma_{Mc}$ = **0.367** N/mm^2
Elastic section modulus of wall;	$Z = t^2 / 6$ = **6016667** mm^3/m
Moment of resistance parallel to bed joints – eq 6.15	$M_{Rd1} = f_{xd1,app} \times Z$ = **1.72** kNm/m
Moment of resistance perpendicular to bed joints – eq 6.15	$M_{Rd2} = f_{xd2} \times Z$ = **2.206** kNm/m

Design moment in panels

Using elastic analysis to determine bending moment coefficients for a vertically spanning panel

Bending moment coefficient;	α = **0.125**
Design moment in wall;	$M_{Ed} = \gamma_{fW} \times \alpha \times W_k \times h^2$ = **1.996** kNm/m

FAIL – Resistance moment is exceeded by design moment

Unreinforced masonry walls subjected to mainly vertical loading – Section 6.1

Partial safety factors for design loads

Partial safety factor for permanent load;	γ_{fG} = **1.35**
Partial safety factor for variable imposed load;	γ_{fQ} = **1.50**

Check vertical loads

Reduction factor for slenderness and eccentricity – Section 6.1.2.2

Design bending moment at top or bottom of wall;	$M_{id} = \gamma_{fG} \times G_k \times e_G + \gamma_{fQ} \times Q_k \times e_Q$ = **0.0** kNm/m
Design vertical load at top or bottom of wall;	$N_{id} = \gamma_{fG} \times G_k + \gamma_{fQ} \times Q_k$ = **49.8** kN/m

Initial eccentricity – cl.5.5.1.1;	$e_{init} = h_{ef} / 450 =$ **7.2** mm
Eccentricity due to horizontal load;	$e_h = M_{Ed} / N_{id} =$ **40.1** mm
Eccentricity at top or bottom of wall – eq 6.5;	$e_i = \max(M_{id} / N_{id} + e_h + e_{init}, 0.05 \times t) =$ **47.3** mm
Reduction factor at top or bottom of wall – eq 6.4;	$\Phi_i = \max(1 - 2 \times e_i / t, 0) =$ **0.502**
Design bending moment at middle of wall;	$M_{md} = \gamma_{fG} \times G_k \times e_G + \gamma_{fQ} \times Q_k \times e_Q =$ **0.0** kNm/m
Design vertical load at middle of wall;	$N_{md} = \gamma_{fG} \times G_k + \gamma_{fQ} \times Q_k + t \times \gamma \times h / 2 =$ **55.8** kN/m
Eccentricity due to horizontal load;	$e_{hm} = M_{Ed} / N_{md} =$ **35.8** mm
Eccentricity at middle of wall due to loads – eq 6.7;	$e_m = M_{md} / N_{md} + e_{hm} + e_{init} =$ **43** mm
Eccentricity at middle of wall due to creep;	$e_k =$ **0** mm
Eccentricity at middle of wall – eq 6.6;	$e_{mk} = \max(e_m + e_k, 0.05 \times t) =$ **43** mm
From eq G.2;	$A_1 = 1 - 2 \times e_{mk} / t =$ **0.547**
Short-term secant modulus of elasticity factor;	$K_E =$ **1000**
Modulus of elasticity – cl.3.7.2;	$E = K_E \times f_k =$ **6170** N/mm^2
Slenderness – eq G.4;	$\lambda = (h_{ef}/t_{ef}) \times \sqrt{[f_k/E]} =$ **0.54**
From eq G.3;	$u = (\lambda - 0.063) / (0.73 - 1.17 \times e_{mk} / t) =$ **1.026**
Reduction factor at middle of wall – eq G.1;	$\Phi_m = \max(A_1 \times e_e^{-(u \times u)/2}, 0) =$ **0.324**
Reduction factor for slenderness and eccentricity;	$\Phi = \min(\Phi_i, \Phi_m) =$ **0.324**

Verification of unreinforced masonry walls subjected to mainly vertical loading – Section 6.1.2

Design value of the vertical load;	$N_{Ed} = \max(N_{id}, N_{md}) =$ **55.785** kN/m
Design compressive strength of masonry;	$f_d = f_k / \gamma_{Mc} =$ **2.057** N/mm^2
Vertical resistance of wall – eq 6.2;	$N_{Rd} = \Phi \times t \times f_d =$ **126.436** kN/m

PASS – Design vertical resistance exceeds applied design vertical load

The Building Regulations have been written in an attempt to apply to buildings wherever they may be in the country, and consequently care should be taken in certain parts of the country where the criteria could be on the edge of the design.

Openings can have an impact on the structural integrity of walls, and Part A of the Building Regulations sets out clear guidance on the extent and positions of openings in relation to walls. See diagram 14 in Part A1/2 of the Building Regulations.

The maximum height and length of walls are prescribed in Part A1/2 of the Building Regulations. The Building Regulations do not deal with walls longer than 12 m, thus specialist advice is required if a wall exceeds 12 m in length measured between walls, chimneys or buttresses.

Clearly, as the height of a building increases, the compressive load increases – this increased load will eventually exceed the compressive strength of the unit employed.

The maximum heights provided in the Building Regulations are limited. For example, condition 2C17 of Part A1/2 of the Building Regulations provides a maximum height of 12 m for which the thicknesses of walls have been calculated. This does not mean that taller structures cannot be built, but that specialist structural calculations will have to be provided in these circumstances since the wall heights are outside the scope of the regulations.

Thus, maximum height-to-width ratios are provided in table 3 of Part A1/2 of the Building Regulations and the maximum is 12 m with a wall thickness of 290 mm from the base for the height of two storeys and 190 mm for the rest of the height.

The maximum height will also depend on the wind loading, which is an important consideration. The tables in diagram 7 of the Building Regulations Part A1/2 provide limits on wall heights considering the wind loading on a building.

For the lateral support to the walls as prescribed under Part A of the Building Regulations to be effective, the maximum storey height is considered to be 2.7 m.

The maximum loading to the walls is the combined dead and imposed load and should not be greater than 70 kN/m as measured at the base of the wall – see regulation 2C24c. This ensures that the height-to-width thicknesses and compressive strengths of the units prescribed are not exceeded.

Buttresses and end restraints

The above criteria in terms of thicknesses and heights of walls are based on effective vertical lateral restraint. This can be provided by buttresses, piers, chimneys and return walls or internal walls, but these need to be fully bonded or securely tied to ensure the structural integrity of the restraint. These restraint options break the wall into smaller lengths, which reduces the effective length of the wall, making it much stronger. However, certain conditions have to be satisfied to ensure this is the case, and one is that the restraining member should provide restraint to the full height of the wall. Other conditions relating to the minimum thicknesses and dimensions of these structures, openings and minimum returns of 665 mm ensure that these act in the manner for which they are employed. Diagrams 12 and 13 of Part A of the Building Regulations provide details on these dimensions and parameters.

Lateral restraint of walls and roofs

Floors and roofs can transfer lateral forces to walls and buttressing members provided they are tied effectively, and this can be seen in Section 2C33 of Part A1/2 of the Building Regulations. The regulations require that lateral restraint is afforded at roof, gable, wall-plate and first-floor level by use of lateral restraint straps conforming to BS EN 845-1. Table 9 and diagrams 15 and 16 of the Building Regulations provide details of these requirements. Essentially this reduces the effective length of the wall and thus reduces the slenderness ratio. Thus, lateral restraint straps can afford considerable stability to a wall by reducing its effective length.

Chapter 16 Basements and Retaining Structures

Structural considerations

The structural basis of a basement wall and a retaining structure are very similar, and essentially the wall is designed in the same manner. Traditionally walls were designed using the Civil Engineering Code of Practice 2, which was superseded in 1994 by BS 8002. More recently walls have been designed to the new Eurocodes, and the code covering retaining walls is Eurocode 7.

There were traditionally three components to consider in the design of a retaining wall, as discussed below.

Sliding

This is where the wall slides forward due to insufficient friction between the base and the ground, or where inadequate passive resistance to the front of the wall is available. In basements, sliding resistance is offered by the opposite wall structure.

Overturning

The resultant thrust must pass through the middle third of the base to ensure that the wall is stable and not subject to overturning. We have discussed the middle-third rule in Chapter 8.

Bearing capacity

The weight on the foundation of the wall and the resultant overturning moments from the thrust will have large bearing pressures on the toe. These have to be less than the bearing capacity of the soil beneath the wall's foundation.

In addition to the above, it was recognised that walls could also fail in the following ways.

Structural failure

Due to poor design or workmanship.

Rotational slip of the surrounding soil

This is prevalent in cohesive soils such as clay, and a slope stability analysis should be undertaken if this is suspected during the design of the wall. It is emphasised that design should be undertaken by a suitably qualified person, since knowledge of soil conditions and any variations thereof can make a difference to the success or failure of such structures.

Structural Design of Buildings, First Edition. Paul Smith.

The more recent Eurocode 7 considers the above failures, but a list of limit states is provided below and these must all be considered in the design of the wall.

- Loss of overall stability.
- Failure of a structural element such as a wall, anchor, etc.
- Combined failure in the ground and in the structural element.
- Failure by hydraulic heave.
- Movement causing collapse or affecting the appearance or efficient use of the structure, or nearby structures or suffices which rely on it.
- Unacceptable leakage through or beneath the wall.
- Unacceptable transport of soil particles through or beneath the wall.
- Unacceptable change in ground water regime.

Safety factors

Under the traditional code of practice, a factor of safety was applied between the values of 2.0 and 3.5 depending on the parameter being considered. Under BS 8002 a factor of safety of 20% was added to the value for the angle of shearing resistance being considered, and allowed the factor of safety to be used throughout the design.

The more recent Eurocode 7 applies various factors of safety depending on the parameter being considered, the design approach being followed and the type of failure under consideration. A different combination of partial design factors is used for the limit state design.

Theory behind the design

The material behind the wall will try to thrust forward in an "active state", that is the material tries to expand and this places the material in tension. The material in front of the wall is in a compressive state, resisting the tensile forces. If there is no movement in the wall at all, the theoretical pressure diagram is as shown in Figure 16.1 and is a triangular distribution. However, if the wall moves it is said to have yielded and the movement can be as a result of sliding or rotation. A rotational effect will result in a triangular pressure diagram but with a smaller triangular distribution, as seen in Figure 16.2. A sliding action will result in a pressure distribution similar to the triangular diagram at the top of the wall, but will move the centre of the pressure higher as seen in Figure 16.3.

If the wall is restricted to a movement equal to 0.5% of its height, then the pressure moves from a third to approximately half the height of the wall. Thus, below this value, the wall is assumed to achieve a triangular pressure distribution (Smith, 1990).

Loading

In accordance with the codes of practice it is recommended that the wall is designed with a minimum surcharge of 10 kN/m^2 to the back of the wall. The surcharge represents stored or accumulated items above the wall, or other loadings such as vehicles. Other

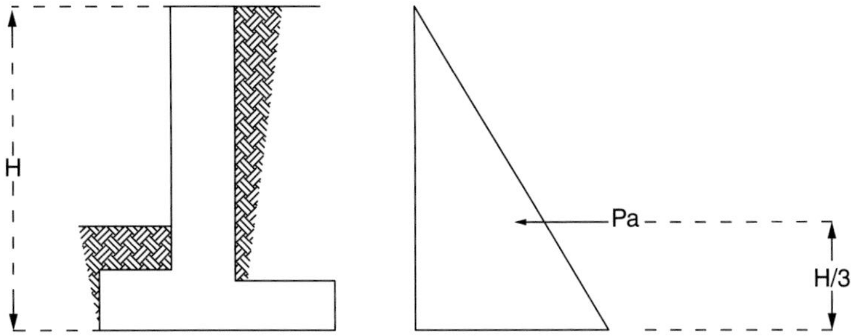

Figure 16.1: Theoretical pressure diagram for a wall unable to move.

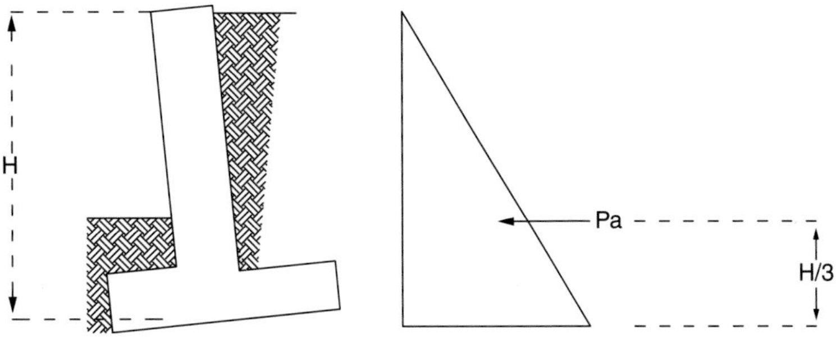

Figure 16.2: Theoretical pressure diagram for a wall subject to rotational effects.

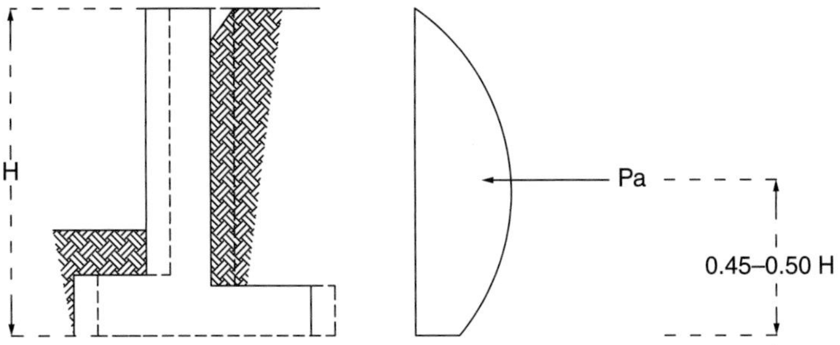

Figure 16.3: Theoretical pressure diagram for a wall subject to sliding effects.

loads such as line loads also need consideration and these are loads that run parallel to the wall, for example a wall set back from the top of the retaining structure.

Axial loads applied to the top of the retaining structure can add to the strength of the wall in terms of rotation and sliding, but will also increase the bearing capacity. The wall may have additional dead and imposed loads – for example where a wall continues above the retained soil height supporting additional storeys of a house. In these circumstances the basement wall supports additional dead and imposed loads and these need to be calculated and added as an axial line load. The eccentricity of these loads should also be considered.

The active pressure is the pressure acting behind the wall from the material retained and is provided by the formula

$$HA = KaQ + Ka\gamma h - 2c\sqrt{Ka} \tag{16.1}$$

where

HA = active pressure
Ka = coefficient of active earth pressure
Q = surcharge
γ = density of soil
h = height of the wall
c = undrained shear strength

with

$$Ka = \frac{1 - \sin\phi}{1 + \sin\phi} \tag{16.2}$$

Φ = angle of shearing resistance

Angle of shearing resistance

This is unique to the soil and if different values are applied to the formula one can see significant changes in the results. Thus, an appreciation and understanding of the soil conditions is required and this should be obtained by undertaking geotechnical tests to determine the parameters of the soil.

Let us consider equation (16.1) and its component parts:

$$HA = KaQ + Ka\gamma h - 2c\sqrt{Ka}$$

This can be written as

$$HA = Hs + Ha \tag{16.3}$$

$$Hs = \text{Pressure due to surcharge} = KaQ \tag{16.4}$$

$$Ha = \text{Pressure due to soil} = Ka\gamma h - 2c\sqrt{Ka} \tag{16.5}$$

It can be seen that the first part of the equation (KaQ) relates to the surcharge, the next part of the equation relates to the pressure due to the retained soil ($Ka\gamma h$). This is reduced by ($2c\sqrt{Ka}$) if cohesion exists in the soil, for example if a clay soil is present.

The pressure diagrams can be seen in Figure 16.4; for the surcharge element of the pressure this is a rectangle and for the soil element of the equation the pressure diagram is represented by a triangle.

If we take the area of this rectangular and triangular pressure diagram then we can calculate the active thrust or load on the wall, which is as follows:

$$\text{Surcharge element } Ps = KaQh\text{; this load acts at a height of } h/2 \tag{16.6}$$

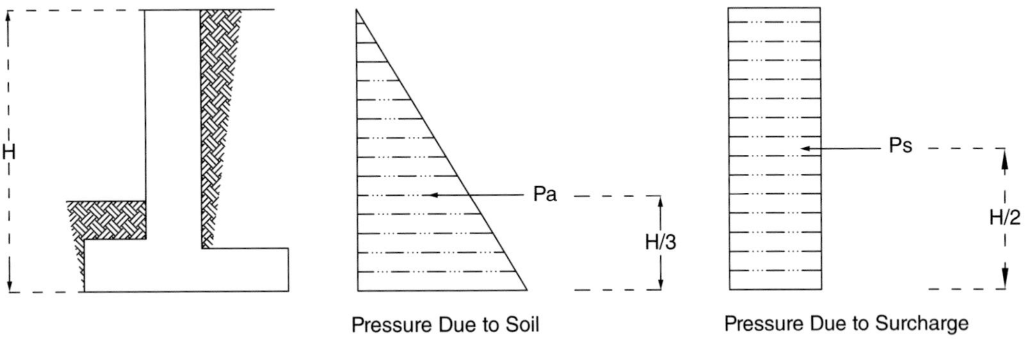

Figure 16.4: Load diagrams on a retaining wall for surcharge and soil.

$$\text{Thrust Pa} = \tfrac{1}{2}(\mathrm{Ka}\gamma \mathrm{h}) \times \mathrm{h} = \tfrac{1}{2}\mathrm{Ka}\gamma \mathrm{h}^2\text{; this load acts at a height of } \mathrm{h}/3 \qquad (16.7)$$

$$\text{Total active thrust PA} = \mathrm{Pa} + \mathrm{Ps} = \mathrm{KaQh} + \tfrac{1}{2}\mathrm{Ka}\gamma \mathrm{h}^2 \qquad (16.8)$$

The position of these can be seen in Figure 16.4.

The passive thrust is provided by the following formula. It is the pressure acting in front of the wall and offers resistance to the active thrust.

$$\mathrm{Pp} = \tfrac{1}{2}\mathrm{Kp}\gamma \mathrm{h}^2 \qquad (16.9)$$

where

$$\mathrm{Kp} = \frac{1 + \sin\phi}{1 - \sin\phi} \qquad (16.10)$$

Effects of water

The effects of water behind a retaining wall can cause significant increases in pressure, as provided by the formula below:

$$\text{Pressure due to water } \mathrm{Hw} = \mathrm{Ka}\gamma_{\mathrm{w}}\mathrm{h} \qquad (16.11)$$

It is with this in mind that effective drainage, through the use of weep holes or a drain behind the wall carrying water to a suitable soakaway or other reciprocal away from the wall, is important to ensure the stability of the wall.

Proportions of walls

Since all the design procedures mentioned analyse a proposed structure rather than providing the dimensions for the wall, it is important to have a realistic starting point otherwise the designer may find they have to design a number of walls until the criteria for the limit state are satisfied. With this in mind, for a cantilever retaining wall we propose the following dimensions:

$$\begin{aligned}&\text{Toe length } h/6 \text{ to } h/8\\&\text{Foundation length } 2/3h\\&\text{Thickness of base } h/10\end{aligned}$$

Thickness of wall h/12, but with a minimum of 300 mm

Here h = height of the retaining structure, including the foundation.

Another method of determining the initial size is provided below (Smith, 2006, p. 258):

$$\begin{aligned}&\text{Toe length } 0.15H\\&\text{Base length } 0.4H \text{ to } 0.7H\\&\text{Thickness of base } 0.1H\end{aligned}$$

where H = height of the retaining structure.

Design example

Let us consider a simple cantilever retaining wall in concrete, with the following parameters (see Figure 16.5):

- Unit weight of soil 18 kN/m^3.
- Bearing capacity of soil 200 kN/m^2.
- Angle of shearing resistance 35°.
- Minimum surcharge 10 kN/m^2.
- The effects of passive resistance ignored.
- Height 4.0 m.
- Cohesionless soil, hence $c' = 0$.

Using the traditional method CP2, from equation (16.2):

$$Ka = \frac{1-\sin\phi}{1+\sin\phi} = \frac{1-\sin 35}{1+\sin 35} = 0.27$$

Active pressure

From equations (16.4) and (16.5):

$$\begin{aligned}Ha &= Ka\gamma h = 0.27 \times 18 \times 4 = 19.44\,kN/m^2\\Hs &= KaQ = 0.27 \times 10 = 2.7\,kN/m^2\end{aligned}$$

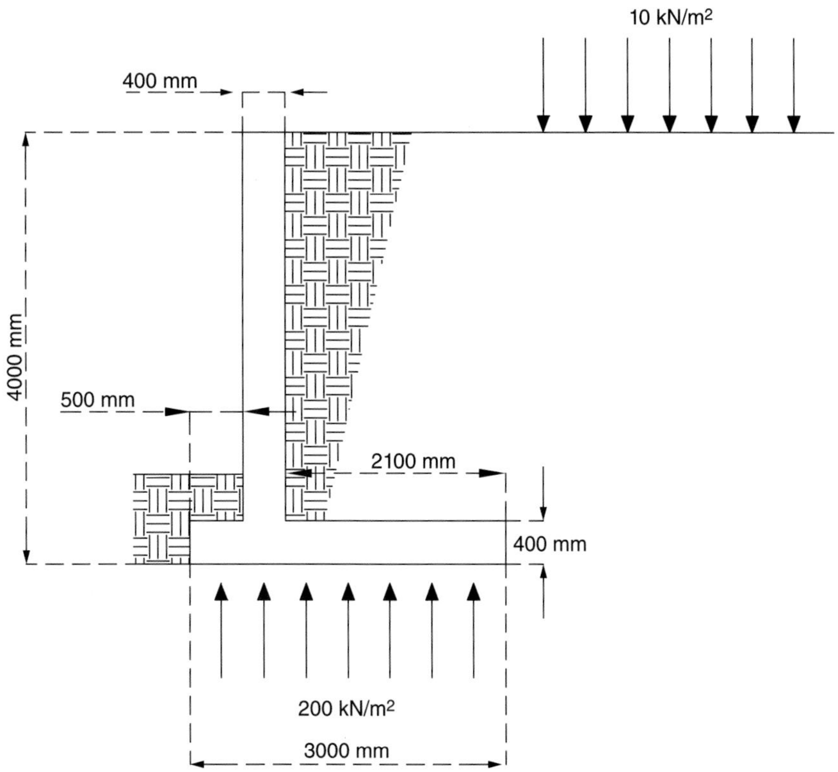

Figure 16.5: Diagram of proposed retaining wall under analysis.

Table 16.1: Vertical loads (Rv) using CP2

Stem of wall	3.6 × 0.4 × 24	34.56
Base	0.4 × 3 × 24	28.8
Soil on heel	18 × 2.1 × 3.6	136.08
Surcharge	2.1 × 10	21
	Total Rv = 220.4 kN	

Active thrust

From equations (16.6) and (16.7):

$$\text{Active thrust from soil Pa} = \tfrac{1}{2}\text{Ka}\gamma\text{h}^2 = \tfrac{1}{2} \times 0.27 \times 18 \times 16 = 38.88\ \text{kN}$$
$$\text{Active thrust from surcharge Ps} = \text{KaQh} = 0.27 \times 10 \times 4 = 10.8\ \text{kN}$$
$$\text{Total active thrust PA} = 49.68\ \text{kN}$$

This is the thrust per metre length of wall.
See Table 16.1 for the vertical loads (Rv) calculated using CP2.

Sliding

Total force causing sliding = Total active thrust = 49.68 kN

$$\text{Force resisting sliding} = \text{Rvtan}\phi \tag{16.12}$$

$$= 220.4 \times \tan 35 = 220.4 \times 0.7 = 154.3\ \text{kN}$$

$$\text{Factor of safety against sliding} = \frac{\text{Force resisting sliding}}{\text{Force causing sliding}} \tag{16.13}$$

$$\text{FOS} = 154.3/49.68 = 3.1$$

Overturning

Taking moments about the toe:

$$\begin{aligned}\text{Overturning moment} &= \text{Pa} \times 4/3 + \text{Pq} \times 4/2 \\ &= 38.88 \times (4/3) + 10.8 \times (4/2) \\ &= 51.84 + 21.6 \\ &= 73.44\ \text{kN·m}\end{aligned}$$

Restoring moment:

Stem	$34.56 \times 0.7 = 24.2$ kN·m
Base	$28.8 \times 1.5 = 43.2$ kN·m
Soil on heel	$136.08 \times 1.95 = 265$ kN·m
Surcharge	$21 \times 1.95 = 41.0$ kN·m

Total restoring moment = 373.8 kN·m

$$\text{Factor of safety against overturning} = \frac{\text{Restoring moment}}{\text{Overturning moment}} \tag{16.14}$$

$$\text{FOS} = 373.8/73.44 = 5.1$$

Bearing capacity

Consider moments about the toe. If we consider that the vertical reaction R acts at a distance y from the toe:

$$\text{Rvy} = \text{Restoring moment} - \text{Overturning moment} \tag{16.15}$$

$$\text{Rvy} = 373.8 - 73.44 = 300.4$$

Thus:

$$y = 300.4/220.4 = \text{Rvy}/\text{Total vertical load Rv} \tag{16.16}$$

y = 1.36 m from the toe

This is within the middle third of the base. From this we can calculate the eccentricity of Rv provided by 3/2 – 1.36 (y calculated above): e = 0.14 m.

$$\text{Bearing pressure} = \frac{\text{Rv}}{\text{B}}\left(1+\frac{6\text{e}}{\text{B}}\right) \tag{16.17}$$

$$= 220.4/3(1+6\times 0.14/3)$$

Thus the bearing pressure is 94 kN/m^2 < 200 kN/m^2, where the factor of safety becomes 200.0/94.0 = 2.12.

Let us now consider the same design using BS 8002. The value for the design angle of shearing resistance is provided by

Using BS 8002:

$$\phi = \frac{\tan 35^\circ}{1.2} \tag{16.18}$$

where 1.2 is the factor of safety. Thus Φ = 29° and from equation (16.2) Ka becomes 0.35.

Active thrust

From equations (16.6) and (16.7):

Active thrust from soil Pa = ½Kaγh^2 = ½ × 0.35 × 18 × 16	= 50.4 kN
Active thrust from surcharge Ps = KaQh = 0.35 × 10 × 4	= 14.0 kN
Total PA	**= 64.4 kN**

Sliding

$$\text{Total force causing sliding} = \text{Total active thrust} = 64.4\,\text{kN}.$$

$$\text{Design}\tan\delta = 0.75(\text{design}\tan 29^\circ) \tag{16.19}$$

Design tan δ = 0.42.
The force resisting sliding, equation (16.12) = Rv tan δ = 220.4 × 0.42 = 92.4.
Since 92.4 > 64.4 kN, the design is satisfactory.

Overturning

Taking moments about the toe:

$$\begin{aligned}\text{Overturning moment} &= \text{Pa}\times 4/3 + \text{Ps}\times 4/2\\ &= 50.4\times(4/3) + 14.0\times(4/2)\\ &= 67.2 + 28.0\\ &= 95.2\,\text{kN·m}\end{aligned}$$

Restoring moment:

Stem	$34.56 \times 0.7 = 24.2\,\text{kN·m}$
Base	$28.8 \times 1.5 = 43.2\,\text{kN·m}$
Soil on heel	$136.1 \times 1.95 = 265.4\,\text{kN·m}$
Surcharge	$21 \times 1.95 = 41.0\,\text{kN·m}$

Total restoring moment = 373.8 kN·m

Now, since 95.2 kN·m (the overturning moment) < 373.8 kN·m (the restoring moment), the design criteria are satisfied.

Bearing capacity

Consider moments about the toe. If we consider that the vertical reaction R acts at a distance y from the toe. From equation (16.14):

$$Rvy = 373.8 - 95.2 = 278.6$$

using equation (16.15). Thus y = 278.6/220.4 = 1.26 m from the toe (see Figure 16.5). This is within the middle third of the base.

From this we can calculate the eccentricity of Rv provided by 3/2 – 1.26 (y calculated as above): e = 0.24 m.

Using equation (16.15) for the bearing capacity:

$$\text{Bearing pressure} = \frac{Rv}{B}\left(1 + \frac{6e}{B}\right) = 220.4/3(1 + 6 \times 0.24/3)$$

Thus the bearing pressure is 108.7 kN/m^2 < 200 kN/m^2 and the design criteria have been met.

Now consider the design using Eurocode 7. First use the EQU limit state design approach.

Using EUROCODE 7:

Overturning

From EC7 we have the following safety parameters:

$$\gamma_{G;dst} = 1.1;\ \ \gamma_{G;stb} = 0.9;\ \ \gamma_{Q;dst} = 1.5;\ \ \gamma_{\phi} = 1.25$$

Design material properties

The value for the design angle of shearing resistance is provided by

$$\phi_d = \frac{\tan 35^\circ}{1.25} \tag{16.20}$$

where 1.25 is the factor of safety. Thus $\Phi = 28^\circ$.

Table 16.2: Vertical loads (Rv) calculated using Eurocode 7

Stem of wall $G_{stem\ d}$	3.6 × 0.4 × 24 × 0.9	31.1
Base $G_{Base\ d}$	0.4 × 3 × 24 × 0.9	25.9
Soil on heel $G_{Heel\ d}$	18 × 2.1 × 3.6 × 0.9	122.5
	Total Rv = 179.5 kN	

From equation (16.2):

$$Ka = \frac{1-\sin\phi_d}{1+\sin\phi_d} = \frac{1-\sin 28}{1+\sin 28} = 0.36$$

Design action

The weights of the wall and soil on the heel are considered a favourable action.

Stem $G_{stem\ d} = 0.4 \times 4.6 \times \gamma$
Base $G_{Base\ d}$
Soil on heal $G_{Heel\ d}$

See Table 16.2.

The thrust from the active earth pressure and the thrust from the surcharge are both considered unfavourable actions.

Using equations (16.6) and (16.7):

$$\text{Active thrust from soil } Pa = \tfrac{1}{2}Ka\gamma h^2\gamma_{G;dst} = \tfrac{1}{2} \times 0.36 \times 18 \times 16 \times 1.1 = 57.0\,\text{kN}$$

$$\text{Active thrust from surcharge } Ps = KaQh\gamma_{Q;dst} = 0.36 \times 10 \times 4 \times 1.5 = 21.6\,\text{kN}$$

Design effects of actions and design resistance

Taking moments about the toe:

$$\begin{aligned}\text{Overturning or destabilising moment} &= Pa \times 4/3 + Pq \times 4/2 \\ &= 57.0 \times (4/3) + 21.6 \times (4/2) \\ &= 76.0 + 43.2 \\ &= 119.2\,\text{kN·m}\end{aligned}$$

Restoring or stabilising moment:

Stem	31.1 × 0.7 = 21.8 kN·m
Base	25.9 × 1.5 = 38.9 kN·m
Soil on heel	122.5 × 1.95 = 238.9 kN·m
Total restoring moment	299.6 kN·m

The EQU limit state requirement has been met, since Mdst < Mstb: 119.2 kN·m < 299.6 kN·m.

Now use the GEO limit state design approach.

Table 16.3: Vertical loads (Rv) for sliding using Eurocode 7

Stem of wall $G_{stem\ d}$	3.6 × 0.4 × 24 × 1.0	34.6
Base $G_{Base\ d}$	0.4 × 3 × 24 × 1.0	28.8
Soil on heel $G_{Heel\ d}$	18 × 2.1 × 3.6 × 1.0	136.1
	Total Rv = 199.5 kN	

Sliding and bearing

$$\text{Combination 1 from EC7 partial factor sets A1} + \text{M1} + \text{R1} \tag{16.21}$$

$$\gamma_{G;dst} = 1.35\gamma_{G;stb} = 1.0\gamma_{Q;dst} = 1.5\gamma_{\phi} = 1.0$$

Material properties

$\phi_d = 35°$. Thus Ka = 0.27, as calculated from equation (16.2).

Consider the design actions

In sliding the weight of the wall is a favourable action, thus the factors of safety are applied as seen in Table 16.3.

The thrust and surcharge on the wall are an unfavourable action.

Applying equations (16.6) and (16.7):

$$\text{Active thrust from soil Pad} = \tfrac{1}{2}\text{Ka}\gamma h^2\gamma_{G;dst} = \tfrac{1}{2} \times 0.27 \times 18 \times 16 \times 1.35 = 52.5\,\text{kN}$$

$$\text{Active thrust from surcharge Psd} = \text{KaQh}\gamma_{Q;dst} = 0.27 \times 10 \times 4 \times 1.5 = 16.2\,\text{kN}$$

Sliding

Using equation (16.8):

Total horizontal thrust = 52.5 + 16.2 = 68.7 kN

Using equations (16.12) and (16.19) to determine tan δ:

Design resistance = Rv tan δ = 199.5 × tan 35 = 139.7 kN

Thus 139.7 kN > 68.7 kN and the design criteria are satisfied.

Bearing

Destabilising moment:

$$\text{Msd} = \text{Mpa} + \text{Msd} \tag{16.22}$$

$$\text{Msd} = 52.5 \times 1.67 + 16.2 \times 2.0 = 87.7 + 32.4 = 120.1\,\text{kN·m}$$

Stabilising moment:

$$\text{Mstb} = \text{Mstem} + \text{Mbase} + \text{Mheal} \tag{16.23}$$

$$\text{Mstb} = 34.6 \times 0.7 + 28.8 \times 1.5 + 136.1 \times 1.95 = 332.8\,\text{kN·m}$$

Applying equation (16.16):

$$\text{Resultant acts at a distance y from the toe Rvdy} = \frac{332.8-120.1}{199.5} = 1.066\ \text{m}$$

This lies within the middle third of the base.

Eccentricity e = 1.5 – 1.066 = 0.43 m

Rv is an unfavourable permanent action: 199.5 × $\gamma_{G;dst}$ (1.35) = 269.3 kN
Applying equation (16.17):

$$\text{Bearing pressure} = \frac{Rv}{B}\left(1+\frac{6e}{B}\right) = 269.3/3(1+6\times 0.43/3)$$

Bearing pressure = 167 kN/m^2 < 200 kN/m^2, therefore the design is satisfactory.

Combination of two partial factor sets A2 + M1 + R1:

$$\gamma_{G;dst} = 1.0\gamma_{G;stb} = 1.0\gamma_{Q;dst} = 1.3\gamma_{\phi} = 125\,\gamma_C = 1.25$$

Material properties

$\phi_d = 35°$. Thus from equation (16.2), Ka = 0.27.

Design actions

In sliding the weight of the wall is a favourable action, thus the factors of safety are applied as already seen in Table 16.3.

The value for the design angle of shearing resistance is provided by $\phi_d = \frac{\tan 35°}{1.25}$ where 1.25 is the factor of safety. Thus Φ = 28°.

Applying equation (16.2):

$$Ka\frac{1-\sin\phi_d}{1+\sin\phi_d} = \frac{1-\sin 28}{1+\sin 28} = 0.36$$

The thrust and surcharge on the wall are an unfavourable action.
Applying equations (16.6) and (16.7):

$$\text{Active thrust from soil Pad} = \tfrac{1}{2}Ka\gamma h^2\gamma_{G;dst} = \tfrac{1}{2}\times 0.36\times 18\times 16\times 1.0 = 51.8\ \text{kN}$$

$$\text{Active thrust from surcharge Psd} = KaQh\,\gamma_{Q;dst} = 0.36\times 10\times 4\times 1.3 = 18.7\ \text{kN}$$

Sliding

Rvd = 199.5 kN
Total horizontal thrust = 51.8 + 18.7 = 70.5 kN
Using equations (16.12) and (16.19):

$$\text{Design resistance} = \text{Rv}\tan\delta = 199.5\times\tan 28 = 106.1\ \text{kN}$$

Thus 106.1 kN > 70.5 kN and the design criteria are satisfied.

Bearing

Destabilising moment using equation (16.22):

$$Msd = Mpa + Msd = 51.8 \times 1.67 + 18.7 \times 2.0 = 86.5 + 37.4 = 123.9\,kN{\cdot}m$$

Stabilising moment using equation (16.23):

$$Mstb = 34.6 \times 0.7 + 28.8 \times 1.5 + 136.1 \times 1.95 = 332.8\,kN{\cdot}m$$

Applying equation (16.16):

$$\text{Resultant acts at a distance y from the toe } Rvdy = \frac{332.8 - 123.9}{199.5} = 1.047\,m$$

This lies within the middle third of the base.
Eccentricity e = 1.5 – 1.047 = 0.453 m
Rv is an unfavourable permanent action 199.5 × $\gamma_{G;dst}$ (1.0) = 199.5 kN
From equation (16.17):

$$\text{Bearing pressure} = \frac{Rv}{B}\left(1 + \frac{6e}{B}\right) = 199.5/3(1 + 6 \times 0.453/3)$$

Bearing pressure = 126.7 kN/m^2 < 200 kN/m^2 therefore the design is adequate.

Specialist advice

The design of retaining structures is a specialist activity, and many of the failures encountered are through poor design or understanding of the parameters that can alter the structural stability of a wall. We have already discussed the impact of water behind retaining walls, but the type of soil also has a serious impact on walls and a poor understanding or a change in soil conditions can mean the difference between the success and failure of a wall.

The design criteria of walls are an important consideration, and there are limiting dimensions for reinforced walls. In the example below we can extract the following elements of the design:

$$\text{Limiting span/Effective depth ratio; } ratio_{max} = \mathbf{18.00} \tag{16.24}$$

$$\text{Actual span/Effective depth ratio; } ratio_{act} = (h_{stem} + d_{stem}/2)/d_{stem} = \mathbf{11.03} \tag{16.25}$$

For example, walls constructed using hollow blocks and in-filled with reinforcement and concrete could easily fail to meet this standard and therefore are prone to failure.

An example of a reinforced block cavity wall can be seen below, using 10 N/mm^2 concrete blocks. The design has been based on the retention of a granular material and passes the checks.

These calculations have been undertaken using structural engineering software and we gratefully acknowledge Tekla (UK) Ltd for their approval in the use of this software.

RETAINING WALL ANALYSIS (BS 8002: 1994).

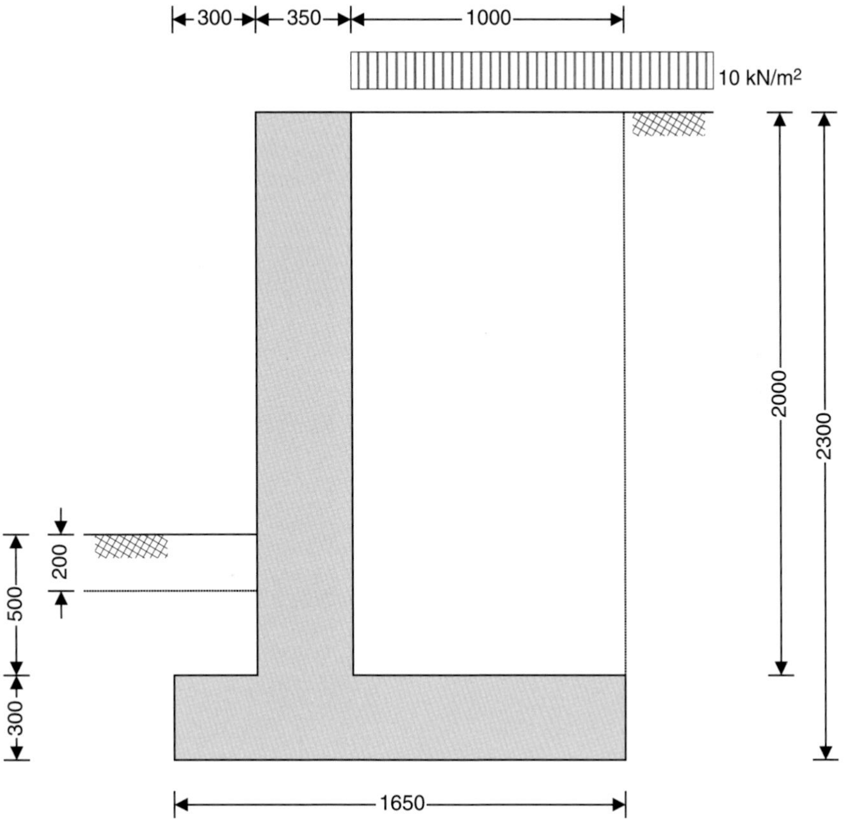

Wall details

Retaining wall type;	**Unpropped cantilever**
Height of retaining wall stem;	h_{stem} = **2000** mm
Thickness of wall stem;	t_{wall} = **350** mm
Length of toe;	l_{toe} = **300** mm
Length of heel;	l_{heel} = **1000** mm
Overall length of base;	$l_{base} = l_{toe} + l_{heel} + t_{wall}$ = **1650** mm
Thickness of base;	t_{base} = **300** mm
Depth of downstand;	d_{ds} = **0** mm
Position of downstand;	l_{ds} = **1200** mm
Thickness of downstand;	t_{ds} = **300** mm
Height of retaining wall;	$h_{wall} = h_{stem} + t_{base} + d_{ds}$ = **2300** mm
Depth of cover in front of wall;	d_{cover} = **500** mm
Depth of unplanned excavation;	d_{exc} = **200** mm
Height of ground water behind wall;	h_{water} = **0** mm
Height of saturated fill above base;	$h_{sat} = \max(h_{water} - t_{base} - d_{ds}, 0\text{ mm})$ = **0** mm
Density of wall construction;	γ_{wall} = **23.6** kN/m^3
Density of base construction;	γ_{base} = **23.6** kN/m^3
Angle of rear face of wall;	α = **90.0** deg
Angle of soil surface behind wall;	β = **0.0** deg
Effective height at virtual back of wall;	$h_{eff} = h_{wall} + l_{heel} \times \tan(\beta)$ = **2300** mm

Retained material details

Mobilisation factor;	M = **1.5**
Moist density of retained material;	γ_m = **18.0** kN/m^3
Saturated density of retained material;	γ_s = **21.0** kN/m^3
Design shear strength;	ϕ' = **24.2** deg
Angle of wall friction;	δ = **0.0** deg

Base material details

Moist density;	γ_{mb} = **18.0** kN/m^3
Design shear strength;	ϕ'_b = **24.2** deg
Design base friction;	δ_b = **18.6** deg
Allowable bearing pressure;	$P_{bearing}$ = **100** kN/m^2

Using Coulomb theory

Active pressure coefficient for retained material

$K_a = \sin(\alpha+\phi')^2/\left(\sin(\alpha)^2 \times \sin(\alpha-\delta) \times [1+\surd(\sin(\phi'+\delta) \times \sin(\phi'-\beta)/(\sin(\alpha-\delta) \times \sin(\alpha+\beta)))]^2\right) = \mathbf{0.419}$

Passive pressure coefficient for base material

$K_p = \sin(90-\phi'_b)^2/\left(\sin(90-\delta_b) \times [1-\surd(\sin(\phi'_b+\delta_b) \times \sin(\phi'_b)/(\sin(90+\delta_b)))]^2\right) = \mathbf{4.187}$

At-rest pressure

At-rest pressure for retained material;	$K_0 = 1 - \sin(\phi') = \mathbf{0.590}$

Loading details

Surcharge load on plan;	Surcharge = **10.0** kN/m^2
Applied vertical dead load on wall;	W_{dead} = **0.0** kN/m
Applied vertical live load on wall;	W_{live} = **0.0** kN/m
Position of applied vertical load on wall;	l_{load} = **0** mm
Applied horizontal dead load on wall;	F_{dead} = **0.0** kN/m
Applied horizontal live load on wall;	F_{live} = **0.0** kN/m
Height of applied horizontal load on wall;	h_{load} = **0** mm

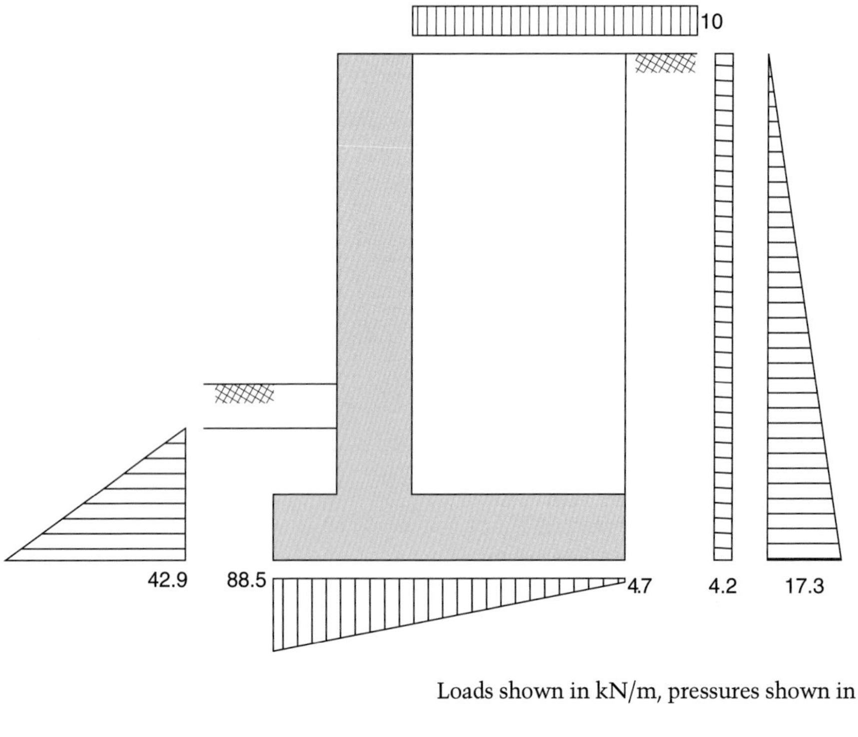

Loads shown in kN/m, pressures shown in kN/m^2

Vertical forces on wall

Wall stem; $w_{wall} = h_{stem} \times t_{wall} \times \gamma_{wall}$ = **16.5** kN/m

Wall base; $w_{base} = l_{base} \times t_{base} \times \gamma_{base}$ = **11.7** kN/m

Surcharge; w_{sur} = Surcharge × l_{heel} = **10** kN/m

Moist backfill to top of wall; $w_{m\_w} = l_{heel} \times (h_{stem} - h_{sat}) \times \gamma_m$ = **36** kN/m

Soil in front of wall; $w_p = l_{toe} \times d_{cover} \times \gamma_{mb}$ = **2.7** kN/m

Total vertical load; $W_{total} = w_{wall} + w_{base} + w_{sur} + w_{m\_w} + w_p$ = **76.9** kN/m

Horizontal forces on wall

Surcharge; $F_{sur} = K_a \times$ Surcharge $\times h_{eff}$ = **9.6** kN/m

Moist backfill above water table; $F_{m\_a} = 0.5 \times K_a \times \gamma_m \times (h_{eff} - h_{water})^2$ = **19.9** kN/m

Total horizontal load; $F_{total} = F_{sur} + F_{m\_a}$ = **29.6** kN/m

Calculate stability against sliding

Passive resistance of soil in front of wall; $F_p = 0.5 \times K_p \times \cos(\delta_b) \times (d_{cover} + t_{base} + d_{ds} - d_{exc})^2 \times \gamma_{mb}$ = **12.9** kN/m

Resistance to sliding; $F_{res} = F_p + (W_{total} - w_{sur} - w_p) \times \tan(\delta_b)$ = **34.5** kN/m

PASS – Resistance force is greater than sliding force

Overturning moments

Surcharge; $M_{sur} = F_{sur} \times (h_{eff} - 2 \times d_{ds})/2$ = **11.1** kNm/m

Moist backfill above water table; $M_{m\_a} = F_{m\_a} \times (h_{eff} + 2 \times h_{water} - 3 \times d_{ds})/3$ = **15.3** kNm/m

Total overturning moment; $M_{ot} = M_{sur} + M_{m\_a}$ = **26.3** kNm/m

Restoring moments

Wall stem; $M_{wall} = w_{wall} \times (l_{toe} + t_{wall}/2)$ = **7.8** kNm/m

Wall base; $M_{base} = w_{base} \times l_{base}/2$ = **9.6** kNm/m

Moist backfill; $M_{m\_r} = (w_{m\_w} \times (l_{base} - l_{heel}/2) + w_{m\_s} \times (l_{base} - l_{heel}/3))$ = **41.4** kNm/m

Total restoring moment; $M_{rest} = M_{wall} + M_{base} + M_{m\_r}$ = **58.9** kNm/m

Check stability against overturning

Total overturning moment; M_{ot} = **26.3** kNm/m

Total restoring moment; M_{rest} = **58.9** kNm/m

PASS – Restoring moment is greater than overturning moment

Check bearing pressure

Surcharge; $M_{sur\_r} = w_{sur} \times (l_{base} - l_{heel}/2)$ = **11.5** kNm/m

Soil in front of wall; $M_{p\_r} = w_p \times l_{toe}/2$ = **0.4** kNm/m

Total moment for bearing; $M_{total} = M_{rest} - M_{ot} + M_{sur\_r} + M_{p\_r}$ = **44.4** kNm/m

Total vertical reaction; $R = W_{total}$ = **76.9** kN/m

Distance to reaction; $x_{bar} = M_{total}/R$ = **578** mm

Eccentricity of reaction; $e = \mathrm{abs}((l_{base}/2) - x_{bar})$ = **247** mm

Reaction acts within middle third of base

Bearing pressure at toe; $p_{toe} = (R/l_{base}) + (6 \times R \times e/l_{base}^2)$ = **88.5** kN/m^2

Bearing pressure at heel; $p_{heel} = (R/l_{base}) - (6 \times R \times e/l_{base}^2)$ = **4.7** kN/m^2

PASS – Maximum bearing pressure is less than allowable bearing pressure

RETAINING WALL DESIGN (BS 8002: 1994)

Ultimate limit state load factors

Dead load factor; γ_{f_d} = **1.4**

Live load factor; γ_{f_l} = **1.6**

Earth and water pressure factor; γ_{f_e} = **1.4**

Factored vertical forces on wall

Wall stem;	$w_{wall\_f} = \gamma_{f\_d} \times h_{stem} \times t_{wall} \times \gamma_{wall}$ = **23.1** kN/m
Wall base;	$w_{base\_f} = \gamma_{f\_d} \times l_{base} \times t_{base} \times \gamma_{base}$ = **16.4** kN/m
Surcharge;	$w_{sur\_f} = \gamma_{f\_l} \times \text{Surcharge} \times l_{heel}$ = **16** kN/m
Moist backfill to top of wall;	$w_{m\_w\_f} = \gamma_{f\_d} \times l_{heel} \times (h_{stem} - h_{sat}) \times \gamma_m$ = **50.4** kN/m
Soil in front of wall;	$w_{p\_f} = \gamma_{f\_d} \times l_{toe} \times d_{cover} \times \gamma_{mb}$ = **3.8** kN/m
Total vertical load;	$W_{total\_f} = w_{wall\_f} + w_{base\_f} + w_{sur\_f} + w_{m\_w\_f} + w_{p\_f}$ = **109.7** kN/m

Factored horizontal at-rest forces on wall

Surcharge;	$F_{sur\_f} = \gamma_{f\_l} \times K_0 \times \text{Surcharge} \times h_{eff}$ = **21.7** kN/m
Moist backfill above water table;	$F_{m\_a\_f} = \gamma_{f\_e} \times 0.5 \times K_0 \times \gamma_m \times (h_{eff} - h_{water})^2$ = **39.3** kN/m
Total horizontal load;	$F_{total\_f} = F_{sur\_f} + F_{m\_a\_f}$ = **61** kN/m
Passive resistance of soil in front of wall;	$F_{p\_f} = \gamma_{f\_e} \times 0.5 \times K_p \times \cos(\delta_b) \times (d_{cover} + t_{base} + d_{ds} - d_{exc})^2 \times \gamma_{mb}$ = **18** kN/m

Factored overturning moments

Surcharge;	$M_{sur\_f} = F_{sur\_f} \times (h_{eff} - 2 \times d_{ds}) / 2$ = **25** kNm/m
Moist backfill above water table;	$M_{m\_a\_f} = F_{m\_a\_f} \times (h_{eff} + 2 \times h_{water} - 3 \times d_{ds}) / 3$ = **30.2** kNm/m
Total overturning moment;	$M_{ot\_f} = M_{sur\_f} + M_{m\_a\_f}$ = **55.1** kNm/m

Restoring moments

Wall stem;	$M_{wall\_f} = w_{wall\_f} \times (l_{toe} + t_{wall} / 2)$ = **11** kNm/m
Wall base;	$M_{base\_f} = w_{base\_f} \times l_{base} / 2$ = **13.5** kNm/m
Surcharge;	$M_{sur\_r\_f} = w_{sur\_f} \times (l_{base} - l_{heel} / 2)$ = **18.4** kNm/m
Moist backfill;	$M_{m\_r\_f} = (w_{m\_w\_f} \times (l_{base} - l_{heel} / 2) + w_{m\_s\_f} \times (l_{base} - l_{heel} / 3))$ = **58** kNm/m
Soil in front of wall;	$M_{p\_r\_f} = w_{p\_f} \times l_{toe} / 2$ = **0.6** kNm/m
Total restoring moment;	$M_{rest\_f} = M_{wall\_f} + M_{base\_f} + M_{sur\_r\_f} + M_{m\_r\_f} + M_{p\_r\_f}$ = **101.4** kNm/m

Check stability against overturning

Total overturning moment;	M_{ot} = **26.3** kNm/m
Total restoring moment;	M_{rest} = **58.9** kNm/m

PASS – Restoring moment is greater than overturning moment

Factored bearing pressure

Total moment for bearing;	$M_{total\_f} = M_{rest\_f} - M_{ot\_f}$ = **46.3** kNm/m
Total vertical reaction;	$R_f = W_{total\_f}$ = **109.7** kN/m
Distance to reaction;	$x_{bar\_f} = M_{total\_f} / R_f$ = **422** mm
Eccentricity of reaction;	$e_f = \text{abs}((l_{base} / 2) - x_{bar\_f})$ = **403** mm

Reaction acts outside middle third of base

Bearing pressure at toe;	$p_{toe\_f} = R_f / (1.5 \times x_{bar\_f})$ = **173.2** kN/m^2
Bearing pressure at heel;	p_{heel_f} = 0 kN/m^2 = **0** kN/m^2
Rate of change of base reaction;	rate = $p_{toe\_f} / (3 \times x_{bar\_f})$ = **136.83** kN/m^2/m
Bearing pressure at stem/toe;	$p_{stem\_toe\_f} = \max(p_{toe\_f} - (\text{rate} \times l_{toe}), 0\ \text{kN/m}^2)$ = **132.2** kN/m^2
Bearing pressure at mid stem;	$p_{stem\_mid\_f} = \max(p_{toe\_f} - (\text{rate} \times (l_{toe} + t_{wall}/2)), 0\ \text{kN/m}^2)$ = **108.2** kN/m^2
Bearing pressure at stem/heel;	$p_{stem\_heel\_f} = \max(p_{toe\_f} - (\text{rate} \times (l_{toe} + t_{wall})), 0\ \text{kN/m}^2)$ = **84.3** kN/m^2

Design of reinforced concrete retaining wall toe (BS 8002: 1994)

Material properties

Characteristic strength of concrete;	f_{cu} = **40** N/mm^2
Characteristic strength of reinforcement;	f_y = **500** N/mm^2

Base details

Minimum area of reinforcement; $k = \mathbf{0.13\%}$

Cover to reinforcement in toe; $c_{toe} = \mathbf{30}$ mm

Calculate shear for toe design

Shear from bearing pressure; $V_{toe\_bear} = (p_{toe\_f} + p_{stem\_toe\_f}) \times l_{toe} / 2 = \mathbf{45.8}$ kN/m

Shear from weight of base; $V_{toe\_wt\_base} = \gamma_{f\_d} \times \gamma_{base} \times l_{toe} \times t_{base} = \mathbf{3}$ kN/m

Shear from weight of soil; $V_{toe\_wt\_soil} = w_{p\_f} - (\gamma_{f\_d} \times \gamma_m \times l_{toe} \times d_{exc}) = \mathbf{2.3}$ kN/m

Total shear for toe design; $V_{toe} = V_{toe\_bear} - V_{toe\_wt\_base} - V_{toe\_wt\_soil} = \mathbf{40.6}$ kN/m

Calculate moment for toe design

Moment from bearing pressure; $M_{toe\_bear} = (2 \times p_{toe\_f} + p_{stem\_mid\_f}) \times (l_{toe} + t_{wall} /2)^2/ 6 = \mathbf{17.1}$ kNm/m

Moment from weight of base; $M_{toe\_wt\_base} = (\gamma_{f\_d} \times \gamma_{base} \times t_{base} \times (l_{toe} + t_{wall} / 2)^2/ 2) = \mathbf{1.1}$ kNm/m

Moment from weight of soil; $M_{toe\_wt\_soil} = (w_{p\_f} - (\gamma_{f\_d} \times \gamma_m \times l_{toe} \times d_{exc})) \times (l_{toe} + t_{wall}) / 2 = \mathbf{0.7}$ kNm/m

Total moment for toe design; $M_{toe} = M_{toe\_bear} - M_{toe\_wt\_base} - M_{toe\_wt\_soil} = \mathbf{15.2}$ kNm/m

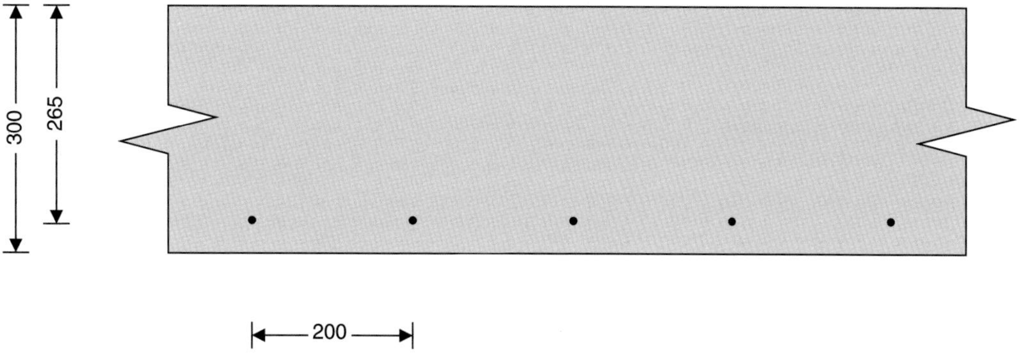

Check toe in bending

Width of toe; $b = \mathbf{1000}$ mm/m

Depth of reinforcement; $d_{toe} = t_{base} - c_{toe} - (\phi_{toe} / 2) = \mathbf{265.0}$ mm

Constant; $K_{toe} = M_{toe} / (b \times d_{toe}^2 \times f_{cu}) = \mathbf{0.005}$

Compression reinforcement is not required

Lever arm;

$z_{toe} = \min(0.5 + \sqrt{(0.25 - (\min(K_{toe}, 0.225)/0.9))}, 0.95) \times d_{toe}$

$z_{toe} = \mathbf{252}$ mm

Area of tension reinforcement required; $A_{s\_toe\_des} = M_{toe} / (0.87 \times f_y \times z_{toe}) = \mathbf{139}$ mm^2/m

Minimum area of tension reinforcement; $A_{s\_toe\_min} = k \times b \times t_{base} = \mathbf{390}$ mm^2/m

Area of tension reinforcement required; $A_{s\_toe\_req} = \mathrm{Max}(A_{s\_toe\_des}, A_{s\_toe\_min}) = \mathbf{390}$ mm^2/m

Reinforcement provided; **A393 mesh**

Area of reinforcement provided; $A_{s\_toe\_prov} = \mathbf{393}$ mm^2/m

PASS – Reinforcement provided at the retaining wall toe is adequate

Check shear resistance at toe

Design shear stress; $v_{toe} = V_{toe} / (b \times d_{toe}) = \mathbf{0.153}$ N/mm^2

Allowable shear stress;

$v_{adm} = \min(0.8 \times \sqrt{(f_{cu}/1\ \mathrm{N/mm^2})}, 5) \times 1\ \mathrm{N/mm^2} = \mathbf{5.000}\ \mathrm{N/mm^2}$

PASS – Design shear stress is less than maximum shear stress

From BS 8110: Part 1: 1997 – Table 3.8

Design concrete shear stress; $v_{c\_toe} = \mathbf{0.434}$ N/mm^2

$v_{toe} < v_{c\_toe}$ – No shear reinforcement required

Design of reinforced concrete retaining wall heel (BS 8002: 1994)

Material properties
Characteristic strength of concrete; f_{cu} = **40** N/mm^2
Characteristic strength of reinforcement; f_y = **500** N/mm^2

Base details
Minimum area of reinforcement; k = **0.13** %
Cover to reinforcement in heel; c_{heel} = **30** mm

Calculate shear for heel design
Shear from bearing pressure; $V_{heel\_bear} = p_{stem\_heel\_f} \times ((3 \times x_{bar\_f}) - l_{toe} - t_{wall})/2$ = **26** kN/m
Shear from weight of base; $V_{heel\_wt\_base} = \gamma_{f\_d} \times \gamma_{base} \times l_{heel} \times t_{base}$ = **9.9** kN/m
Shear from weight of moist backfill; $V_{heel\_wt\_m} = w_{m\_w\_f}$ = **50.4** kN/m
Shear from surcharge; $V_{heel\_sur} = w_{sur\_f}$ = **16** kN/m
Total shear for heel design; $V_{heel} = - V_{heel\_bear} + V_{heel\_wt\_base} + V_{heel\_wt\_m} + V_{heel\_sur}$ = **50.3** kN/m

Calculate moment for heel design
Moment from bearing pressure; $M_{heel\_bear} = p_{stem\_mid\_f} \times ((3 \times x_{bar\_f}) - l_{toe} - t_{wall}/2)^2/6$ = **11.3** kNm/m
Moment from weight of base; $M_{heel\_wt\_base} = (\gamma_{f\_d} \times \gamma_{base} \times t_{base} \times (l_{heel} + t_{wall}/2)^2/2)$ = **6.8** kNm/m
Moment from weight of moist backfill; $M_{heel\_wt\_m} = w_{m\_w\_f} \times (l_{heel} + t_{wall})/2$ = **34** kNm/m
Moment from surcharge; $M_{heel\_sur} = w_{sur\_f} \times (l_{heel} + t_{wall})/2$ = **10.8** kNm/m
Total moment for heel design; $M_{heel} = - M_{heel\_bear} + M_{heel\_wt\_base} + M_{heel\_wt\_m} + M_{heel\_sur}$ = **40.4** kNm/m

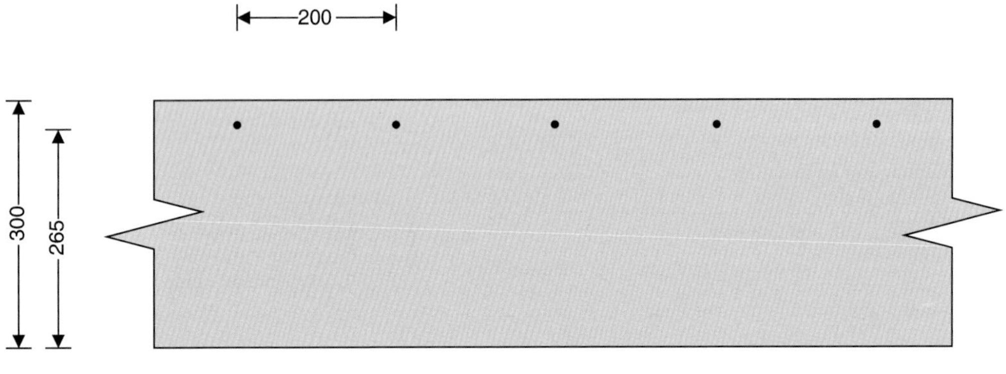

Check heel in bending
Width of heel; b = **1000** mm/m
Depth of reinforcement; $d_{heel} = t_{base} - c_{heel} - (\phi_{heel}/2)$ = **265.0** mm
Constant; $K_{heel} = M_{heel}/(b \times d_{heel}^2 \times f_{cu})$ = **0.014**
Compression reinforcement is not required

Lever arm;
$z_{heel} = \min(0.5 + \sqrt{(0.25 - (\min(K_{heel}, 0.225)/0.9))}, 0.95) \times d_{heel}$
z_{heel} = **252** mm
Area of tension reinforcement required; $A_{s\_heel\_des} = M_{heel}/(0.87 \times f_y \times z_{heel})$ = **369** mm^2/m
Minimum area of tension reinforcement; $A_{s\_heel\_min} = k \times b \times t_{base}$ = **390** mm^2/m
Area of tension reinforcement required; $A_{s\_heel\_req} = \mathrm{Max}(A_{s\_heel\_des}, A_{s\_heel\_min})$ = **390** mm^2/m
Reinforcement provided; **A393 mesh**
Area of reinforcement provided; $A_{s\_heel\_prov}$ = **393** mm^2/m
PASS – Reinforcement provided at the retaining wall heel is adequate

Check shear resistance at heel

Design shear stress; $v_{heel} = V_{heel} / (b \times d_{heel}) =$ **0.190** N/mm^2

Allowable shear stress;

$v_{adm} = \min(0.8 \times \sqrt{(f_{cu}/1\,N/mm^2)}, 5) \times 1\,N/mm^2 = \mathbf{5.000}\,N/mm^2$

PASS – Design shear stress is less than maximum shear stress

From BS 8110: Part 1: 1997 – Table 3.8

Design concrete shear stress; $v_{c\_heel} =$ **0.434** N/mm^2

$v_{heel} < v_{c\_heel}$ – No shear reinforcement required

Design of cavity reinforced masonry retaining wall stem – BS 5628-2: 2000

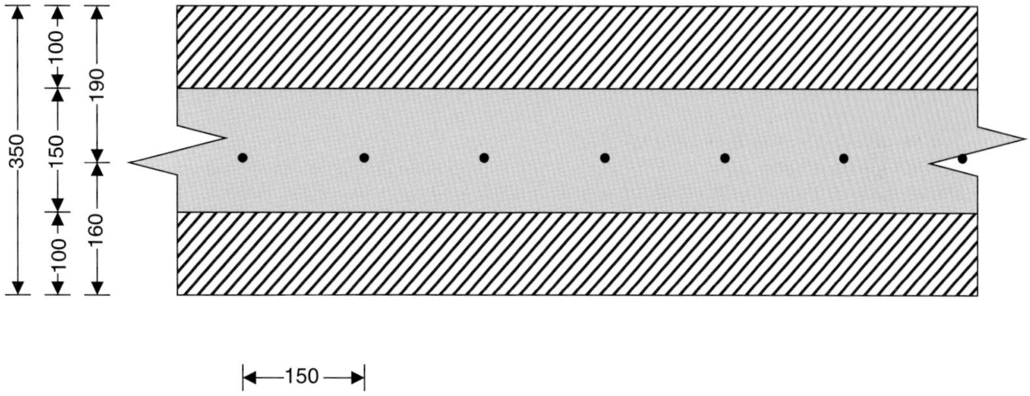

Wall details

Thickness of outer leaf of wall; $t_{outer} =$ **100** mm

Thickness of inner leaf of wall; $t_{inner} =$ **100** mm

Thickness of reinforced cavity; $t_{cavity} = t_{wall} - t_{outer} - t_{inner} =$ **150** mm

Depth of stem reinforcement; $d_{stem} =$ **190** mm

Masonry details

Masonry type; **Aggregate concrete blocks no voids**

Compressive strength of units; $p_{unit} =$ **10.0** N/mm^2

Mortar designation; **(ii)**

Category of manufactoring control of units; **Category II**

Partial safety factor for material strength; $\gamma_{mm} =$ **2.3**

Characteristic compressive strength of masonry

Least horizontal dimension of masonry units; $b_{unit} =$ **100.0** mm

Height of masonry units; $h_{unit} =$ **215.0** mm

Ratio of height to least horizontal dimension; ratio $= h_{unit}/b_{unit} =$ **2.2**

From BS 5628: 2 – Table 3d, mortar (ii)

Characteristic compressive strength; $f_k =$ **8.1** N/mm^2

Factored horizontal at-rest forces on stem

Surcharge; $F_{s\_sur\_f} = \gamma_{f\_l} \times K_0 \times Surcharge \times (h_{eff} - t_{base} - d_{ds}) =$ **18.9** kN/m

Moist backfill above water table; $F_{s\_m\_a\_f} = 0.5 \times \gamma_{f\_e} \times K_0 \times \gamma_m \times (h_{eff} - t_{base} - d_{ds} - h_{sat})^2$ = **29.7** kN/m

Calculate shear for stem design

Shear at base of stem; $V_{stem} = F_{s\_sur\_f} + F_{s\_m\_a\_f} =$ **48.6** kN/m

Calculate moment for stem design

Surcharge; $M_{s\_sur} = F_{s\_sur\_f} \times (h_{stem} + t_{base})/2 =$ **21.7** kNm/m

Moist backfill above water table; $M_{s\_m\_a} = F_{s\_m\_a\_f} \times (2 \times h_{sat} + h_{eff} - d_{ds} + t_{base} / 2)/3 =$ **24.3** kNm/m

Total moment for stem design; $M_{stem} = M_{s\_sur} + M_{s\_m\_a} =$ **46** kNm/m

Check maximum design moment for wall stem

Width of wall; $b = \mathbf{1000}$ mm/m

Maximum design moment; $M_{d\_stem} = 0.4 \times f_k \times b \times d_{stem}^2 / \gamma_{mm} = \mathbf{51.1}$ kNm/m

PASS – Applied moment is less than maximum design moment

Check wall stem in bending

Moment of resistance factor; $Q = M_{stem} / d_{stem}^2 = \mathbf{1.274}$ N/mm²

$Q = 2 \times c \times (1 - c) \times f_k / \gamma_{mm}$

Lever arm factor; $c = \mathbf{0.765}$

Lever arm; $z_{stem} = \min(0.95, c) \times d_{stem} = \mathbf{145.3}$ mm

Area of tension reinforcement required; $A_{s\_stem\_des} = M_{stem} \times \gamma_{ms} / (f_y \times z_{stem}) = \mathbf{728}$ mm²/m

Minimum area of tension reinforcement; $A_{s\_stem\_min} = k \times b \times t_{wall} = \mathbf{455}$ mm²/m

Area of tension reinforcement required; $A_{s\_stem\_req} = \mathrm{Max}(A_{s\_stem\_des}, A_{s\_stem\_min}) = \mathbf{728}$ mm²/m

Reinforcement provided; **12 mm dia.bars @ 150 mm centres**

Area of reinforcement provided; $A_{s\_stem\_prov} = \mathbf{754}$ mm²/m

PASS – Reinforcement provided at the retaining wall stem is adequate

Check shear resistance at wall stem

Design shear stress; $v_{stem} = V_{stem} / (b \times d_{stem}) = \mathbf{0.256}$ N/mm²

Basic characteristic shear strength of masonry; $f_{vbas} = \min[0.35 + (17.5 \times A_{s\_stem\_prov} / (b \times d_{stem})), 0.7] \times 1$ N/mm²

$f_{vbas} = \mathbf{0.419}$ N/mm²

Shear span; $a = M_{stem} / V_{stem} = \mathbf{946.1}$ mm

Characteristic shear strength of masonry; $f_v = \mathrm{Min}(f_{vbas} \times \max(2.5 - 0.25 \times (a / d_{stem}), 1), 1.75\text{N/mm}^2)$

$f_v = \mathbf{0.526}$ N/mm²

Allowable shear stress; $v_{adm} = f_v / \gamma_{mv} = \mathbf{0.263}$ N/mm²

PASS – Design shear stress is less than maximum shear stress

Check limiting dimensions

Limiting span/effective depth ratio; $ratio_{max} = \mathbf{18.00}$

Actual span/effective depth ratio; $ratio_{act} = (h_{stem} + d_{stem} / 2) / d_{stem} = \mathbf{11.03}$

PASS – Span to depth ratio is acceptable

Axial load check

Factored axial load on wall; $N_{wall} = ([t_{wall} \times h_{stem} \times \gamma_{wall} + W_{dead}] \times \gamma_{f\_d}) + (W_{live} \times \gamma_{f\_l}) = \mathbf{23.1}$ kN/m

Limiting axial load; $N_{limit} = 0.1 \times f_k \times t_{wall} = \mathbf{285.0}$ kN/m

Applied axial load may be ignored – calculations valid

Indicative retaining wall reinforcement diagram

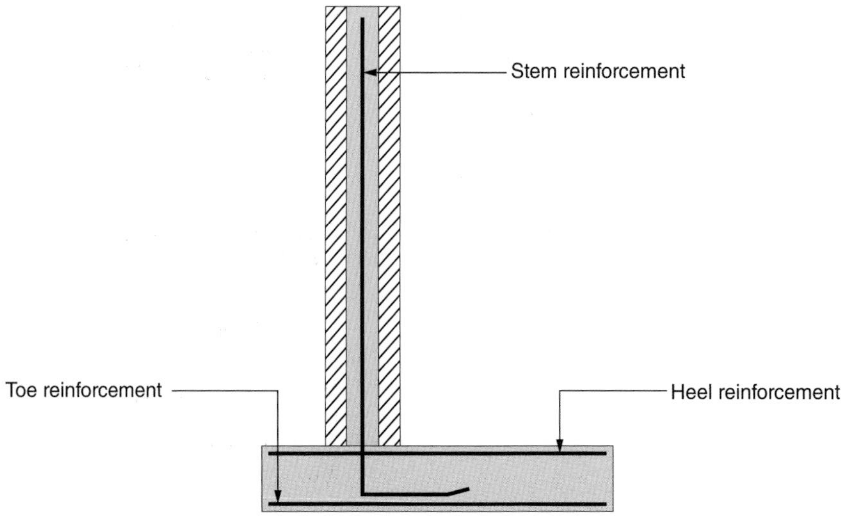

Toe mesh - A393 (393 mm^2/m)
Heel mesh - A393 (393 mm^2/m)
Stem bars - 12 mm dia.@ 150 mm centres (754 mm^2/m)

Let us consider the design of the same wall and make some changes. Assume that the wall retains a saturated material to a height of 1.2 m and that the angle of shearing resistance has decreased to 25° due to the material being different from what was first thought – that is a sandy gravel.

These calculations have been undertaken using structural engineering software and we gratefully acknowledge Tekla (UK) Ltd for their approval in the use of this software.

RETAINING WALL ANALYSIS (BS 8002: 1994).

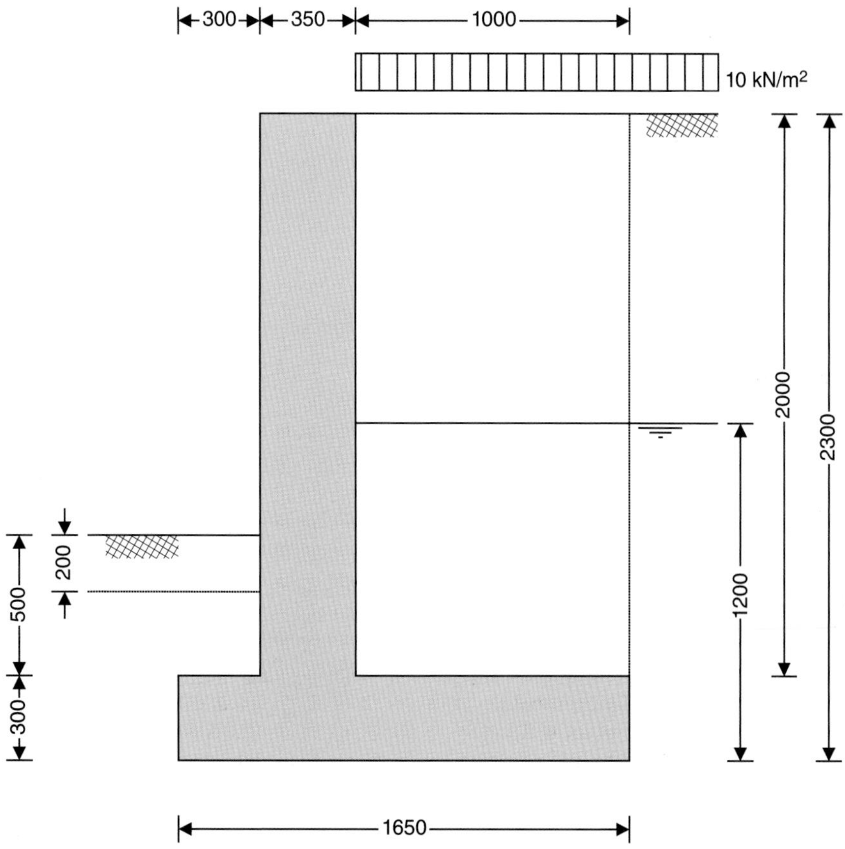

Wall details

Retaining wall type;	**Unpropped cantilever**
Height of retaining wall stem;	h_{stem} = **2000** mm
Thickness of wall stem;	t_{wall} = **350** mm
Length of toe;	l_{toe} = **300** mm
Length of heel;	l_{heel} = **1000** mm
Overall length of base;	$l_{base} = l_{toe} + l_{heel} + t_{wall}$ = **1650** mm
Thickness of base;	t_{base} = **300** mm
Depth of downstand;	d_{ds} = **0** mm
Position of downstand;	l_{ds} = **1200** mm
Thickness of downstand;	t_{ds} = **300** mm
Height of retaining wall;	$h_{wall} = h_{stem} + t_{base} + d_{ds}$ = **2300** mm
Depth of cover in front of wall;	d_{cover} = **500** mm
Depth of unplanned excavation;	d_{exc} = **200** mm
Height of ground water behind wall;	h_{water} = **1200** mm
Height of saturated fill above base;	$h_{sat} = \max(h_{water} - t_{base} - d_{ds}, 0\text{ mm})$ = **900** mm
Density of wall construction;	γ_{wall} = **23.6** kN/m^3
Density of base construction;	γ_{base} = **23.6** kN/m^3
Angle of rear face of wall;	α = **90.0** deg
Angle of soil surface behind wall;	β = **0.0** deg
Effective height at virtual back of wall;	$h_{eff} = h_{wall} + l_{heel} \times \tan(\beta)$ = **2300** mm

Retained material details

Mobilisation factor;	M = **1.5**
Moist density of retained material;	γ_m = **21.0** kN/m^3
Saturated density of retained material;	γ_s = **23.0** kN/m^3
Design shear strength;	ϕ' = **17.3** deg
Angle of wall friction;	δ = **13.1** deg

Base material details

Moist density;	γ_{mb} = **18.0** kN/m^3
Design shear strength;	ϕ'_b = **24.2** deg
Design base friction;	δ_b = **18.6** deg
Allowable bearing pressure;	$P_{bearing}$ = **100** kN/m^2

Using Coulomb theory

Active pressure coefficient for retained material

$$K_a = \sin(\alpha+\phi')^2/\left(\sin(\alpha)^2 \times \sin(\alpha-\delta) \times \left[1+\sqrt{(}\sin(\phi'+\delta) \times \sin(\phi'-\beta)/(\sin(\alpha-\delta) \times \sin(\alpha+\beta)))\right]^2\right) = \mathbf{0.482}$$

Passive pressure coefficient for base material

$$K_p = \sin(90-\phi'_b b)^2/\left(\sin(90-\delta_b) \times \left[1-\sqrt{(}\sin(\phi'_b+\delta_b) \times \sin(\phi'_b)/(\sin(90+\delta_b)))\right]^2\right) = \mathbf{4.187}$$

At-rest pressure

At-rest pressure for retained material;	$K_0 = 1 - \sin(\phi')$ = **0.703**

Loading details

Surcharge load on plan;	Surcharge = **10.0** kN/m^2
Applied vertical dead load on wall;	W_{dead} = **0.0** kN/m
Applied vertical live load on wall;	W_{live} = **0.0** kN/m
Position of applied vertical load on wall;	l_{load} = **0** mm
Applied horizontal dead load on wall;	F_{dead} = **0.0** kN/m
Applied horizontal live load on wall;	F_{live} = **0.0** kN/m
Height of applied horizontal load on wall;	h_{load} = **0** mm

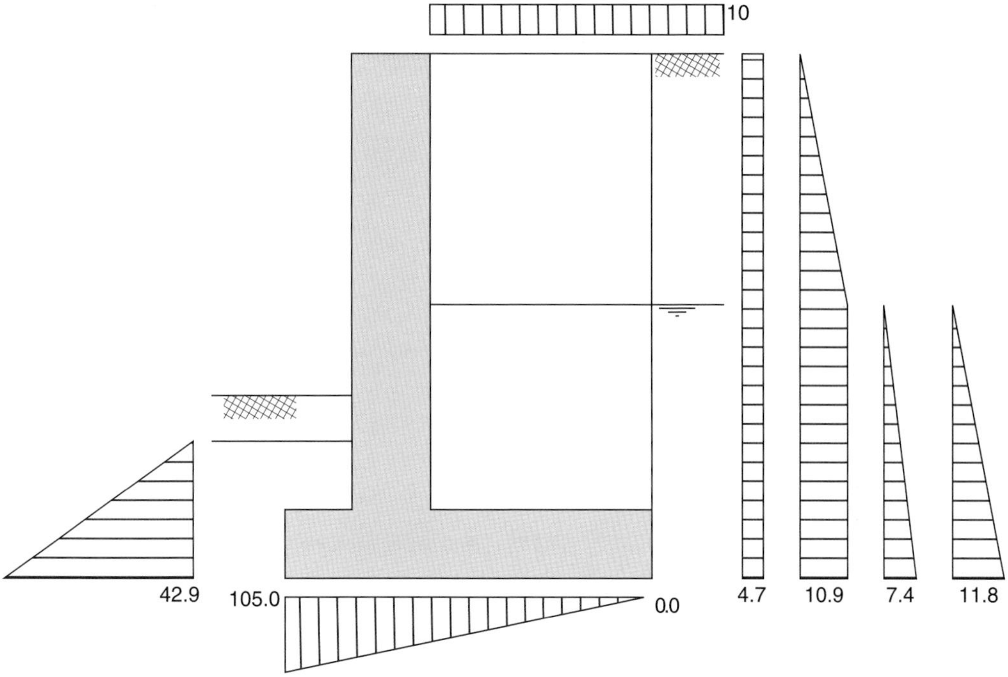

Loads shown in kN/m, pressures shown in kN/m^2

Vertical forces on wall

Wall stem;	$w_{wall} = h_{stem} \times t_{wall} \times \gamma_{wall} =$ **16.5** kN/m
Wall base;	$w_{base} = l_{base} \times t_{base} \times \gamma_{base} =$ **11.7** kN/m
Surcharge;	$w_{sur} = \text{Surcharge} \times l_{heel} =$ **10** kN/m
Moist backfill to top of wall;	$w_{m\_w} = l_{heel} \times (h_{stem} - h_{sat}) \times \gamma_m =$ **23.1** kN/m
Saturated backfill;	$w_s = l_{heel} \times h_{sat} \times \gamma_s =$ **20.7** kN/m
Soil in front of wall;	$w_p = l_{toe} \times d_{cover} \times \gamma_{mb} =$ **2.7** kN/m
Total vertical load;	$W_{total} = w_{wall} + w_{base} + w_{sur} + w_{m\_w} + w_s + w_p$ = **84.7** kN/m

Horizontal forces on wall

Surcharge;	$F_{sur} = K_a \times \cos(90 - \alpha + \delta) \times \text{Surcharge} \times h_{eff} =$ **10.8** kN/m
Moist backfill above water table;	$F_{m\_a} = 0.5 \times K_a \times \cos(90 - \alpha + \delta) \times \gamma_m \times (h_{eff} - h_{water})^2$ = **6** kN/m
Moist backfill below water table;	$F_{m\_b} = K_a \times \cos(90 - \alpha + \delta) \times \gamma_m \times (h_{eff} - h_{water}) \times h_{water} =$ **13** kN/m
Saturated backfill;	$F_s = 0.5 \times K_a \times \cos(90 - \alpha + \delta) \times (\gamma_s - \gamma_{water}) \times h_{water}^2$ = **4.5** kN/m
Water;	$F_{water} = 0.5 \times h_{water}^2 \times \gamma_{water} =$ **7.1** kN/m
Total horizontal load;	$F_{total} = F_{sur} + F_{m\_a} + F_{m\_b} + F_s + F_{water} =$ **41.3** kN/m

Calculate stability against sliding

Passive resistance of soil in front of wall;	$F_p = 0.5 \times K_p \times \cos(\delta_b) \times (d_{cover} + t_{base} + d_{ds} - d_{exc})^2 \times \gamma_{mb} =$ **12.9** kN/m
Resistance to sliding;	$F_{res} = F_p + (W_{total} - w_{sur} - w_p) \times \tan(\delta_b) =$ **37.1** kN/m

FAIL – Sliding force is greater than resisting force

Overturning moments

Surcharge;	$M_{sur} = F_{sur} \times (h_{eff} - 2 \times d_{ds})/2$ = **12.4** kNm/m
Moist backfill above water table;	$M_{m\_a} = F_{m\_a} \times (h_{eff} + 2 \times h_{water} - 3 \times d_{ds})/3$ = **9.3** kNm/m
Moist backfill below water table;	$M_{m\_b} = F_{m\_b} \times (h_{water} - 2 \times d_{ds})/2$ = **7.8** kNm/m
Saturated backfill;	$M_s = F_s \times (h_{water} - 3 \times d_{ds})/3$ = **1.8** kNm/m
Water;	$M_{water} = F_{water} \times (h_{water} - 3 \times d_{ds})/3$ = **2.8** kNm/m
Total overturning moment;	$M_{ot} = M_{sur} + M_{m\_a} + M_{m\_b} + M_s + M_{water}$ = **34.2** kNm/m

Restoring moments

Wall stem;	$M_{wall} = w_{wall} \times (l_{toe} + t_{wall}/2)$ = **7.8** kNm/m
Wall base;	$M_{base} = w_{base} \times l_{base}/2$ = **9.6** kNm/m
Moist backfill;	$M_{m\_r} = (w_{m\_w} \times (l_{base} - l_{heel}/2) + w_{m\_s} \times (l_{base} - l_{heel}/3))$ = **26.6** kNm/m
Saturated backfill;	$M_{s\_r} = w_s \times (l_{base} - l_{heel}/2)$ = **23.8** kNm/m
Total restoring moment;	$M_{rest} = M_{wall} + M_{base} + M_{m\_r} + M_{s\_r}$ = **67.9** kNm/m

Check stability against overturning

Total overturning moment;	M_{ot} = **34.2** kNm/m
Total restoring moment;	M_{rest} = **67.9** kNm/m

PASS – Restoring moment is greater than overturning moment

Check bearing pressure

Surcharge;	$M_{sur\_r} = w_{sur} \times (l_{base} - l_{heel}/2)$ = **11.5** kNm/m
Soil in front of wall;	$M_{p\_r} = w_p \times l_{toe}/2$ = **0.4** kNm/m
Total moment for bearing;	$M_{total} = M_{rest} - M_{ot} + M_{sur\_r} + M_{p\_r}$ = **45.6** kNm/m
Total vertical reaction;	$R = W_{total}$ = **84.7** kN/m
Distance to reaction;	$x_{bar} = M_{total}/R$ = **538** mm
Eccentricity of reaction;	$e = abs((l_{base}/2) - x_{bar})$ = **287** mm

Reaction acts outside middle third of base

Bearing pressure at toe;	$p_{toe} = R / (1.5 \times x_{bar})$ = **105** kN/m^2
Bearing pressure at heel;	p_{heel} = 0 kN/m^2 = **0** kN/m^2

FAIL – Maximum bearing pressure exceeds allowable bearing pressure

RETAINING WALL DESIGN (BS 8002: 1994)

Ultimate limit state load factors

Dead load factor;	γ_{f_d} = **1.4**
Live load factor;	γ_{f_l} = **1.6**
Earth and water pressure factor;	γ_{f_e} = **1.4**

Factored vertical forces on wall

Wall stem;	$w_{wall\_f} = \gamma_{f\_d} \times h_{stem} \times t_{wall} \times \gamma_{wall}$ = **23.1** kN/m
Wall base;	$w_{base\_f} = \gamma_{f\_d} \times l_{base} \times t_{base} \times \gamma_{base}$ = **16.4** kN/m
Surcharge;	$w_{sur\_f} = \gamma_{f\_l} \times$ Surcharge $\times l_{heel}$ = **16** kN/m
Moist backfill to top of wall;	$w_{m\_w\_f} = \gamma_{f\_d} \times l_{heel} \times (h_{stem} - h_{sat}) \times \gamma_m$ = **32.3** kN/m
Saturated backfill;	$w_{s\_f} = \gamma_{f\_d} \times l_{heel} \times h_{sat} \times \gamma_s$ = **29** kN/m
Soil in front of wall;	$w_{p\_f} = \gamma_{f\_d} \times l_{toe} \times d_{cover} \times \gamma_{mb}$ = **3.8** kN/m
Total vertical load;	$W_{total\_f} = w_{wall\_f} + w_{base\_f} + w_{sur\_f} + w_{m\_w\_f} + w_{s\_f} + w_{p\_f}$ = **120.6** kN/m

Factored horizontal at-rest forces on wall

Surcharge;	$F_{sur\_f} = \gamma_{f\_l} \times K_0 \times$ Surcharge $\times h_{eff}$ = **25.9** kN/m
Moist backfill above water table;	$F_{m\_a\_f} = \gamma_{f\_e} \times 0.5 \times K_0 \times \gamma_m \times (h_{eff} - h_{water})^2$ = **12.5** kN/m
Moist backfill below water table;	$F_{m\_b\_f} = \gamma_{f\_e} \times K_0 \times \gamma_m \times (h_{eff} - h_{water}) \times h_{water}$ = **27.3** kN/m
Saturated backfill;	$F_{s\_f} = \gamma_{f\_e} \times 0.5 \times K_0 \times (\gamma_s - \gamma_{water}) \times h_{water}^2$ = **9.3** kN/m

Water; $F_{water\_f} = \gamma_{f\_e} \times 0.5 \times h_{water}^2 \times \gamma_{water} = \mathbf{9.9}$ kN/m

Total horizontal load; $F_{total\_f} = F_{sur\_f} + F_{m\_a\_f} + F_{m\_b\_f} + F_{s\_f} + F_{water\_f} = \mathbf{84.9}$ kN/m

Passive resistance of soil in front of wall; $F_{p\_f} = \gamma_{f\_e} \times 0.5 \times K_p \times \cos(\delta_b) \times (d_{cover} + t_{base} + d_{ds} - d_{exc})^2 \times \gamma_{mb} = \mathbf{18}$ kN/m

Factored overturning moments

Surcharge; $M_{sur\_f} = F_{sur\_f} \times (h_{eff} - 2 \times d_{ds})/2 = \mathbf{29.7}$ kNm/m

Moist backfill above water table; $M_{m\_a\_f} = F_{m\_a\_f} \times (h_{eff} + 2 \times h_{water} - 3 \times d_{ds})/3 = \mathbf{19.6}$ kNm/m

Moist backfill below water table; $M_{m\_b\_f} = F_{m\_b\_f} \times (h_{water} - 2 \times d_{ds})/2 = \mathbf{16.4}$ kNm/m

Saturated backfill; $M_{s\_f} = F_{s\_f} \times (h_{water} - 3 \times d_{ds})/3 = \mathbf{3.7}$ kNm/m

Water; $M_{water\_f} = F_{water\_f} \times (h_{water} - 3 \times d_{ds})/3 = \mathbf{4}$ kNm/m

Total overturning moment; $M_{ot\_f} = M_{sur\_f} + M_{m\_a\_f} + M_{m\_b\_f} + M_{s\_f} + M_{water\_f} = \mathbf{73.4}$ kNm/m

Restoring moments

Wall stem; $M_{wall\_f} = w_{wall\_f} \times (l_{toe} + t_{wall}/2) = \mathbf{11}$ kNm/m

Wall base; $M_{base\_f} = w_{base\_f} \times l_{base}/2 = \mathbf{13.5}$ kNm/m

Surcharge; $M_{sur\_r\_f} = w_{sur\_f} \times (l_{base} - l_{heel}/2) = \mathbf{18.4}$ kNm/m

Moist backfill; $M_{m\_r\_f} = (w_{m\_w\_f} \times (l_{base} - l_{heel}/2) + w_{m\_s\_f} \times (l_{base} - l_{heel}/3)) = \mathbf{37.2}$ kNm/m

Saturated backfill; $M_{s\_r\_f} = w_{s\_f} \times (l_{base} - l_{heel}/2) = \mathbf{33.3}$ kNm/m

Soil in front of wall; $M_{p\_r\_f} = w_{p\_f} \times l_{toe}/2 = \mathbf{0.6}$ kNm/m

Total restoring moment; $M_{rest\_f} = M_{wall\_f} + M_{base\_f} + M_{sur\_r\_f} + M_{m\_r\_f} + M_{s\_r\_f} + M_{p\_r\_f} = \mathbf{114}$ kNm/m

Check stability against overturning

Total overturning moment; $M_{ot} = \mathbf{34.2}$ kNm/m

Total restoring moment; $M_{rest} = \mathbf{67.9}$ kNm/m

PASS – Restoring moment is greater than overturning moment

Factored bearing pressure

Total moment for bearing; $M_{total\_f} = M_{rest\_f} - M_{ot\_f} = \mathbf{40.6}$ kNm/m

Total vertical reaction; $R_f = W_{total\_f} = \mathbf{120.6}$ kN/m

Distance to reaction; $x_{bar\_f} = M_{total\_f}/R_f = \mathbf{337}$ mm

Eccentricity of reaction; $e_f = \text{abs}((l_{base}/2) - x_{bar\_f}) = \mathbf{488}$ mm

Reaction acts outside middle third of base

Bearing pressure at toe; $p_{toe\_f} = R_f/(1.5 \times x_{bar\_f}) = \mathbf{238.8}$ kN/m^2

Bearing pressure at heel; $p_{heel\_f} = 0$ kN/m^2 = **0** kN/m^2

Rate of change of base reaction; $rate = p_{toe\_f}/(3 \times x_{bar\_f}) = \mathbf{236.41}$ kN/m^2/m

Bearing pressure at stem/toe; $p_{stem\_toe\_f} = \max(p_{toe\_f} - (rate \times l_{toe}), 0\ \text{kN/m}^2) = \mathbf{167.9}$ kN/m^2

Bearing pressure at mid stem; $p_{stem\_mid\_f} = \max(p_{toe\_f} - (rate \times (l_{toe} + t_{wall}/2)), 0\ \text{kN/m}^2) = \mathbf{126.5}$ kN/m^2

Bearing pressure at stem/heel; $p_{stem\_heel\_f} = \max(p_{toe\_f} - (rate \times (l_{toe} + t_{wall})), 0\ \text{kN/m}^2) = \mathbf{85.1}$ kN/m^2

Design of reinforced concrete retaining wall toe (BS 8002: 1994)

Material properties

Characteristic strength of concrete; $f_{cu} = \mathbf{40}$ N/mm^2

Characteristic strength of reinforcement; $f_y = \mathbf{500}$ N/mm^2

Base details

Minimum area of reinforcement; $k = \mathbf{0.13}$ %

Cover to reinforcement in toe; $c_{toe} = \mathbf{30}$ mm

Calculate shear for toe design

Shear from bearing pressure; $V_{toe\_bear} = (p_{toe\_f} + p_{stem\_toe\_f}) \times l_{toe}/2 = \mathbf{61}$ kN/m

Shear from weight of base; $V_{toe\_wt\_base} = \gamma_{f\_d} \times \gamma_{base} \times l_{toe} \times t_{base} = \mathbf{3}$ kN/m

Shear from weight of soil; $V_{toe\_wt\_soil} = w_{p\_f} - (\gamma_{f\_d} \times \gamma_m \times l_{toe} \times d_{exc}) = \mathbf{2}$ kN/m

Total shear for toe design; $V_{toe} = V_{toe\_bear} - V_{toe\_wt\_base} - V_{toe\_wt\_soil} = \mathbf{56}$ kN/m

Calculate moment for toe design

Moment from bearing pressure; $M_{toe\_bear} = (2 \times p_{toe\_f} + p_{stem\_mid\_f}) \times (l_{toe} + t_{wall}/2)^2/6 = \mathbf{22.7}$ kNm/m

Moment from weight of base; $M_{toe\_wt\_base} = (\gamma_{f\_d} \times \gamma_{base} \times t_{base} \times (l_{toe} + t_{wall}/2)^2/2) = \mathbf{1.1}$ kNm/m

Moment from weight of soil; $M_{toe\_wt\_soil} = (w_{p\_f} - (\gamma_{f\_d} \times \gamma_m \times l_{toe} \times d_{exc})) \times (l_{toe} + t_{wall})/2 = \mathbf{0.7}$ kNm/m

Total moment for toe design; $M_{toe} = M_{toe\_bear} - M_{toe\_wt\_base} - M_{toe\_wt\_soil} = \mathbf{20.9}$ kNm/m

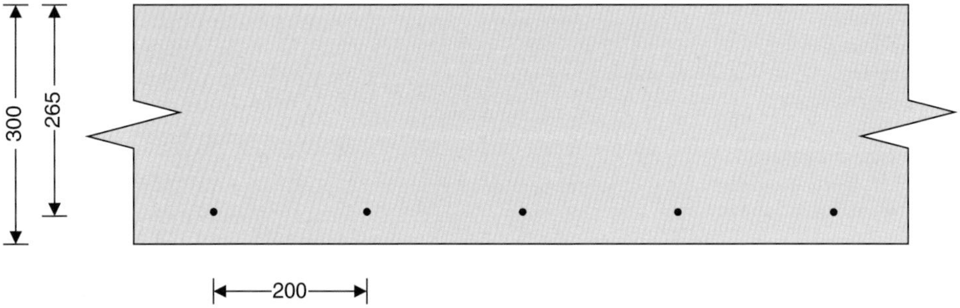

Check toe in bending

Width of toe; $b = \mathbf{1000}$ mm/m

Depth of reinforcement; $d_{toe} = t_{base} - c_{toe} - (\phi_{toe}/2) = \mathbf{265.0}$ mm

Constant; $K_{toe} = M_{toe}/(b \times d_{toe}^2 \times f_{cu}) = \mathbf{0.007}$

Compression reinforcement is not required

Lever arm;

$z_{toe} = \min(0.5 + \sqrt{(0.25 - (\min(K_{toe}, 0.225)/0.9))}, 0.95) \times d_{toe}$

$z_{toe} = \mathbf{252}$ mm

Area of tension reinforcement required; $A_{s\_toe\_des} = M_{toe}/(0.87 \times f_y \times z_{toe}) = \mathbf{191}$ mm²/m

Minimum area of tension reinforcement; $A_{s\_toe\_min} = k \times b \times t_{base} = \mathbf{390}$ mm²/m

Area of tension reinforcement required; $A_{s\_toe\_req} = \mathrm{Max}(A_{s\_toe\_des}, A_{s\_toe\_min}) = \mathbf{390}$ mm²/m

Reinforcement provided; **A393 mesh**

Area of reinforcement provided; $A_{s\_toe\_prov} = \mathbf{393}$ mm²/m

PASS – Reinforcement provided at the retaining wall toe is adequate

Check shear resistance at toe

Design shear stress; $v_{toe} = V_{toe}/(b \times d_{toe}) = \mathbf{0.211}$ N/mm²

Allowable shear stress;

$v_{adm} = \min(0.8 \times \sqrt{(f_{cu}/1\ \mathrm{N/mm^2})}, 5) \times 1\ \mathrm{N/mm^2} = \mathbf{5.000}$ N/mm²

PASS – Design shear stress is less than maximum shear stress

From BS 8110: Part 1: 1997 – Table 3.8

Design concrete shear stress; $v_{c\_toe} = \mathbf{0.434}$ N/mm²

$v_{toe} < v_{c\_toe}$ – No shear reinforcement required

Design of reinforced concrete retaining wall heel (BS 8002: 1994)

Material properties

Characteristic strength of concrete; $f_{cu} = \mathbf{40}$ N/mm²

Characteristic strength of reinforcement; $f_y = \mathbf{500}$ N/mm²

Base details

Minimum area of reinforcement; $k = \mathbf{0.13}\%$

Cover to reinforcement in heel; $c_{heel} = \mathbf{30}$ mm

Calculate shear for heel design

Shear from bearing pressure; $V_{heel\_bear} = p_{stem\_heel\_f} \times ((3 \times x_{bar\_f}) - l_{toe} - t_{wall})/2 = \mathbf{15.3}$ kN/m

Shear from weight of base; $V_{heel\_wt\_base} = \gamma_{f\_d} \times \gamma_{base} \times l_{heel} \times t_{base} = \mathbf{9.9}$ kN/m

Shear from weight of moist backfill; $V_{heel\_wt\_m} = w_{m\_w\_f} = \mathbf{32.3}$ kN/m

Shear from weight of saturated backfill; $V_{heel\_wt\_s} = w_{s\_f} = \mathbf{29}$ kN/m

Shear from surcharge; $V_{heel\_sur} = w_{sur\_f} = \mathbf{16}$ kN/m

Total shear for heel design; $V_{heel} = - V_{heel\_bear} + V_{heel\_wt\_base} + V_{heel\_wt\_m} + V_{heel\_wt\_s} + V_{heel\_sur} = \mathbf{71.9}$ kN/m

Calculate moment for heel design

Moment from bearing pressure; $M_{heel\_bear} = p_{stem\_mid\_f} \times ((3 \times x_{bar\_f}) - l_{toe} - t_{wall}/2)^2/6 = \mathbf{6}$ kNm/m

Moment from weight of base; $M_{heel\_wt\_base} = (\gamma_{f\_d} \times \gamma_{base} \times t_{base} \times (l_{heel} + t_{wall}/2)^2/2) = \mathbf{6.8}$ kNm/m

Moment from weight of moist backfill; $M_{heel\_wt\_m} = w_{m\_w\_f} \times (l_{heel} + t_{wall})/2 = \mathbf{21.8}$ kNm/m

Moment from weight of saturated backfill; $M_{heel\_wt\_s} = w_{s\_f} \times (l_{heel} + t_{wall})/2 = \mathbf{19.6}$ kNm/m

Moment from surcharge; $M_{heel\_sur} = w_{sur\_f} \times (l_{heel} + t_{wall})/2 = \mathbf{10.8}$ kNm/m

Total moment for heel design; $M_{heel} = - M_{heel\_bear} + M_{heel\_wt\_base} + M_{heel\_wt\_m} + M_{heel\_wt\_s} + M_{heel\_sur} = \mathbf{53}$ kNm/m

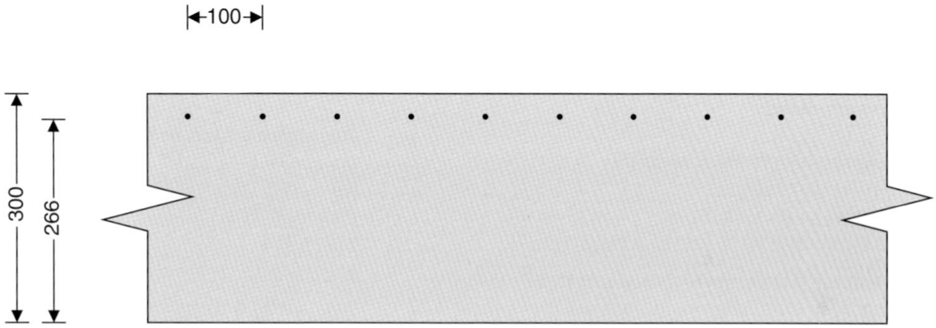

Check heel in bending

Width of heel; $b = \mathbf{1000}$ mm/m

Depth of reinforcement; $d_{heel} = t_{base} - c_{heel} - (\phi_{heel}/2) = \mathbf{266.0}$ mm

Constant; $K_{heel} = M_{heel}/(b \times d_{heel}^2 \times f_{cu}) = \mathbf{0.019}$

Compression reinforcement is not required

Lever arm;

$z_{heel} = \min(0.5 + \sqrt{(0.25 - (\min(K_{heel}, 0.225)/0.9))}, 0.95) \times d_{heel}$

$z_{heel} = \mathbf{253}$ mm

Area of tension reinforcement required; $A_{s\_heel\_des} = M_{heel}/(0.87 \times f_y \times z_{heel}) = \mathbf{482}$ mm^2/m

Minimum area of tension reinforcement; $A_{s\_heel\_min} = k \times b \times t_{base} = \mathbf{390}$ mm^2/m

Area of tension reinforcement required; $A_{s\_heel\_req} = \mathrm{Max}(A_{s\_heel\_des}, A_{s\_heel\_min}) = \mathbf{482}$ mm^2/m

Reinforcement provided; **B503 mesh**

Area of reinforcement provided; $A_{s\_heel\_prov} = \mathbf{503}$ mm^2/m

PASS – Reinforcement provided at the retaining wall heel is adequate

Check shear resistance at heel

Design shear stress; $v_{heel} = V_{heel}/(b \times d_{heel}) = \mathbf{0.270}$ N/mm^2

Allowable shear stress;

$v_{adm} = \min(0.8 \times \sqrt{(f_{cu}/1\ \mathrm{N/mm^2})}, 5) \times 1\ \mathrm{N/mm^2} = \mathbf{5.000}$ N/mm^2

PASS – Design shear stress is less than maximum shear stress

From BS 8110: Part 1: 1997 – Table 3.8

Design concrete shear stress; $v_{c\_heel} = \mathbf{0.470}$ N/mm^2

$v_{heel} < v_{c\_heel}$ – No shear reinforcement required

Design of cavity reinforced masonry retaining wall stem – BS 5628-2: 2000

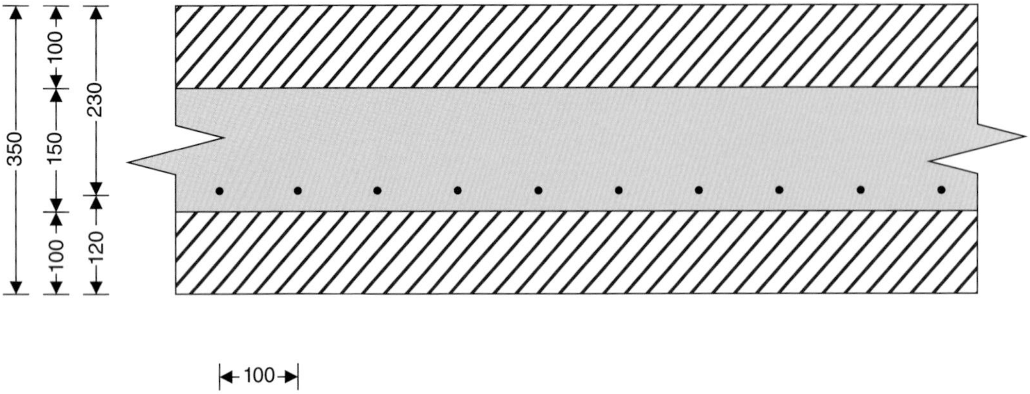

Wall details

Thickness of outer leaf of wall; t_{outer} = **100** mm

Thickness of inner leaf of wall; t_{inner} = **100** mm

Thickness of reinforced cavity; $t_{cavity} = t_{wall} - t_{outer} - t_{inner}$ = **150** mm

Depth of stem reinforcement; d_{stem} = **230** mm

Masonry details

Masonry type; **Aggregate concrete blocks no voids**

Compressive strength of units; p_{unit} = **10.0** N/mm^2

Mortar designation; **(ii)**

Category of manufactoring control of units; **Category II**

Partial safety factor for material strength; γ_{mm} = **2.3**

Characteristic compressive strength of masonry

Least horizontal dimension of masonry units; b_{unit} = **100.0** mm

Height of masonry units; h_{unit} = **215.0** mm

Ratio of height to least horizontal dimension; ratio = h_{unit}/b_{unit} = **2.2**

From BS 5628: 2 Table 3d, mortar (ii)

Characteristic compressive strength; f_k = **8.1** N/mm^2

Factored horizontal at-rest forces on stem

Surcharge; $F_{s\_sur\_f} = \gamma_{f\_l} \times K_0 \times \text{Surcharge} \times (h_{eff} - t_{base} - d_{ds})$ = **22.5** kN/m

Moist backfill above water table; $F_{s\_m\_a\_f} = 0.5 \times \gamma_{f\_e} \times K_0 \times \gamma_m \times (h_{eff} - t_{base} - d_{ds} - h_{sat})^2$ = **12.5** kN/m

Moist backfill below water table; $F_{s\_m\_b\_f} = \gamma_{f\_e} \times K_0 \times \gamma_m \times (h_{eff} - t_{base} - d_{ds} - h_{sat}) \times h_{sat}$ = **20.5** kN/m

Saturated backfill; $F_{s\_s\_f} = 0.5 \times \gamma_{f\_e} \times K_0 \times (\gamma_s - \gamma_{water}) \times h_{sat}^2$ = **5.3** kN/m

Water; $F_{s\_water\_f} = 0.5 \times \gamma_{f\_e} \times \gamma_{water} \times h_{sat}^2$ = **5.6** kN/m

Calculate shear for stem design

Shear at base of stem; $V_{stem} = F_{s\_sur\_f} + F_{s\_m\_a\_f} + F_{s\_m\_b\_f} + F_{s\_s\_f} + F_{s\_water\_f}$ = **66.2** kN/m

Calculate moment for stem design

Surcharge; $M_{s\_sur} = F_{s\_sur\_f} \times (h_{stem} + t_{base})/2$ = **25.9** kNm/m

Moist backfill above water table; $M_{s\_m\_a} = F_{s\_m\_a\_f} \times (2 \times h_{sat} + h_{eff} - d_{ds} + t_{base}/2)/3$ = **17.7** kNm/m

Moist backfill below water table; $M_{s\_m\_b} = F_{s\_m\_b\_f} \times h_{sat}/2$ = **9.2** kNm/m

Saturated backfill; $M_{s\_s} = F_{s\_s\_f} \times h_{sat}/3$ = **1.6** kNm/m

Water; $M_{s\_water} = F_{s\_water\_f} \times h_{sat}/3$ = **1.7** kNm/m

Total moment for stem design; $M_{stem} = M_{s\_sur} + M_{s\_m\_a} + M_{s\_m\_b} + M_{s\_s} + M_{s\_water}$ = **56** kNm/m

Check maximum design moment for wall stem

Width of wall; $b = \mathbf{1000}$ mm/m

Maximum design moment; $M_{d\_stem} = 0.4 \times f_k \times b \times d_{stem}{}^2/\gamma_{mm} = \mathbf{74.9}$ kNm/m

PASS – Applied moment is less than maximum design moment

Check wall stem in bending

Moment of resistance factor; $Q = M_{stem}/d_{stem}{}^2 = \mathbf{1.059}$ N/mm^2

$Q = 2 \times c \times (1 - c) \times f_k/\gamma_{mm}$

Lever arm factor; $c = \mathbf{0.817}$

Lever arm; $z_{stem} = \min(0.95, c) \times d_{stem} = \mathbf{187.9}$ mm

Area of tension reinforcement required; $A_{s\_stem\_des} = M_{stem} \times \gamma_{ms}/(f_y \times z_{stem}) = \mathbf{686}$ mm^2/m

Minimum area of tension reinforcement; $A_{s\_stem\_min} = k \times b \times t_{wall} = \mathbf{455}$ mm^2/m

Area of tension reinforcement required; $A_{s\_stem\_req} = \text{Max}(A_{s\_stem\_des}, A_{s\_stem\_min}) = \mathbf{686}$ mm^2/m

Reinforcement provided; **B785 mesh**

Area of reinforcement provided; $A_{s\_stem\_prov} = \mathbf{785}$ mm^2/m

PASS – Reinforcement provided at the retaining wall stem is adequate

Check shear resistance at wall stem

Design shear stress; $v_{stem} = V_{stem}/(b \times d_{stem}) = \mathbf{0.288}$ N/mm^2

Basic characteristic shear strength of masonry; $f_{vbas} = \min[0.35 + (17.5 \times A_{s\_stem\_prov}/(b \times d_{stem})), 0.7] \times 1$ N/mm^2

$f_{vbas} = \mathbf{0.410}$ N/mm^2

Shear span; $a = M_{stem}/V_{stem} = \mathbf{845.4}$ mm

Characteristic shear strength of masonry; $f_v = \text{Min}(f_{vbas} \times \max(2.5 - 0.25 \times (a/d_{stem}), 1), 1.75\text{N/mm}^2)$

$f_v = \mathbf{0.648}$ N/mm^2

Allowable shear stress; $v_{adm} = f_v/\gamma_{mv} = \mathbf{0.324}$ N/mm^2

PASS – Design shear stress is less than maximum shear stress

Check limiting dimensions

Limiting span/effective depth ratio; $\text{ratio}_{max} = \mathbf{18.00}$

Actual span/effective depth ratio; $\text{ratio}_{act} = (h_{stem} + d_{stem}/2)/d_{stem} = \mathbf{9.20}$

PASS – Span to depth ratio is acceptable

Axial load check

Factored axial load on wall; $N_{wall} = ([t_{wall} \times h_{stem} \times \gamma_{wall} + W_{dead}] \times \gamma_{f\_d}) + (W_{live} \times \gamma_{f\_l}) = \mathbf{23.1}$ kN/m

Limiting axial load; $N_{limit} = 0.1 \times f_k \times t_{wall} = \mathbf{285.0}$ kN/m

Applied axial load may be ignored – calculations valid

Indicative retaining wall reinforcement diagram

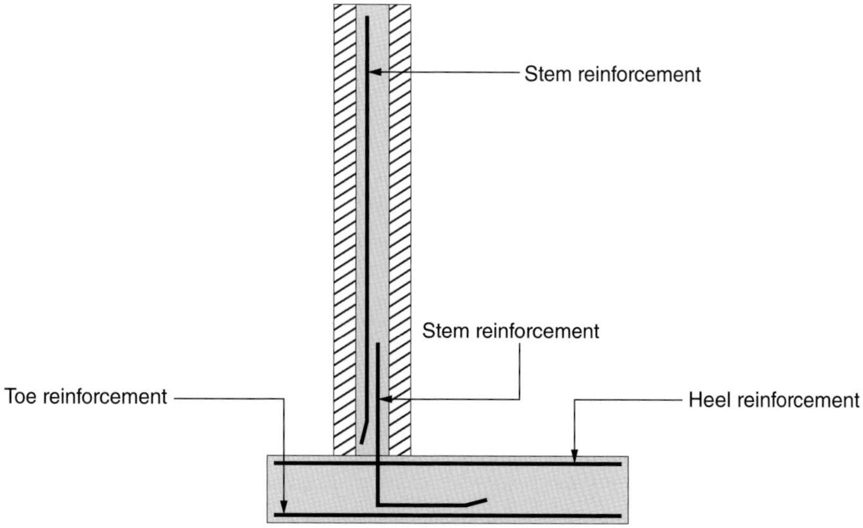

Toe mesh - A393 (393 mm^2/m)
Heel mesh - B503 (503 mm^2/m)
Stem mesh - B785 (785 mm^2/m)

We have changed the reinforcement to comply with the design, but as can be seen the overall stability of the wall has now been compromised in respect of bearing pressure and sliding. These parameters which have been changed can easily be overlooked by someone undertaking the constriction or design of a wall without knowledge of the soil and ground conditions. With this in mind, it is strongly recommended that retaining walls are designed by suitably qualified professionals.

Types of wall

The most commonly used retaining structures are listed as follows.

Solid masonry

These are mass construction gravity walls and rely on their own weight to resist the lateral pressures placed on the wall. They are effective for relatively low retaining heights since they become uneconomic at greater heights. The wall is unable to sustain tensile stresses and will fail if these are induced.

Reinforced cavity

Reinforced cavity walls can be constructed from brick, concrete block or both and have a cavity with a reinforced concrete in-fill. The tensile stresses in the base and stem are absorbed by the reinforcement and consequently can be constructed to retain greater heights than solid gravity walls (up to 6–7.0 m). The base and foundation must act as one monolithic structure to enable the tensile forces to be absorbed.

Reinforced concrete

These types of wall follow the same principle as reinforced cavity walls and usually have a backward taper to accommodate any movements or forward tilting.

Crib walls

A photograph of a crib wall can be seen in Figure 16.6.

The wall comprises cribs constructed using timber, concrete or steel and filled with granular material.

This type of wall is a mass gravity retaining structure but capable of sustaining small movements and water behind the wall has the ability to pass through the granular material, thus removing the hydrostatic pressure behind the wall. The walls are quick to construct and can be constructed piecemeal.

The width of the wall is typically 0.5–1.0 H, where H is the height of the retained material. These types of wall are efficient to heights similar to mass concrete and brick retaining walls, 6–7.0 m.

Gabion walls

A photograph of a gabion wall can be seen in Figure 16.7.

The wall is a mass gravity retaining structure and comprises a series of baskets – usually of metal – which have a granular in-fill. The mesh to the baskets can be protected

Figure 16.6: Photograph showing a typical crib wall construction.

Figure 16.7: Photograph showing a gabion wall.

using plastic to resist corrosive environments. The laying of the stone will play an important part in the strength of the wall, and if material is dropped in or badly placed then the mass of the wall will not be as compact as if the material had been placed by hand. The visual impact of a hand-filled basket is also much more aesthetically pleasing than one where material has been in-filled by machine.

The base of the gabion wall is approximately 0.5 H, where H is the height of the retained material.

Reinforced earth or anchored walls

This type of wall is similar to a cantilever wall but has additional stability provided by rows of ties fixed to the wall and extending into the retained soil. The action of the soil on the ties has the positive effect of preventing overturning and sliding.

Embedded walls

Embedded walls extend into the ground to a greater depth than a cantilever wall and rely on the passive resistance of the material in front of the wall. In some circumstances anchors can be used to provide additional support.

Figure 16.8: Showing the commencement of a basement structure.

Basements

An example of the commencement of a basement structure can be seen in Figure 16.8, which is a photograph of an excavated basement with a proposed reinforced base. The reinforcement has been placed prior to the concrete pour. The base will be connected to a reinforced cavity wall at the sides of the base to retain the soil.

The walls will be suitably designed to ensure water and moisture does not penetrate the walls and cause dampness in the structure. Careful consideration has to be given to the height of the water table, and if this is likely to rise over the seasons then a sump to collect the water and a submersible pump will have to be designed to discharge any rising levels of water.

If the basement is to be a habitable room then other matters will need to be afforded careful consideration – for example the means of escape in the event of a fire. The basement structure will have to meet structural considerations and also comply with any other Building Regulation requirements – for example insulation, means of escape and radon protection if required.

Chapter 17 Structural Alterations

Preliminary considerations

Before undertaking any structural alterations in a building, the necessary approvals need to be sought. These may include planning approvals, Building Regulation approvals and if the property is listed, then Listed Building Consent will also be required. The works should be designed and undertaken by a competent person, who fully understands the load paths and the extent of the work to be carried out.

Removal of walls

The first thing to ascertain before removing a wall is whether it is load bearing. A note of caution at this point: just because a wall is made of timber stud does not necessarily mean that it is not load bearing. Professional advice should be sought at an early stage. Other factors that may affect the removal of internal walls are the lateral stability of the external walls if the wall to be removed is acting as a shear wall or buttress.

If the wall is load bearing then support will have to be afforded to the floors and walls above before commencing work. All services should be removed or isolated from the subject wall. A method statement including the method of temporary support should be undertaken, which includes a risk assessment of the work. In many cases works are commenced and the wall supported before it is realised how the steel will be inserted. For example, if the wall is supported by props and pins and the steel is not in position, it may be found impractical to place the steel without removing some of the props, which will become difficult if the wall has been removed. A method statement and risk assessment will assist in resolving these issues prior to commencement. In addition to this, an appreciation of the loads to be supported will be necessary to determine the amount of temporary support required. For example, the load on an individual acrow prop will need to be assessed and the capacity and number of acrow props calculated. The loads sustained by acrow props are available from the suppliers, and these will vary depending on the length due to the slenderness ratio of the prop changing as the length increases.

The wall can be supported on acrow props, strong boys, pins and in some cases the floor joists can assist in the support of the wall if they penetrate sufficiently into the wall. Thus, a prop placed under a board – which in turn is placed under the joists – can afford a certain degree of support.

Some consideration will have to be given to the wall construction. For example, a weaker wall such as one constructed in lime mortar can be prone to progressive

Structural Design of Buildings, First Edition. Paul Smith.

Table 17.1: Typical beam sizes for openings

Opening (m)	Beam size
3.0	203 × 102 × 23 U.B.
3.5	254 × 102 × 28 U.B.
4.0	305 × 127 × 37 U.B.
4.5	305 × 127 × 42 U.B.

migration of loose masonry if sufficient support is not provided. It may be found that in such walls the amount of support required is more frequent along the length of the wall than for a similar wall constructed in a cement mortar, which has a greater bond.

Table 17.1 provides typical beam section sizes for a single-skin wall supporting a floor either side and a wall 2.7 m above. The beams show different opening sizes from 2.5 m to 4.5 m in intervals of 0.5 m for a floor span not exceeding 4.0 m on each side of the wall. We have made no allowance for roof load, but have allowed for a 4.0 m ceiling bearing onto the wall either side. The imposed loads used are 1.5 kN/m^2 for the floor and 0.25 kN/m^2 in the ceiling for storage. We have undertaken the design using BS 5950 and assumed that the floor joists offer no intermediate lateral restraint to the beam.

Dead loads

Floor	55 × 10/100 × 4 = 2.20 kN/m
Wall	55 × 4 × 10/1000 × 2.7 = 5.9 kN/m
Plaster	22 × 2 × 10/1000 × 2.7 = 1.19 kN/m
Ceiling	33 × 10/1000 × 4 = 1.28 kN/m
Total	**10.61kN/m**

Imposed loads

Floor loading	1.5 × 4 = 6.0 kN/m
Ceiling/loft storage	0.25 × 4 = 1.0 kN/m
Total	**7 kN/m**

The above loads are unfactored loads

There is a practical aspect to the design, to ensure that the beam width is similar to the wall width – particularly if it is bearing onto a wall running parallel to the beam, otherwise the edges of the beam will be overhanging the edge of the wall. However, on the longer spans it has been necessary to opt for a wider beam, which can be hidden in the plaster to the sides. Interestingly, wider beams can sustain higher loads and for the 3.5 m span a 203 × 133 × 30 U.B. would suffice, which is a wider but shallower beam.

It is also interesting to note that if we considered the floor to offer full lateral restraint (not permitted under the code, but a number of engineers do allow this at their own discretion) then a 203 × 102 × 23 U.B. would be seen to work up to a span of 4.0 m. This is a much more economical section than the 305 × 127 × 37 U.B., but strictly under the code would not be permitted.

The bearing is also an essential part of the design, and padstones are sometimes necessary to spread the load of the beam over a wider area of masonry. There are many factors affecting the design and length of the bearing and padstone. Some considerations are provided below:

- What is the material of the wall on which the beam bears – for example brick, concrete block, aerated block? These have different compressive strengths that will affect the bearing.
- Is the bearing running parallel or perpendicular to the wall?

Finally, one has to consider the load to the foundations since a concentrated load may now be directed via the load path to the base of the remaining wall, thus making a large load on a smaller area of foundation. This will have an impact on the bearing pressure which may surpass the bearing capacity of the subsoil beneath the foundation.

The above is for guidance only and the design should be undertaken by competent and qualified structural engineers who will provide beam sizes, bearings and padstone sizes depending on the particular circumstances.

Alterations to timbers and trusses

It is important to understand why structural members are employed and their importance in the load path before commencing on any alterations. Roof frames are particularly vulnerable to alterations; if timber members are considered inconvenient or obstructive there is a tendency to remove or adjust these without any concern for the consequences. Purlins are sometimes cut to facilitate dormers, which affect the structural integrity of the rafters they support. Traditional trusses are amended to facilitate door openings in loft conversions, sometimes without a full understanding of the consequences. Moving a tie member and placing it higher up the truss could result in the structural tie falling outside the tension zone and becoming a compression member, thus leaving the truss vulnerable to distortion and roof spread.

Many such alterations have been undertaken, particularly in older properties and arguably not more so than in traditional timber frame buildings. This was particularly prevalent in the 1970s, when many alterations were made and grants were readily available for the modernisation of older properties.

Window openings in older stone cottages were widened to allow more light without necessarily specifying the correct lintels. Alterations trapped damp in timber lintels by the advent of modern mortars rather than the lime mortars traditionally used in these buildings. Thus, over time, the timber lintels suffer wet rot and deteriorate. It is ironic that cottages which have stood the test of time for over a hundred years suddenly suffer and deteriorate in modern times. Alterations made without consideration of a building's properties have caused such buildings to suffer serious damage in less than a generation of the alterations being undertaken.

A roof frame is an intricate structure and alterations undertaken must be based on the guidance of a suitably qualified person. Trusses have a series of members that act in compression and tension, forming an integral structure. The removal or alteration of one member can distribute loads to other members that are not able to sustain these additional loads. A pegged joint in a traditional timber truss is reputed to be able to carry a load of approximately one tonne, but if this is overloaded this capacity can soon be exceeded. In modern truss design the connection may be found to be the weakest structural element, thus the alteration of the members of a truss may leave the timbers able to accommodate the additional loading but the connections may fail.

If alterations to a cut roof are being undertaken, then the order in which the alterations are made is probably as important as the strengthening of the remaining members. This ensures that the load paths of the building are maintained and the structural integrity of the roof frame is not compromised.

Steel plates and flitch connections can be used to repair timber members, but this should be undertaken sympathetically, particularly if the building has significant architectural merit or heritage. An example of a steel plate to a purlin and a plated connection at the base of a truss which has been cut can be seen in the photographs in Figures 17.1 and 17.2, respectively.

Figure 17.1: Showing steel plate to strengthen purlin.

Figure 17.2: Showing steel plate to strengthen a truss connection.

People owning listed buildings need to appreciate that they are custodians of the building; repairs should be undertaken for future generations to appreciate it, and in accordance with the historic context of the building.

Alterations to roof structures for dormers

We will consider loft conversions separately, but in this section we examine the alteration of roof structures to facilitate dormers or the introduction of dormer windows. Bungalows constructed in the 1960s have considerable accommodation in the roof space and some of these have been constructed with dormer windows, see Figure 1.12 in Chapter 1. In some cases these have been extended, and in other cases the dormer has been introduced retrospectively. Care must be taken in executing this type of work, and the dormers employed can be quite large. Properties of this period are invariably constructed using a cut roof with rafters spanning over purlins.

The purlin will be cut or shortened if a dormer is added or extended, and the support to the purlin end either side of the dormer needs to be properly assessed. The load path needs to be considered and the purlin end supported on a load-bearing wall carrying the load to the ground or via appropriate beams at ceiling level on the ground floor. In some circumstances the purlin is supported from floor joists which will require strengthening. Without strengthening the floor joists, the ceiling on the ground floor will bulge due to excessive deflections (or in the worst-case scenario, fail).

Consideration also has to be given to the beam supporting the dormer roof and the rafter ends from the main roof. The beam is normally supported by a trimmer rafter to the sides of the dormer, which also supports the side walls to the dormer. The design of this rafter needs to be considered, and in many circumstances these rafters are a multiple of the single rafter size (possibly only two or three), but in reality much more strength is required – particularly for large-span dormers.

When examining such properties these considerations need to be assessed. Other issues worthy of consideration are to examine if the roof or side walls have a change in material from the original design and if these have a significant weight difference which will impact on the structural components of the dormer.

Loft conversions

There are a number of considerations when undertaking a loft conversion, not only structural considerations but also height requirements and insulation requirements. The first consideration should be the height obtainable in the converted space, and as a general rule the roof should provide a head height of at least 2.15 m for half of its width. This height has to be assessed in recognition of the following.

Will the rafters need to be strengthened with respect to the additional load from plasterboard and insulation? Typically a concrete tile will weigh 50 kg/m^2 and the felt and batons will be 10 kg/m^2, thus the total dead load in this example will be 60 kg/m^2. Now, considering the insulation which can typically weigh 5.67 kg/m^2 and the plasterboard which can weigh 11 kg m^2, this equates to an additional load of 17 kg/m^2 which is an increase of just over 28%.

In addition to this, the thickness of the rafters may be determined by the type of felt lining used. A breathable felt will require a 25 mm air gap between the underside of the felt and the insulation, whereas a bituminous felt will require a 50 mm air gap. This means that a minimum 125 mm or 150 mm deep rafter section will be needed to meet this requirement, assuming 100 mm of insulation between the rafters. New rafters can be placed alongside depending on space requirements, or the existing rafters may be strengthened.

The purlin will also require some structural assessment to ensure it can sustain the additional load and be strengthened as necessary. Dormer windows may necessitate the removal or cutting of a purlin, and one option is to support the purlin from a beam at ceiling level and create a small load-bearing stud wall between the purlin and the beam to transmit the loads to the new beam.

The ceiling joists are unlikely to be able to sustain the additional loads from the proposed usage. The ceiling joists will have been designed to carry an imposed load of 0.25–0.5 kN/m^2, but a proposed conversion will lead to an imposed floor load which under residential use will be 1.5 kN/m^2. Thus the ceiling joists will need to be strengthened, which may be difficult and cause damage to the plaster below if ceiling binders are to be removed. Consequently, it is normal practice to install new beams spanning between the new steel beams which support the purlin. A diagram of such an arrangement can be seen in Figure 17.3.

A modern trussed roof is more difficult to convert, and expert advice should be obtained prior to undertaking any alterations to such a structure.

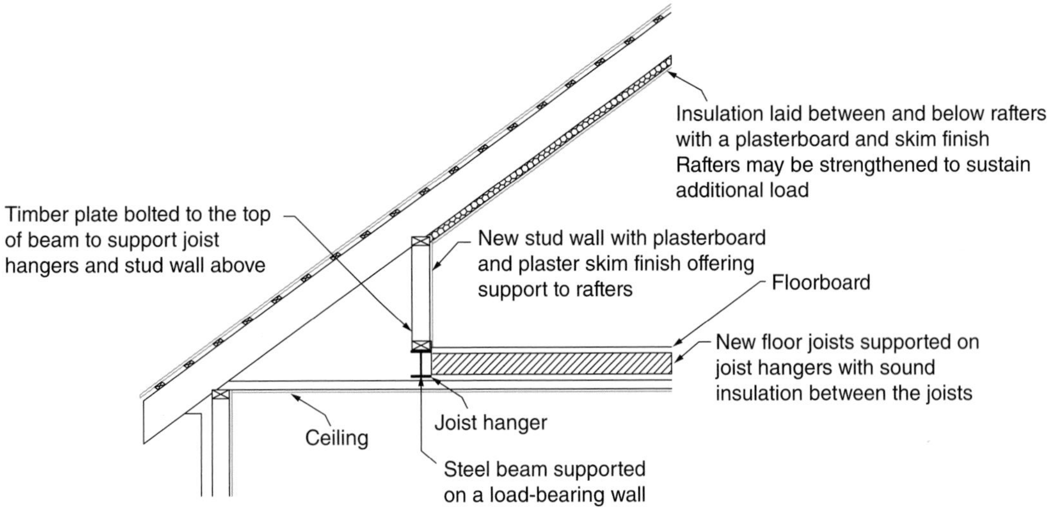

Figure 17.3: Typical arrangement of roof and ceiling joist strengthening for a loft conversion.

Table 17.2: Analysis results using a timber beam

	Permissible	Applied	Utilisation
Bearing stress (N/mm^2)	2.4	0.408	0.170
Bending stress (N/mm^2)	7.842	5.357	0.683
Shear stress (N/mm^2)	0.710	0.306	0.431
Deflection (mm)	10.5	9.971	0.950

Flitch beams

A flitch beam is a beam comprising timber and steel plates between or at the side of a single or a number of timber members. These types of beam can be used in their own right or as a repair to existing beams. The introduction of a steel plate to a timber section can make it much stronger. The analysis below considers a 3.5 m beam loaded with a dead load of 3 kN/m^2 and an imposed load of 1.5 kN/m^2. Firstly, consider a timber beam 200 mm × 200 mm in Grade C24 timber. The results can be seen in Table 17.2.

Now consider a flitch beam comprising two Grade C24 timber beams 75 mm × 200 mm with a central steel plate sandwiched between, 8 mm × 195 mm. The M12 bolts pass through the timber, through the steel plate and are spaced at 500 mm centres. Let us now analyse the results using the same loading. See Table 17.3.

It can clearly be seen that the flitch beam is much stronger and the beam is actually smaller, measuring 158 mm × 200 mm. It is therefore smaller in cross-sectional area than its equivalent section in timber. Any smaller section in timber would initially fail in the deflection criterion.

Table 17.3: Analysis results using a flitch beam

	Permissible	Applied	Utilisation
Bearing stress (N/mm^2)	2.4	0.553	0.231
Bending stress (N/mm^2)	7.842	3.747	0.478
Shear stress (N/mm^2)	0.710	0.214	0.302
Deflection (mm)	10.5	4.858	0.463

The analysis using a flitch beam is seen below. These calculations have been undertaken using structural engineering software and we gratefully acknowledge Tekla (UK) Ltd for their approval in the use of this software.

FLITCH BEAM ANALYSIS & DESIGN TO BS 5268-2: 2002.

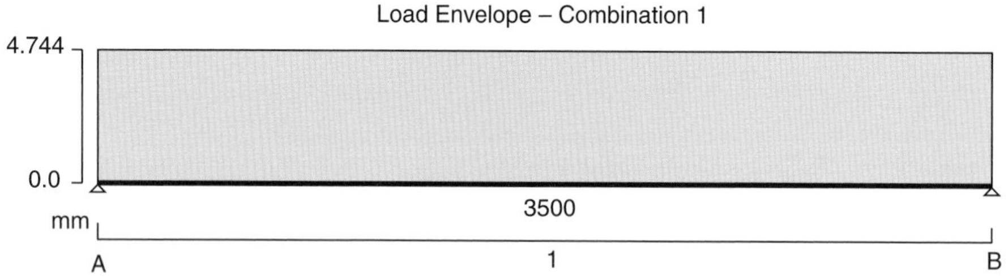

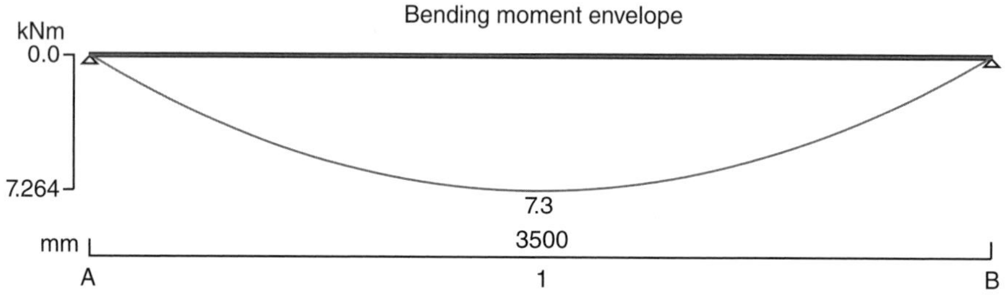

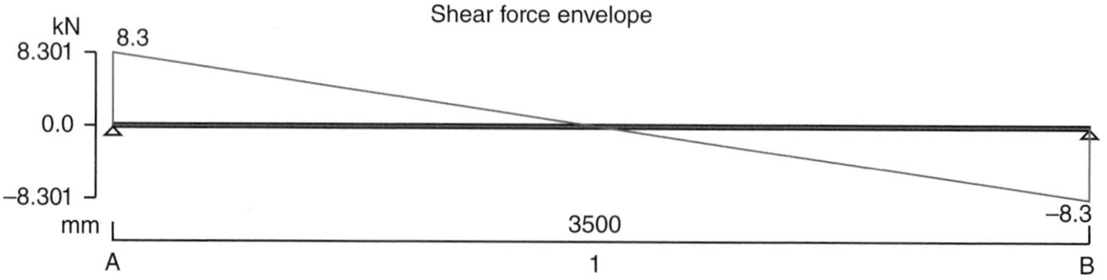

Applied loading
Beam loads

Dead full UDL 3.000 kN/m
Imposed full UDL 1.500 kN/m
Dead self-weight of beam × 1

Load combinations

Load combination 1	Support A	Dead × 1.00 Imposed × 1.00
	Span 1	Dead × 1.00 Imposed × 1.00
	Support B	Dead × 1.00 Imposed × 1.00

Analysis results

Maximum moment;	M_{max} = **7.264** kNm;	M_{min} = **0.000** kNm
Design moment;	M = max(abs(M_{max}),abs(M_{min})) = **7.264** kNm	
Maximum shear;	F_{max} = **8.301** kN;	F_{min} = **-8.301** kN
Design shear;	F = max(abs(F_{max}),abs(F_{min})) = **8.301** kN	
Total load on beam;	W_{tot} = **16.603** kN	
Reactions at support A;	R_{A_max} = **8.301** kN;	R_{A_min} = **8.301** kN
Unfactored dead load reaction at support A;	R_{A_Dead} = **5.676** kN	
Unfactored imposed load reaction at support A;	$R_{A\_Imposed}$ = **2.625** kN	
Reactions at support B;	R_{B_max} = **8.301** kN;	R_{B_min} = **8.301** kN
Unfactored dead load reaction at support B;	R_{B_Dead} = **5.676** kN	
Unfactored imposed load reaction at support B;	$R_{B\_Imposed}$ = **2.625** kN	

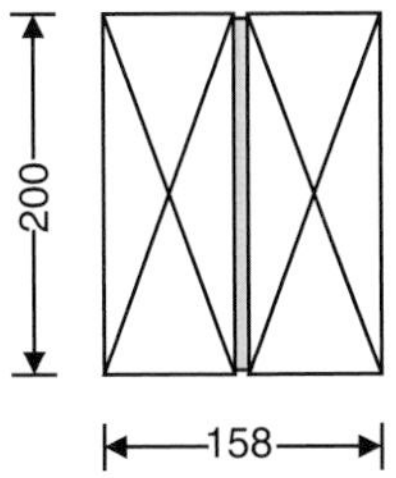

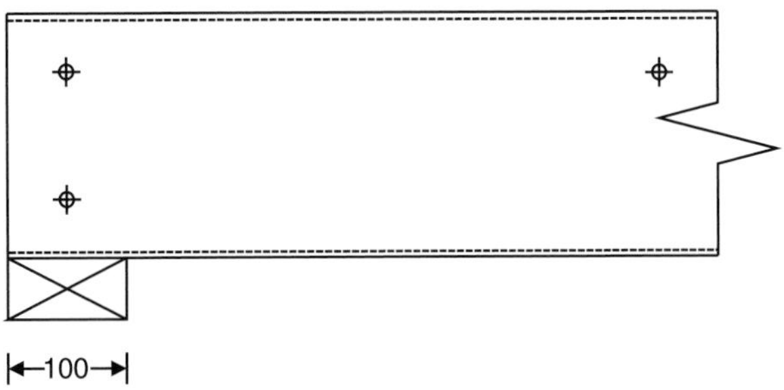

Timber section details

Breadth of timber sections;	b = **75** mm
Depth of timber sections;	h = **200** mm
Number of timber sections in member;	N = **2**
Timber strength class;	**C24**

Steel section details

Breadth of steel plate;	b_s = **8** mm
Depth of steel plate;	h_s = **195** mm
Number of steel plates in beam;	N_s = **1**
Steel stress;	p_y = **165** N/mm^2
Bolt diameter;	ϕ_b = **12** mm

Member details

Service class of timber;	**1**
Load duration;	**Long term**
Length of bearing;	L_b = **100** mm

Section properties

Cross-sectional area of beam;	$A = N \times b \times h =$ **30000** mm^2
Timber section modulus;	$Z_{xt} = N \times b \times h^2/6 =$ **1000000** mm^3
Steel section modulus;	$Z_{xs} = N_s \times b_s \times h_s^2/6 =$ **50700** mm^3
Second moment of area of timber;	$I_{xt} = N \times b \times h^3/12 =$ **100000000** mm^4
Second moment of area of steel;	$I_{xs} = N_s \times b_s \times h_s^3/12 =$ **4943250** mm^4

Load proportions
Instant deflection under permanent actions; u_{instG} = **3.028** mm
Instant deflection under principal variable action; u_{instQ1} = **1.400** mm
k_{def} = **0.6**
ψ_2 = **0.3**
Final minimum modulus of elasticity $E_{min,fin} = E_{min} \times (u_{instG} + u_{instQ1})/(u_{instG} + u_{instQ1} + k_{def} \times (u_{instG} + \psi_2 \times u_{instQ1}))$ = **4907** N/mm^2
Proportion of applied load in timber; $k_t = E_{mean} \times I_{xt} / (E_{mean} \times I_{xt} + E_{S5950} \times I_{xs})$ = **0.516**
Proportion of applied load in steel; $k_s = 1.1 \times E_{S5950} \times I_{xs} / (E_{min,fin} \times I_{xt} + E_{S5950} \times I_{xs})$ = **0.741**

Modification factors
Duration of loading – Table 17; K_3 = **1.00**
Bearing stress – Table 18; K_4 = **1.00**
Total depth of member – cl.2.10.6; $K_7 = (300\ \text{mm}/h)^{0.11}$ = **1.05**
Load sharing – cl.2.9; K_8 = **1.00**

Lateral support – cl.2.10.8
No lateral support
Permissible depth-to-breadth ratio – Table 19; **2.00**
Actual depth-to-breadth ratio; $h/(N \times b + N_s \times b_s)$ = **1.27**

PASS – Lateral support is adequate

Compression perpendicular to grain
Permissible bearing stress (no wane); $\sigma_{c\_adm} = \sigma_{cp1} \times K_3 \times K_4 \times K_8$ = **2.400** N/mm^2
Applied bearing stress; $\sigma_{c\_a} = R_{A\_max}/(N \times b \times L_b)$ = **0.553** N/mm^2
$\sigma_{c\_a}/\sigma_{c\_adm}$ = **0.231**

PASS – Applied compressive stress is less than permissible compressive stress at bearing

Bending parallel to grain
Permissible bending stress; $\sigma_{m\_adm} = \sigma_m \times K_3 \times K_7 \times K_8$ = **7.842** N/mm^2
Applied timber bending stress; $\sigma_{m\_a} = k_t \times M/Z_{xt}$ = **3.747** N/mm^2
$\sigma_{m\_a}/\sigma_{m\_adm}$ = **0.478**

PASS – Timber bending stress is less than permissible timber bending stress

Applied steel bending stress; $\sigma_{m\_a\_s} = k_s \times M/Z_{xs}$ = **106.178** N/mm^2
$\sigma_{m\_a\_s}/p_y$ = **0.644**

PASS – Steel bending stress is less than permissible steel bending stress

Check beam in shear
Permissible shear stress; $\tau_{adm} = \tau \times K_{2s} \times K_3 \times K_8$ = **0.710** N/mm^2
Applied shear stress; $\tau_a = 3 \times k_t \times F/(2 \times A)$ = **0.214** N/mm^2
τ_a/τ_{adm} = **0.302**

PASS – Shear stress within permissible limits

Deflection
Modulus of elasticity for deflection; $E = E_{mean}$ = **10800** N/mm^2
Permissible deflection; $\delta_{adm} = \min(14\ \text{mm}, 0.003 \times L_{s1})$ = **10.500** mm
Bending deflection; δ_{b_s1} = **4.428** mm
Shear deflection; δ_{v_s1} = **0.430** mm
Total deflection; $\delta_a = \delta_{b\_s1} + \delta_{v\_s1}$ = **4.858** mm
δ_a/δ_{adm} = **0.463**

PASS – Total deflection is less than permissible deflection

Flitch plate bolting requirements
Total load on beam; W_{tot} = **16.603** kN
Total load taken by steel; $W_s = k_s \times W_{tot}$ = **12.304** kN
Basic bolt shear load – Table 71; v_{90} = **2.692** kN
Number of interfaces; $N_{int} = (N + N_s) - 1$ = **2**
Number of bolts required at supports; $N_{be} = \max(k_s \times R_{A\_max} / (N_{int} \times v_{90}), 2)$ = **2**

Limiting bolt spacing;	S_{limit} = min(2.5 × h, 600 mm) = **500** mm
Maximum bolt spacing;	S_{max} = **500** mm
Minimum number of bolts along length of beam;	$N_{bl} = W_s/(N_{int} \times v_{90})$ = **2.286**

- Provide a minimum of 2 No.12 mm diameter bolts at each support
- Provide 12 mm diameter bolts at maximum 500 mm centres staggered 50 mm alternately above and below the centre line

Minimum bolt spacings

Minimum end spacing;	$S_{end} = 4 \times \phi_b$ = **48** mm
Minimum edge spacing;	$S_{edge} = 4 \times \phi_b$ = **48** mm
Minimum bolt spacing;	$S_{bolt} = 4 \times \phi_b$ = **48** mm
Minimum washer diameter;	$\phi_w = 3 \times \phi_b$ = **36** mm
Minimum washer thickness;	$t_w = 0.25 \times \phi_b$ = **3** mm

Lintels and openings

Openings in walls can be undertaken using lintels – such as steel lintels or precast concrete lintels. Suppliers of such lintels provide load tables for various lintels depending on the span of the opening, and these are available upon request.

Steel lintels will vary in width and style depending on the wall construction. For example, different types of lintel will be required for external solid walls, internal single-leaf walls and cavity walls. In the case of cavity walls the width of the cavity will vary depending on the date of construction, and it is important to ensure that the correct width of lintel is purchased. The load transmitted to the lintels is also important, and this needs to be calculated prior to purchasing a lintel. In the case of steel lintels these can be obtained as Normal or Standard Load, Heavy Load and Extra Heavy Load lintels depending on the loads to be sustained. For a typical 1.5 m lintel in a cavity wall of 100 mm cavity width the uniformly distributed loads sustained can vary between 14 kN, 40 kN and 70 kN depending on the type of lintel used. The specification assumes that the load is distributed evenly along the length of the lintel.

The load ratio between the front and rear flange is also critical in determining the lintel type and load distribution. Furthermore, lintels of this type cannot sustain point loads and if the lintel is required to support a beam then the lintel manufacturer should be consulted. The bearing for steel lintels is usually a minimum of 150 mm.

Precast concrete lintels also have specific load capacities depending on the manufacturer, and these should be available from the supplier upon request. The size of the reinforcement contained within the precast section will be dependent on the load. The fire resistance and chemical exposure may be another consideration that will assist in the determination of the lintel. Typically, a 100 mm × 70 mm deep lintel spanning 1000 mm will typically sustain a load of 8.8 kN, but a doubly reinforced concrete lintel 100 mm × 110 mm deep will sustain a load of 19 kN, which is a considerable difference.

The bearing requirements will also be specified by the supplier, and this is important in the application of such products. Bearings can range from 100 mm to 150 mm depending on the span and type of lintel.

All structural alterations require the consent of Building Control and will need to be inspected as the work progresses. It is essential that suitably qualified professionals are consulted over any structural alterations undertaken on a property.

Chapter 18 Structural Defects in Buildings

Structural defects

There are many publications available covering this subject matter, and there are a number of publications that cover this in detail and dedicate volumes to the defects in buildings. In this chapter we hope to examine a few of the more common structural defects likely to be encountered in buildings, and discuss their likely causes and solutions.

Let us first examine the types of failure that are possible and the characteristic cracking nature and style.

Compression

This is when forces act in alignment and push against each other, causing crushing. The compression force usually results in a defined area of squeezed failure, crushing or crumbling within the structure.

Most materials in building construction are strong in compression and this type of failure, although rare, can be found in places where concentrated loads have been directed down a structure – for example a narrow area of masonry between large windows.

Tension

This is where two forces act against each other, pulling in opposite directions but in alignment, thus this type of failure tends to pull the structure apart and consequently the failure crack is typically vertical.

Shear

This is similar to tension failure, except the forces are not in alignment and consequently the failure crack is diagonal; typical shear failure has cracking at 45°.

Random cracking

Usually found in concrete, this type of failure is due to tension and results from the material not being the same thickness (homogeneous).

Structural Design of Buildings, First Edition. Paul Smith.

Location of cracking

The weakest part of a building is over the doors and windows, or through existing joints. The location of cracking will determine the most likely cause, and diagonal cracking will normally lie over the problem, with the cause being under the diagonal line of the crack.

Roof spread

Roof spread occurs when there is little or no lateral restraint of the roof, and the roof thrusts outwards at wall-plate level. Lateral restraint is usually provided by the ceiling collars or joists connecting to the rafters, creating a triangulation of the roof frame and thus ensuring that the lateral thrust forces on the roof are resisted.

Roof spread can be evidenced by the walls bowing out just below the eaves, and often there has been lateral movement on the wall plate which may have been pushed over. The rafters may also have slipped from their bearing and parted at the ridge. Other indicators are that the skirting board may have moved, and this may be evidenced by the original paint mark on the floorboards being seen away from the skirting board.

Solution

Retrospective longitudinal ties between the rafters can be employed to ensure the triangulation of the roof frame. The connection is usually via a bolt or plate connector.

Lateral restraint straps are a requirement on all new buildings to reduce the likelihood of such a failure. These can be provided retrospectively by using bat straps tied at eaves and first-floor level to prevent or reduce the possibility of roof spread. The bat straps or lateral restraint straps are galvanised straps measuring 5 mm × 30 mm × 1200 mm, and Part A of the Building Regulations recommends these be employed at 2.0 m centres.

Helifix™ ties can be provided, which pass through the wall and screw into floor joists either longitudinally or laterally with noggins placed between the joists. These have to be installed in accordance with the suppliers' and manufacturers' instructions.

A ridge beam can be provided to prevent the roof from progressing vertically downwards, and essentially supports some of the weight of the roof.

Full-length ties can be provided between the floor joists or notched over the floor joists with plates visible on the side walls. The tie member needs to be tight and the plates need to touch the walls, otherwise they are rendered useless. Thus the tension tie usually has a swivel joint in the centre to tighten the rod or bar, ensuring it is kept in tension.

Settlement

Settlement usually occurs within 10–20 years of construction, and although often referred to or mistaken as subsidence it is distinctly different. Bonshor and Bonshor (1996, p. 75) offer the following definition of settlement: "Settlement – the compression of the ground below the foundation resulting from the loads applied by the building." This can be as a result of consolidation of the subsoil material, such as gravel, and usually occurs relatively quickly. The consolidation of clays, in contrast, can occur over a number of years.

Settlement does not cause cracking but differential settlement does, and this can occur if there is a variation in the material below the building and the building compresses or settles by a differential amount.

Solution

Usually settlement occurs and is self-stabilising – although the cracking can be quite extensive, the cracks can be sealed with no further likelihood of movement.

Shrinkage due to thermal and moisture movements

We have already discussed the effects of thermal and moisture changes in Chapter 8. This type of problem can be identified by tension cracks normally occurring between windows and doors, but will not extend below the damp-proof course. This type of cracking is generally straight and vertical, although it may step around the bricks. The cracking is characterised as being the same thickness top and bottom, and generally no more than 2.0 mm in width. The Building Research Establishment proposes that cracks resulting from thermal and moisture changes are between 0.5 and 5 mm wide.

Moreover, the cracks will be at regular intervals along the wall or floor – for example at quarter, half or third points. These cracks will occur where expansion or movement joints would naturally fall, and the filling of such joints with cement may result in the crack reoccurring. Other areas which are prone to this type of cracking are between changes in section, for example where a structure changes thickness from say a 100 mm wall to a 200 mm wall.

It should also be noted that where walls are constructed from different materials – such as a brick outer skin and a block inner skin – the symptoms will be different due to the exposure temperature, the materials having different coefficients of thermal expansion and differential resistance to changes due to moisture.

Differential expansion can occur between the internal and external skins of a wall, and this is particularly so when the internal skin of material is maintained at a constant temperature and humidity. This can result in the bulging of a wall and in some cases, for example in Victorian dwellings constructed of solid brick, the bulging is such that it separates the wall.

In the manufacture of the brick and following the oven-drying process the brick will shrink, then when exposed to moisture it will expand. This effect can be caused by leaking guttering, rain or other sources of moisture. Concrete, however, shrinks when exposed to moisture. Thus, if a concrete material is placed against a brick material and exposed to moisture, tension cracks can result due to the opposing forces. This may be a long-term effect and in the case of brickwork, where the material can expand by up to 1.0 mm/m, this can result in a typical house being 8.0 mm longer following construction.

Timber is particularly susceptible to shrinkage across the grain, and typically this can be 2.0 mm across a 100 mm section. Green timber is even more susceptible to shrinkage, and the effects of this on timber frame structures using green oak for example can be considerable. Reversible changes are also possible if the humidity and temperature increase.

Notably in cavity walls differential movement can occur between the blockwork and brickwork, and this can result in the cavity becoming larger at corners. If brick ties are placed too close to the corner, this will restrain the wall and result in cracking.

Shrinkage cracking normally occurs within the first 12 months after construction.

Solution

The best solution in masonry is to provide a silicone or mastic sealant to the joint to allow future expansion and movement, since the crack will occur at a natural expansion joint.

Bulging walls can be stabilised using Helifix™ ties inserted through the masonry walls and power driven through the floor joists. The tie is then resin fixed to the wall.

Movement of brickwork along the damp-proof course

Movement of brickwork along the damp-proof course can occur, particularly in long walls, where the brickwork can over-sail the corner of the wall. This is a result of the irreversible moisture changes in the brickwork above the damp-proof course, unlike the brickwork below the damp-proof course which is usually at constant moisture content provided by the ground moisture and consequently not subjected to such changes. When the bricks are delivered from the kiln an initial shrinkage will occur. However, as the bricks absorb moisture from the surrounding air the brick will expand and continue to do so long term. The movement can continue for a number of years following manufacture.

Solution

Movement joints can be installed, but there is little that can be done to overcome this type of problem.

Subsidence

Subsidence is due to the downward movement of part of a foundation, caused by activity in the ground. The damage is caused when part of the building remains in its original position and part of it experiences downward movement, thus causing a shear movement. The following criteria must be met:

- By nature of the shear failure the resulting cracking will be diagonal and most likely follow the weakest part of the building, through the doors and windows.
- The cracking generally passes through the damp-proof course to ground level.
- The cracking will be seen on both sides of the wall, since one side of the wall cannot subside without the other.
- The crack will tend to cease at the corner of a building or at the eaves. Cracking that ceases and commences in the centre of a wall is not as a result of subsidence.
- The cracking will be tapered, but this does not necessarily mean that a tapered crack is only attributable to subsidence. The angle of the taper is important, because the problem causing the subsidence usually lies under the taper.

Figure 18.1: A tapered subsidence joint between the line of an extension.

Horizontal cracking is possible, but this will be tapered and will follow a line of weakness in the wall – such as a damp-proof membrane or between openings that are close together.

Vertical joints are not generally associated with subsidence unless they follow a construction joint, for example vertically between the line of a more recent extension the cracking will be tapered and extend the full height of the structure. See Figure 18.1.

The other exception to a vertical crack indicating subsidence is where the cracking is between two openings, such as window openings on the first and ground floor. In this situation the cracking will be either side of the openings, and one of the cracks will be thicker at the top and the other at the bottom.

There are a number of reasons why buildings subside, but the list below provides some examples.

Trees

The extraction of moisture from clay soils results in volumetric changes, and this can cause subsidence. The introduction of moisture to such subsoils can also result in heave, which has similar characteristics. This is the most common cause of subsidence on clay soils, which are susceptible to volumetric changes. Essentially the trees extract moisture from the clay subsoil, causing the soil to contract or shrink, thus resulting in a downward movement of the foundation. This effect is exasperated during times of drought. In the winter months, when the ground becomes more saturated with moisture, the clay will swell and return to its original volume – thus the property may experience movement across the seasons due to swelling and contraction of the clay subsoil.

Moisture

Leaking drains can wash away smaller particles contained in sandy gravels and as the smaller particles are washed away by the flowing water, the volume of the soil will reduce, thus leading to a downward movement of the foundation.

Moisture can also reduce the bearing capacity of any soil, including clays, making them susceptible to subsidence. The photograph in Figure 18.2 shows a diagonal crack on the corner of a building. The diagonal line tells us that the problem lies under the crack, and it can be seen that the location of a drain on the corner of the building is the most likely cause. It would therefore be advantageous to excavate around the drain to ensure it is not leaking.

Voids

Voids caused by mines, caves, sinkholes, disused wells or collapsed drains in close proximity to a house will cause downward movement of the foundation. A photograph of a building over a recently discovered well can be seen in Figure 18.3. In this case the well was capped using a reinforced slab.

Filled ground

Filled ground using building material, unconsolidated fill or waste will result in subsidence, since the overburden load from the property causes the material to compress. If the ground contains vegetation, which continues to degrade over time, then this will result in further subsidence.

Solution

The extent of the problem usually determines the solution, but possible remedies include underpinning and micro-piles. An underpinning schedule can be seen in Chapter 8. Other

Figure 18.2: Photograph showing diagonal subsidence crack in a building.

Figure 18.3: Photograph of a building constructed over a well.

solutions include the use of pillar insertion systems, such as Uretex™ who manufacture PowerPile® polymer pillars. Essentially the product is an expansive polymer, which is inserted into voids or an expansive casing under the foundation. This can stabilise and in some cases lift a slab or pavement by filling the voids and improving the bearing capacity of the subsoil. In the case of the expansive casing this acts like a pile, and the casing is inserted into the ground via a 50 mm diameter hole. The expansive casing is inserted and filled with the geopolymer, which expands the casing up to 400 mm diameter.

Chemical reactions

Sulphate attack

Sulphate is found in mortar and in bricks, and will react with water resulting in a chemical reaction. The following soluble sulphates are found in cement: sodium, potassium, magnesium and tricalcium aluminate. For the reaction to take place, significant water is required and this can be provided by natural ground water, rain or leaking pipes and drains. When the reaction takes place the cement attempts to return to its original constituents, which have a greater volume than the cement mixture, thus the material expands.

Hot flue gases pass up the chimney and condense higher up, particularly where the chimney exits the property and becomes exposed and consequently cooler. The condensation leads to the formation of water, which migrates into the wall. This can result in the deposition of sulphate salts, which attack the mortar causing it to expand. This leads to a lean in the chimney, usually on the cooler side.

In some cases, if the driving rain is constant and from one side of the chimney, the moisture can cause sulphate attack on this side of the chimney.

In walls the effect is determined by the restraint, and will expand on the side exposed to rain or moisture. If the wall is restrained then the wall will bow and horizontal cracking will be seen. This is similar to, and often mistaken for, wall tie failure. If the wall is unrestrained, the wall will expand upwards.

In concrete floors, sulphate attack leads to a doming of the floor slab and in extreme cases this can lift the walls. Another indicator is cracking between the parquet blocks. With the advent of damp-proof membranes in the 1970s/1980s, protection was afforded to the concrete since these prevented sulphates being in contact with it.

There are materials which are known to trigger sulphate attacks. Some examples are Red Shale (a by-product of coal mining), which was used as a hardcore beneath the concrete floor slab and Black Ash (a by-product from power stations, particularly plentiful in the 1950s). Plaster also triggers sulphate attacks, and if builders' rubbish has been placed under a floor unprotected this can result in sulphate attack.

Solution

To confirm sulphate attack the material has to be analysed. However, the solution is to remove concrete floors and relay using a damp-proof membrane.

Chimneys may have to be reconstructed, depending on the severity of the defect, but a liner should be employed to protect the masonry.

High alumina cement (HAC)

This was an additive introduced to concrete to allow it to get very strong quickly, but unfortunately it then gets weaker! HAC was introduced into the building trade in the 1950s and continued to be used until the 1970s; it was predominantly used in pre-stressed concrete beams. By this time most concrete of this type will have reached its minimum strength, and thus if it is still standing arguably it should not be a problem.

The other problem with this type of structural element is that the structural components for which it was used generally had very small sections of reinforcement, typically 6 mm diameter bars, and very little cover to the reinforcement. Thus, if corrosion occurs there is not much reinforcement to corrode before failure can occur. This is particularly concerning where these structural elements have been employed in damp conditions.

Wall tie failure

Traditionally wall ties were constructed using galvanised iron or steel, and when exposed to moisture the wall ties were susceptible to corrosion. This causes the outer wall to become detached from the inner wall and can lead to collapse. Also, due to the expanded metal, the cement joints become forced apart. The symptoms of this type of failure are that cracking is evident along the horizontal joints – either continuously or at regular intervals and four courses apart. The cracking coincides with the spacing of the wall ties.

Invasive tests can be undertaken using an endoscope or by removing a brick to locate a wall tie and examine if corrosion is present.

Solution

One solution is to prevent rain penetration by the addition of a cladding, render or silicone moisture sealant. Helifix™ provides dry fix wall ties that are power driven into both leaves of masonry walls via a small pilot hole and resin bonded to tie the leaves together. However, it may be necessary to remove and reconstruct the areas of wall affected and insert new wall ties. In extreme cases it may prove necessary to reconstruct large areas of wall.

Damp

Damp is a problem within buildings and the structural components are affected also. Concrete will shrink and brick will expand in moisture, thus opposing forces will result in tension and cracking in a wall containing both materials.

Steel will rust from exposure to moisture due to oxidation. This will cause expansion and steel will expand up to four to ten times its original volume, thus causing distortion and cracking in a structure. Consequently, in steel frame structures the rusting at the base of a steel column may cause enough tension to pull the column from its holding bolts.

Steel frame buildings are prone to corrosion due to the ingress of moisture, and most mortgage companies require a structural assessment of these buildings to ascertain if they are structurally sound. In such cases columns are examined on exposed elevations and under kitchen or bathroom windows to check the steel frame for corrosion.

Damp can also affect reinforced concrete and if the reinforcement is allowed to corrode due to the ingress of moisture, this can cause spalling of the concrete and a weakening of the structural element.

Damp also weakens the subsoil under foundations and can cause localised downward movements of the foundation if the bearing pressure exceeds the bearing capacity of the soil, thus leading to subsidence.

Moisture ingress near timber lintels can lead to wet and dry rot and in extreme cases the bearing may be compromised and the lintel fail. Figure 18.4 shows a photograph of a house constructed using stone rubble walls, which has been pointed using cement pointing. Penetrating damp has been trapped in the wall through the use of the cement pointing. If the re-pointing had been undertaken using lime rather than cement, this would have allowed the moisture to escape and the wall to breathe. Consequently, the cement mortar has led to wet-rot damage to a beam and collapse over the window.

Overloading

The overloading of a section or part of a building can result in movement on the foundation or the compressive strength of the supporting material being exceeded. This is common in buildings with large openings, where the loads from the floors and roof are directed down a small masonry panel acting as a column. The material forming the column may become overloaded and the excessive compression causes it to fail, but more likely the small foundation on which the column or small length of wall bears

Figure 18.4: Photograph showing the loss of bearing of a beam subject to wet-rot deterioration.

Figure 18.5: Photograph showing overloaded column of brickwork which has failed in compression.

Figure 18.6: Photograph of a deflected purlin.

is inadequate to support the load. Consequently the bearing pressure exceeds the bearing capacity of the soil and failure can occur.

Figure 18.5 is a photograph showing an overloaded element of brickwork between windows in a Georgian property after one of the windows has been relocated adjacent to an original window.

In other cases an overloaded member may suffer deflection, which is not a structural defect in itself but can cause unsightly deflections in roof structures or cracking in brittle finishes such as plaster. Figure 18.6 is a photograph of a purlin which has suffered deflection due to being undersized for the loads imposed on it.

Solution

In the case of an overloaded wall section the solution is to strengthen the wall panel by thickening the cross-sectional area and/or increasing the size of the foundation to ensure the bearing pressure is less than the bearing capacity of the soil on which it rests.

Where timber members such as roof or floor members and beams are overloaded, the structural sections can be strengthened using the addition of steel plates, providing new members or thickening of existing members.

Professional advice

It is important to determine the cause of structural failures to ensure that the symptom not the cause is being addressed. This may involve exploratory and invasive tests undertaken by specialists. Professional advice should be sought if structural defects are identified, not only to address the problem but also to ensure that there are no other ongoing problems or knock-on effects. It may also be a requirement under any insurance policy taken out on the building to inform the insurance company concerned, and they may wish to appoint their own surveyor or loss adjuster. Failure to notify insurance companies of such defects may compromise future insurance cover and any future claims.

Chapter 19 The Ancient Use of Sign and Geometry in the Setting Out of Buildings

Daisy wheel

The daisy wheel is composed of six-point geometry with diameter D and radius R, by placing the compass at point 1 and drawing an arc of radius R as shown in Figure 19.1. The procedure is repeated where the arc touches the circle, and you continue to draw the arcs as seen.

The pattern that emerges can be called a number of names, including the daisy wheel, seed of life, hex and rose petal, and has significant meaning for the Knights Templar.

There are many theories concerning daisy wheels, including good-luck symbols and signs/warnings to drive away evil spirits – mainly to protect harvests. Sometimes they are seen at the entrance to churches and have a religious meaning. However, there is evidence to suggest that these signs were used by Medieval carpenters to set out timber frames for houses, and indeed by masons to set out the geometry of cathedrals.

Simple geometrical techniques were used by carpenters and masons to set out buildings and architectural features. These techniques were quite secretive until 1776, when Gaspard Mongue published *Descriptive Geometry, or the Art of Science of Masonic Symbolism.*

Daisy wheel symbols can sometimes be found in timber frame houses, etched out in the timber. The overlap of the daisy wheel and the number of petals determined the design of the external and internal structure of the building. By joining the petals using a series of lines we can obtain a hexagon, an equilateral triangle and a star of David.

If we examine triangle ABC, the shape provides the angles 30°, 60° and 90° and for the petals, the 15° angle as seen in Figure 19.2. It can also be seen that if we join the side petals then a rectangle results which has a ratio of 1 for the shortest side to 2 for the diagonal.

The golden number or golden mean

This is a significant ratio represented by the Greek letter ϕ. This is the ratio where the length of an object is divided in a particular place along its length, with the ratio of the shorter length to the longer length being equal to the ratio of the longer length to the whole length. The diagram in Figure 19.3 demonstrates this.

If a rectangle is drawn using the divisions as shown in Figure 19.3, we can create a golden rectangle. A curve is subsequently drawn at the point of intersection of the golden mean ratio and the rectangle further divided using the golden rectangle along its width.

Structural Design of Buildings, First Edition. Paul Smith.

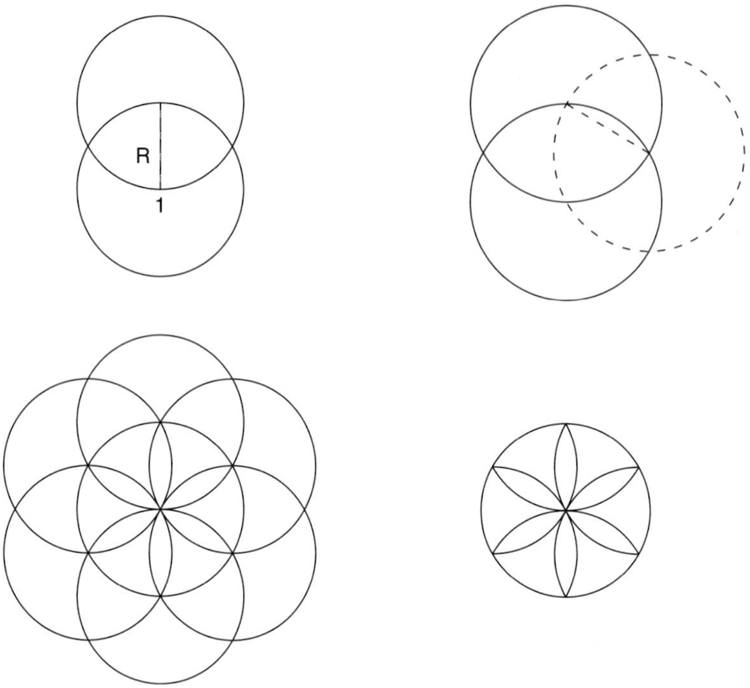

Figure 19.1: The construction of a daisy wheel.

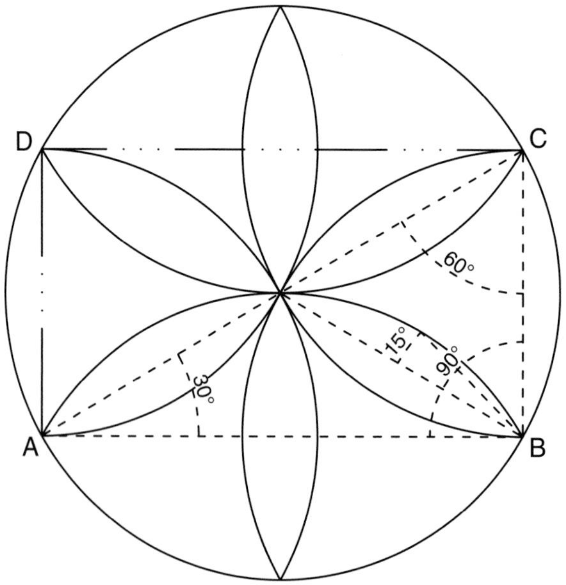

Figure 19.2: Construction of a geometric triangle using the daisy wheel.

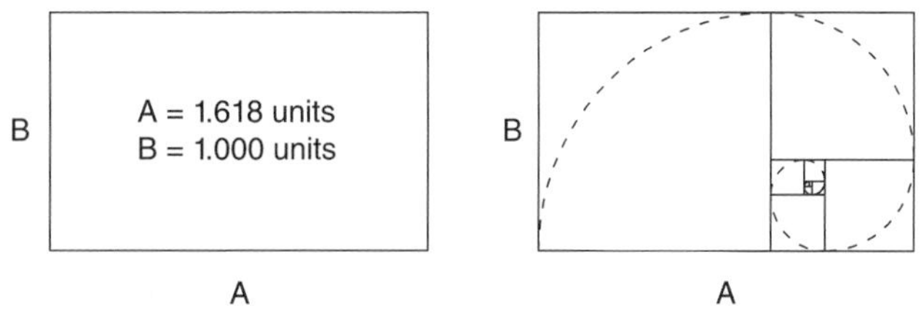

Figure 19.3: Representation of the golden number.

By repeating this process between the remaining widths of the remaining rectangle and drawing a curve to the point of intersection, one can see that a spiral emerges.

The golden number or golden mean has significance in a number of areas, such as the geometry of the human body.

Pythagoras

Let us examine an equal-sided square and take a line between the corners to form a triangle of equal sides. If the equal sides have a dimension of one unit, then by Pythagoras the hypotenuse will be equal to the root of two units.

Consider a square of equal sides 1 unit.

The diagonal across the square is equal to

$$\sqrt{(1^2 + 1^2)} = \sqrt{2} \tag{19.1}$$

This ratio of 1 unit to $\sqrt{2}$ units was used by early masons in the setting out of churches and cathedrals. The length and width of such buildings were set out using the geometry of squares and their diagonals. For example, the width of a church may have equated to one unit of the side of a square, whereby the length of the entire building is the diagonal or square root of the diagonal. It can be seen in some cases, such as Norwich Cathedral, that the cloister is set out as a square and by using the side of the square as equal to 1 unit the diagonal of the square is $\sqrt{2}$ units. The nave has then been constructed to this length of $\sqrt{2}$ units. If we take the length of the nave and make this the square, then the diagonal of this square is the length of the entire church.

Masonic markings

The Freemasons are a society of secrets rather than a secret society, and there are mystical emblems and symbols found in stonework and buildings throughout the world that have significant meaning to the Masons. In churches and cathedrals there are many examples of Masonic carvings, and these include quite prominent carvings. There are also other markings of which the exact purpose is unknown, although assumptions can be made about their significance.

The masons employed a marking system as a method to establish how much work or stone had been laid, as a means of receiving payment, and these marks were carved in the

Figure 19.4: Photograph showing a Masonic mark.

stone. This also established which mason was responsible for a given section of wall, and was a signature.

Other assumptions are that the markings are layout or construction marks, which identify the position or order in which a stone should be laid. The Masons in Medieval times were most likely illiterate, but could understand a system of signs used for the assembly of buildings and the order in which the stone components were to be constructed.

Finally, there are also assumptions that the marks identify the location or source of the stone.

Masons still use these marks, but in the repair of cathedrals and churches they have to be hidden and can occur on the internal face of the repaired stone masonry. The markings vary in size – sometimes as small as 25 mm and in other examples as large as 150 mm. The markings can be simple scratches or significant markings, and may include symbols or numbers. The markings can include crosses, stars or intersecting lines at different angles. Examples of these types of mark can be found throughout the world, and an example can be seen in the photograph in Figure 19.4, taken in a church in Herefordshire.

Ordnance datum bench marks

Bench marks are established on permanent structures and can be either fundamental bench marks or bench marks. Fundamental bench marks record the height above an ordnance datum, usually taken as a mean sea level. From the fundamental bench mark, a network of bench marks extend out providing a reference for heights across the country. The horizontal carved bench provides a datum on which the surveyor can rest a plate and

Figure 19.5: Photograph showing an ordnance datum bench mark.

levelling staff. The level of this datum can be obtained from the relevant government body responsible. In the UK the body responsible is the Ordnance Survey, and using this information the relative heights and positions of points can be ascertained or established across a particular site. A photograph of such a marking can be seen in Figure 19.5.

This chapter has described some of the symbols found on historic buildings and structures across the country. Some of these are surrounded with mystery and others are used for mathematical purposes. Whatever their purpose, the mystery and identification of such symbols can be both rewarding and exciting.

References

Bonshor, R.B. and Bonshor, L.L. (1996) *Cracking in Buildings*. Construction Research Communications Ltd: Watford.
Brunskill, R.W. (1994) *Timber Building in Britain*, 2nd edn. Yale University Press: New Haven, CT.
Middleton, G.A.T. (1905) *Building Materials: Their Nature, Properties and Manufacture*. B.T. Batsford: London.
Mitchell, C.F. (1930) *Building Construction*. B.T. Batsford: London.
Smith, G.N. (1990) *Lateral Earth Pressure*, 6th edn. BSP Professional Books: Oxford.
Smith, I. (2006) *Smith's Elements of Soil Mechanics*. Blackwell: Oxford.
Statham, I. (1977) *Earth Surface Sediment Transport*. Oxford University Press: Oxford.
Taylor, R. (2003) *Church Building and Furniture, How to Read a Church*. Random House: New York.

Further Reading

Barnbrook, G. *et al.* (1975) *Concrete Practice*. Cement and Concrete Association: Slough.
Brunskill, R.W. (2004) *Traditional Buildings of Britain*. Cassell: London.
Case, J. and Chilver, A.H. (1971) *Strength of Materials and Structures*. Edward Arnold: London.
Chanakya, A. (2003) *Design of Structural Elements*. Spons Press: London.
Department of Scientific and Industrial Research (1959) *Principles of Modern Building*. HMSO: London.
Gwynne, A. (2013) *Guide to Building Control*. John Wiley: Chichester.
Handyside, C.C. (1950) *Building Materials Science and Practice*. The Architectural Press: London.
Harrison, H.W. and Trotman, P.M. (2002) *BRE Building Elements: Foundations, Basements and External Works*. Construction Research Communications Ltd: London.
Heyman, J. (1997) *The Stone Skeleton*. Cambridge University Press: Cambridge.
Heyman, J. (1999) *The Science of Structural Engineering*. Imperial College Press: London.
Hilson, B. (1993) *Basic Structural Behaviour*. Thomas Telford Services Ltd: London.
Hunt, R. and Suhr, M. (2008) *Old House Handbook*. Frances Lincon Ltd: London.
McKenzie, W.M.C. (2001) *Design of Structural Masonry*. Palgrave: New York.
McKenzie, W.M.C. (2013) *Design of Structural Elements to Eurocodes*. Palgrave Macmillan: Basingstoke.
Moseley, W.H., Bungey, J.H. and Hulse, R. (1999) *Reinforced Concrete Design*. Palgrave: New York.
Powys, A.R. (1929) *Repair of Ancient Buildings*. J.M. Dent: London.
Shirley, D.E. (1985) *Introduction to Concrete*. Cement and Concrete Association: Slough.
Yorke, T. (2008) *British Architectural Styles*. Countryside Books: Berkshire.
Yorke, T. (2010) *Timber Framed Buildings Explained*. Countryside Books: Berkshire.
Yorke, T. (2011) *Tracing the History of Houses*. Countryside Books: Berkshire.

Structural Design of Buildings, First Edition. Paul Smith.

Index

Structural Design of Buildings, First Edition. Paul Smith.